Machine Learning for Protein Science and Engineering

A subject collection from *Cold Spring Harbor Perspectives in Biology*

Machine Learning for Protein Science and Engineering

A subject collection from *Cold Spring Harbor Perspectives in Biology*

EDITED BY

Peter K. Koo
Cold Spring Harbor Laboratory

Christian Dallago
Duke University
NVIDIA Corp.

Ananthan Nambiar
University of Illinois Urbana-Champaign

Kevin K. Yang
Microsoft Research

COLD SPRING HARBOR LABORATORY PRESS
Cold Spring Harbor, New York • www.cshlpress.org

Machine Learning for Protein Science and Engineering
A subject collection from *Cold Spring Harbor Perspectives in Biology*
Articles online at www.cshperspectives.org

Executive Editor	Richard Sever
Project Supervisor	Barbara Acosta
Editorial Assistant	Danett Gil
Permissions Administrator	Carol Brown
Production Editor	Diane Schubach
Production Manager/Cover Designer	Denise Weiss
Publisher	John Inglis

Front cover artwork: Illustration of actinidin protein. Image © Science RF from the photo agency Science Photo Library (SPL) (adobe stock 699492068).

Library of Congress Cataloging-in-Publication Data

Names: Koo, Peter, editor. | Dallago, Christian, editor. | Nambiar, Ananthan, editor. | Yang, Kevin K., editor.
Title: Machine learning for protein science and engineering / edited by Peter K. Koo, Cold Spring Harbor Laboratory, Christian Dallago, Technical University of Munich. Ananthan Nambiar, University of Illinois Urbana-Champaign and Kevin K. Yang, Microsoft Research New England.
Description: Cold Spring Harbor, New York : Cold Spring Harbor Laboratory Press, [2025] | "A subject collection from Cold Spring Harbor perspectives in medicine"-- Title page. | Includes bibliographical references and index. | Summary: "Tremendous advances in machine learning are enabling scientists to make computational predictions about numerous biological phenomena. The volume will examine how machine learning is being used to help understand the structure and function of proteins"--Provided by publisher.
Identifiers: LCCN 2024049729 (print) | LCCN 2024049730 (ebook) | ISBN 9781621824800 (hardcover) | ISBN 9781621824817 (epub)
Subjects: LCSH: Protein engineering--Data processing. | Proteins--Analysis--Data processing.
Classification: LCC TP248.65.P76 C66 1994 (print) | LCC TP248.65.P76 (ebook) | DDC 660.6/3--dc23/eng/20250116
LC record available at https://lccn.loc.gov/2024049729
LC ebook record available at https://lccn.loc.gov/2024049730

All World Wide Web addresses are accurate to the best of our knowledge at the time of printing.

For a complete catalog of all Cold Spring Harbor Laboratory Press publications, visit our website at www.cshlpress.org.

Contents

v

Preface

PROTEIN SCIENCE HAS MADE TREMENDOUS PROGRESS since the early studies in the mid-twentieth century. The landmark paper on the structure of myoglobin by John Kendrew, followed by the structure of hemoglobin published by Max Perutz, opened the doors to structural biology and deepened our understanding of protein function. Around the same time, Frederick Sanger's sequencing of insulin provided a glimpse into the primary structure of a protein. Together, these discoveries laid the foundation for protein science, establishing the importance of both protein structure and amino acid sequence, and sparking fundamental questions about the relationship between the two.

In the decades that followed, advances in sequencing technologies, structural biology, and bioinformatics allowed researchers to investigate proteins at an unprecedented scale. Large sequence and structure databases, high-throughput experimental methods, and computational modeling techniques transformed protein science into a data-rich field. With this influx of data came both the opportunity and challenge of extracting meaningful patterns, predicting function from sequence, and engineering new proteins with desired functions. The past few years have seen particularly impactful breakthroughs at the intersection of machine learning and protein science, beginning with the introduction of transformer-based protein language models in 2019 and 2020. These models, inspired by natural language processing, demonstrated an ability to learn meaningful representations of proteins from large-scale protein sequence data, capturing evolutionary, functional, and structural signals. This was soon followed by the introduction of AlphaFold 2, which set a new standard for protein structure prediction. These developments have fundamentally changed the field of protein science, with applications in protein function annotation, variant effect prediction, and generative protein design.

This book explores the rapidly evolving intersection of machine learning and protein science. The first chapters (Chapters 1 and 2) introduce machine learning approaches for learning representations of proteins, including applications to antibody comprehension. Subsequent chapters cover statistical models of coevolution (Chapter 3) and large-scale homology searches (Chapter 4), which have implications for protein structure prediction. The middle chapters examine machine learning applications in functional annotation and evolution, including variant effect prediction (Chapter 5) and the fundamental question of whether protein novelty is predictable (Chapter 6). We then explore generative models for both protein sequence and structure (Chapters 7–9). The final chapter (Chapter 10) reflects on the environmental impact of applying large-scale machine learning in protein science and engineering, acknowledging the need to balance technological advancement with sustainable computational practices.

PETER K. KOO
CHRISTIAN DALLAGO
ANANTHAN NAMBIAR
KEVIN K. YANG

Artificial Intelligence Learns Protein Prediction

Michael Heinzinger[1] and Burkhard Rost[1,2,3,4]

[1]Technical University of Munich (TUM) School of School of Computation, Information and Technology (CIT), Bioinformatics and Computational Biology - i12, 85748 Garching/Munich, Germany

[2]Institute for Advanced Study (TUM-IAS), 85748 Garching/Munich, Germany

[3]TUM School of Life Sciences Weihenstephan (WZW), 85354 Freising, Germany

[4]Department of Biochemistry and Molecular Biophysics, Columbia University, New York, New York 10032, USA

Correspondence: mheinzinger@rostlab.org

From *AlphaGO* over *StableDiffusion* to *ChatGPT*, the recent decade of exponential advances in artificial intelligence (AI) has been altering life. In parallel, advances in computational biology are beginning to decode the language of life: *AlphaFold2* leaped forward in protein structure prediction, and protein language models (pLMs) replaced expertise and evolutionary information from multiple sequence alignments with information learned from reoccurring patterns in databases of billions of proteins without experimental annotations other than the amino acid sequences. None of those tools could have been developed 10 years ago; all will increase the wealth of experimental data and speed up the cycle from idea to proof. AI is affecting molecular and medical biology at giant steps, and the most important might be the leap toward more powerful protein design.

SCIENCE FICTION OR FUTURE SCIENCE?

Walking her dog, Dr. Elena decides to engineer a bacterium efficiently gobbling up all those painkillers she had to swallow after her recent tooth extraction. She hopes to immerse those bugs into a wastewater facility. On her phone, she begins collecting a few dozen enzymes known to catalyze reactions similar to those needed to digest the environmentally toxic ingredients of that medicine. She downloads the bug's sequences and predicts the three-dimensional (3D) structures for all proteins (Box 1). Before reaching home, she has already created the most likely functional 3D scaffolds relevant for the proteins to bind the toxins, has applied a protein language model (pLM) to generate millions of new sequences that might have similar 3D structures, has selected a few tens of top candidates for experimental testing. She has sent those sequences to her laboratory robot. When she reaches the laboratory, 2 hours, later the robot has already gone through the first round of optimization with results being fed back directly to the pLM for further refinement of the top candidates. All is ready for more detailed experimental analysis thanks to the advances in artificial intelligence (AI).

EVOLUTIONARY INFORMATION POWER CHARGES PROTEIN PREDICTION

Secondary Structure Prediction Jumped by Combining AI and Alignments

The application of advanced machine learning (here for simplicity coined AI) to protein prediction began 35 years ago, with simple artificial feedforward neural networks (ANNs) predicting protein secondary structure (Bohr et al. 1988; Qian and Sejnowski 1988). Although these, along with subsequent publications, provided the proof-of-principle for a powerful new technique, its breakthrough came by combining AI and evolutionary information (EI) as derived from multiple sequence alignments (MSAs) (Rost and Sander 1992, 1993). Secondary structure had been predicted by both AI (Bohr et al. 1988; Qian and Sejnowski 1988) and EI (Zvelebil et al. 1987); the successful novelty was the combination of both. This succeeded because ANNs captured long-range information (sequence separation between residues i and j such that, e.g., $|i - j| > 15$) much better than other statistical analyses of MSAs. The formula AI + EI was so successful that performance rose above what had been published in many textbooks as the theoretical limit, namely, a three-state per-residue accuracy of Q3 = 65% (Fig. 1; purple dashed horizontal line). The first method, dubbed PHD, surprisingly reached above 72% (Rost 1993, 1996), which pushed the advance more than the three decades of improvements and data collection before (Fig. 1). The trick was to put more complex information (replace single sequences by protein families described by EI) into advanced learning methods capable to mine this complexity. Fine-grained control of the tool (balanced training and stacking of simple ANNs into constructs that resembled some aspects of deep learning a decade before its introduction) allowed to also fix other aspects of the problem not reflected in the simple three-state per-residue accuracy (Rost 1993).

AI + EI Recipe Boosts Other Aspects of Structure Prediction

The successful combination of AI + EI was expanded to other features of protein 1D structure, including the prediction of solvent accessibility and membrane regions (Rost 1996). The initial objective had been to predict interresidue distance maps (2D structure; Rost 1993), but the complexity of this objective required another decade (Punta and Rost 2005a). Although successful in many ways (Punta and Rost 2005b; Schlessinger et al. 2007), such methods did not suffice to generalize from 2D to 3D structure predictions. Overall, AI + EI broke through many ceilings, including predicting molecular function (Rost et al. 2003), but appeared to fail accurately predicting 2D or 3D structure.

In fact, the major advance toward 2D predictions sufficient for the generation of 3D structure

Cite this article as *Cold Spring Harb Perspect Biol* doi: 10.1101/cshperspect.a041458

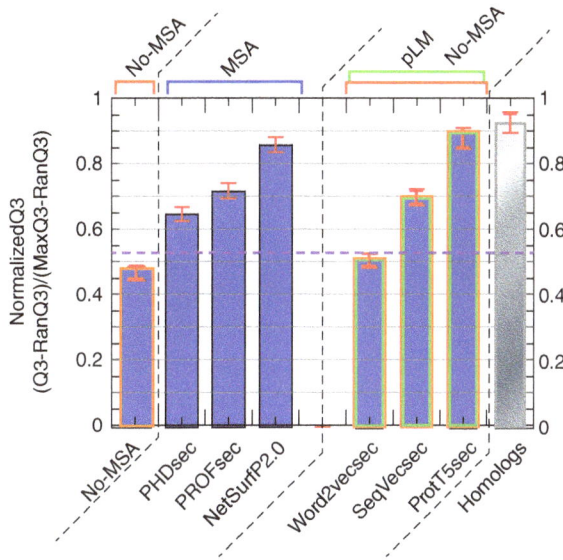

Figure 1. Rise in secondary structure prediction. Protein secondary structure prediction might be the simplest and best-understood aspect of structure prediction. Therefore, we use it as a proxy to compare different methods. The y-axis shows the performance in terms of normalized Q3 (Rost et al. 1994). This is defined as follows: NormalizedQ3 = (Q3-RandomQ3)/(MaxQ3-RandomQ3); Q3: three-state per-residue accuracy (helix, strand, other); MaxQ3 = 92% approximates the secondary structure string agreement between alternate experimental structures for the same protein (Andersen et al. 2002); and minimal performance, namely, random, RandomQ3 = 35% (Rost et al. 1994). By definition, NormalizedQ3 ranges from 0 (random) to 1 for predictions reaching the experimental resolution. Methods: *No-MSA* refers to simple statistical models or simple artificial feedforward neural networks (ANNs) not using evolutionary information (EI) from multiple sequence alignments (MSAs), *MSA-based*: PHD_{sec} marks the first stacked system of ANNs combining artificial intelligence (AI) and EI (Rost and Sander 1992), $PROF_{sec}$ (Rost 2001) uses richer MSAs (from PSI-BLAST rather than BLAST; Altschul et al. 1997), $NetSurfP2$ (Klausen et al. 2019) marks the top recent MSA-based predictions; *pLM-based* (protein language model methods): $Word2vec_{sec}$ (Heinzinger et al. 2019) is the context-independent first generation of language models (LMs) (bag-of-words; Mikolov et al. 2013), $SeqVec_{sec}$ (Heinzinger et al. 2019) uses the LM called ELMo (Peters et al. 2018), and $ProtT5_{sec}$ (Elnaggar et al. 2021) is based on transformers; *Homologs* marks the ~88% agreement of secondary structure between proteins with similar sequences (Rost et al. 1994). The red-arrowed error bars roughly approximate a comparison between methods developed over the course of three decades and assessed on a diversity of data sets. The horizontal dashed purple line (marking Q3 ~ 65%) has been considered the top reachable for many years. In fact, this was true before AI + EI. The noncontext-aware Word2vec value confirms this level. On a side note, the rise for MSA-based solutions from PHD_{sec} (1992) to NetSurfP2.0 (2019; *middle* set of blue bars) required about three orders of magnitude larger databases combined with much advanced AI (from long short-term memories [LSTMs] to convolutional neural networks [CNNs]) and took almost three decades. The comparable advance for pLM-based solutions from Word2vec to ProtT5 took a little more than 3 years.

originated from a combination of advanced processing of EI in the form of evolutionary couplings (Marks et al. 2011). The successful signal-to-noise filtering that turned those couplings into a breakthrough solution required statistical models (Lapedes et al. 1999; Weigt et al. 2009; Balakrishnan et al. 2011; Marks et al. 2011; Jones et al. 2012; Seemayer et al. 2014). This advance was orthogonal to another set of tools largely dominating CASP (critical assessment of protein structure prediction) (Moult et al. 1995, 1999, 2007; Kryshtafovych et al. 2007), namely, programs using more or less directly comparative modeling (Baker and Sali 2001; Bonneau et al. 2001; Pieper et al. 2011; Biasini et al. 2014). Although comparative modeling and evolutionary coupling-based advances remained AI-free, the next step, once again, combined the simpler statistical models with AI-based models (Wang et al. 2017; Yang et al. 2020). This approach peaked in *AlphaFold1* (Senior et al. 2020) (officially *AlphaFold*), the last method still on the way toward accurate 3D prediction.

ALPHAFOLDOLOGY IS ALL THERE IS FROM NOW ON?

Leap in 3D Prediction by AI Just in Time

In December 2020, *AlphaFold2* (Jumper et al. 2021) broke through in protein structure prediction at CASP14 (Kryshtafovych et al. 2021). Like all top structure predictions since PHD_{sec} (Fig. 1), *AlphaFold2* succeeded through using EI from MSAs. Why did it take 28 years to get to this point? Simply put, the breakthrough could not have happened any earlier because it was rooted in three crucial advances coinciding at that point in time: (1) software (i.e., the advanced layered architectures) (deep neural networks—especially, the flexibility offered by *Transformers* paired with geometry-aware attention operations rendering the input 3D structure invariant to global rotations and translations), (2) hardware (advanced GPUs and TPUs), and (3) database sizes (*BFD*; Steinegger et al. 2019) is 10-times *UniProt* (The UniProt Consortium 2021). That the tool of the year (Marx 2022) came at all in 2020 required many successful novel solutions cleverly engineered by a large team of advanced experts fueled with sufficient resources (Jumper et al. 2021). Among many other inventions, the novelties included learning explicitly to predict (1) reliability (in the form of the so-called *pLDDT* score); (2) particular shapes (no method had ever implemented this explicitly); (3) to iterate over and thereby refine its own predictions; and (4) to increase data set size and diversity by training on its own predictions (which in turn needed a method as successful as *AlphaFold2*).

All these advances required ingenious AI engineering coding physical and geometrical constraints (inductive biases) directly into the components of the models, or more precisely: into its architecture. Joining these components with learning features in a way known as end-to-end (i.e., by backpropagating the gradient from the predicted 3D structure to the MSA) allowed the model to directly learn features from the wealth of experimental 3D structures deposited in the PDB (Burley et al. 2023). This solution was in stark contrast to previous approaches that either relied on expertly crafted features or required external tools for computing input (evolutionary couplings) or targets (3D structures).

AlphaFold2 also leveraged the concept of distograms (Rost 1993) introduced unsuccessfully 30 years earlier. The DeepMind engineers made it work. One way to showcase the jump: While *AlphaFold1* outperformed any single method at CASP13 in 2018, it was not the best for any protein. In contrast, *AlphaFold2* at CASP14 in 2020 clearly outperformed each method for every protein, and immediately helped experimental structure determination (Millán et al. 2021). At CASP15 in 2022, *AlphaFold2* became the method to judge others by. Although many methods have reached performance levels that would have stunned the world 30 months ago, none has consistently reached the top, yet. So far, this seems as true for methods attempting to fully reengineer *AlphaFold2* as for those seeking "smarter" ways to improve.

AlphaFold2 Changes Experimental and Structural Biology

Molecular biologists have reacted by using the new tool (e.g., to advance experimental high-resolution determination of protein 3D structure) (Millán et al. 2021; Akdel et al. 2022; Bryant et al. 2022; Laurents 2022; Thorn 2022) to optimize docking and complement molecular dynamics (Guo et al. 2022; Laurents 2022; Tsaban et al. 2022). Many colleagues use the method successfully even for tasks for which it was not designed, such as predicting regions of intrinsically disordered proteins (Bryant et al. 2022; Guo et al. 2022; Ilzhöefer et al. 2022), or of permanent protein–protein interactions (PPIs) (Evans et al. 2021; Bryant et al. 2022; Johansson-Åkhe and Wallner 2022).

Although apparently successful for modeling the interaction between permanently bound

Cite this article as *Cold Spring Harb Perspect Biol* doi: 10.1101/cshperspect.a041458

constituents of proteins, neither *AlphaFold2* nor the interaction specialist *AlphaFold-Multimer* (Evans et al. 2021) appear to rise up to the challenges of predicting binding of transient, physical PPIs for experimentally uncharacterized protein pairs (Burke et al. 2023; L Kaindl and B Rost, unpubl.). In fact, most existing methods either infer or predict such PPIs through the simple annotation transfer referred to as comparative modeling or homology-based modeling/inference applying the simple logic: if the sequence similarity between two proteins Q and A exceeds an empirically established threshold T, copy annotation of A to Q. PPI prediction is an even tougher nut to crack than 3D structure prediction, and the appropriate assessment of methods is easier to get wrong than right (Park and Marcotte 2012; Hamp and Rost 2015). Another aspect that appears not fully covered by *AlphaFold2* is the prediction of the effect of sequence variation (Weissenow et al. 2022a,b). The reason is that *AlphaFold2* reaches its peak performance by generating a family average rather than a protein-specific prediction. We might expect that this limitation could be bypassed because *AlphaFold2* has internal components that allow it to weigh the query sequence against the MSA, thereby moving between protein-specific and family-average. However, there is way too little experimental data on the mutational effect on 3D structure to leverage this effect. Thirty-one years ago, when the successful combination of AI and EI won the day (Rost and Sander 1992), the distinction between the two never became relevant due to lower performance (Fig. 1: no bar reaches the level of Homologs to the very right; although we have not assessed this, we assume that *AlphaFold2* would approach the experimental error, i.e., approach NormalizedQ3~1). Despite undisputed success, there are limits—even for systems such as *AlphaFold2*.

Overall, very atypical for AI applications, the performance of *AlphaFold2* may even exceed the hype it created at CASP14 (Kryshtafovych et al. 2021). The *AlphaFoldDB* database now (June 2023) holds over 217 million 3D predictions (Tunyasuvunakool et al. 2021), and if you want to run the method on your set of sequences without requiring too much computing resources, *Colab-*

Fold (Mirdita et al. 2022) offers easy and fast access to the tool (essentially gaining speed by flexibly adjusting the number of iterations and by more efficient MSA generation).

Better Resolution of Structure Space

How much do the millions of accurate 3D structure predictions change our perception of structure space (i.e., the *CATH* [Sillitoe et al. 2021] or *SCOP* [Andreeva et al. 2020] classification of proteins by their 3D structure)? A recent analysis of 370,000 confident *AlphaFold2* 3D predictions could assign more than 90% of these to one of the known CATH domain superfamilies (Bordin et al. 2023). Expert analysis of the nonmatching human proteins revealed 25 novel superfamilies. Overall, the 130,000 predicted structures increased the number of "known folds" (i.e., compact 3D structural scaffolds typically shared between many distantly related proteins by more than 30%). An impressive tool for viewing protein structure space, which incidentally relies on *FoldSeek* (van Kempen et al. 2024) for comparing structures (below), has recently been made available by the developers of *ESMFold* (Lin et al. 2023). The confluence of two tools on the opposite side of the spectrum of data analysis, namely, on the side of great detail, the CATH-based analysis having experts visually compare the similarities and differences in detail for about a thousand proteins and on the side of large numbers the ESMFold-based threshold-driven automatic comparisons of 700 million proteins. The resulting message is so strong because it originates from both ends (detailed AlphaFold2/CATH and coarse-grained ESMFold perspective): The most important gain from 3D structure predictions is in refining known superfamilies, in linking previously isolated proteins to known groups, and providing some evidence for where to hunt for new "folds."

New Solutions Boost the Power of *AlphaFold2*

Methods in computational and experimental biology that benefit from the results of *AlphaFold2* are mushrooming. In fact, just 2 years after the publication of *AlphaFold2*, their number and diversity are already beyond the scope of this perspective. Nevertheless, we want to highlight what we consider possibly the most important impact from accurate 3D structure predictions of the day, namely, *Foldseek* (van Kempen et al. 2024). For three decades, the most important criterion for the development of methods comparing protein 3D structures was accuracy (Taylor and Orengo 1989; Kolodny et al. 2005): The number of known 3D structures added at any time point were so small that computing resources were only of minor concern. With *AlphaFoldDB* (Tunyasuvunakool et al. 2021) and *ESMFold* (Lin et al. 2022) making hundreds of millions 3D structure predictions available, the field needed another revolution: Reliable 3D structure comparisons are three orders of magnitude faster than existing tools to make the gigantic amount of new data searchable and thereby useable. *FoldSeek* found a genius solution toward this end by mapping 3D structure through a so-called vector quantised-variational autoencoder (VQ-VAE) (van den Oord et al. 2017) onto an alphabet of 20 different letters. The resulting 3Di "states" can be imagined as configurations in the backbone angles most informative for known protein structures (possibly reminding more experienced scholars of the informative 3D motifs used early on to predict structure; Bystroff and Baker 1998). The 3Di states are described by 20 letters that can be used to tap into the amazing accelerations realized for sequence comparisons through the blazingly fast *MMseqs2* (Mirdita et al. 2019). This engineering marvel enables *Foldseek* to reliably compare 3D structures (predictions or observations) at the speed of the faster ever reliable sequence comparisons. Just like *AlphaFold2*, *Foldseek* is a solution that would have been impossible 7 years ago, and it could never have been as useful as after the publication of *AlphaFoldDB* (Tunyasuvunakool et al. 2021). No matter how big this story in itself may be, it seeds an even bigger rift between past and future: Ultimately, *Foldseek* is the beginning of the end for sequence comparisons as they have been perfected for more than 42 years (Smith and Waterman 1981). Foldseek substitutes 1D sequence comparisons of proteins by 3D structure comparisons (although technically projected onto 1D sequences). Comparisons using 3D predic-

tions already outperform sequence-based alignment methods in many ways (Heinzinger et al. 2022; Schütze et al. 2022; van Kempen et al. 2024), even for relatively inaccurate 3D prediction methods (Weissenow et al. 2022b).

The term *AlphaFoldology* has been used half-jokingly by colleagues to capture some of the results of the amount of change brought about by the breakthrough *AlphaFold2*: From here on, all that is needed in structure prediction is to understand what we can and cannot do with *AlphaFold2*, to actually spread and dive into the word, reason, or discourse signifying the Greek word λόγος. Clearly, this has described many activities over the last 2 years. Another flurry of activity has been dedicated toward inventing tools that provide simpler access to Alpha-Fold2-level predictions, such as *ColabFold*, or that render the results from *AlphaFold2* even more useful, such as *Foldseek*. Undoubtedly, we will witness many more improvements.

The idea that chaining the tools AlphaFold2/ColabFold-Foldseek will replace traditional ways to generate MSAs leads to an interesting circle: AlphaFold2 is so successful because it uses large MSAs. How can we maintain the success when removing the foundation? Today, we may or may not quite get there yet. However, thanks to pLMs, tomorrow we might easily accomplish the seemingly impossible.

PROTEIN LANGUAGE MODELS (pLMs) TRIGGER PARADIGM SHIFTS

Protein Language Models (pLMs): Learning the Language of Life

Advances in natural language processing (NLP) spawned pLMs (Alley et al. 2019; Bepler and Berger 2019, 2021; Heinzinger et al. 2019; Rives et al. 2021; Elnaggar et al. 2022). pLMs leverage protein sequence databases that have outgrown computers for 28 years; they require no annotation (*label* in AI-jargon) except for the amino acid sequence. Through transfer learning, pLMs bridge the sequence-annotation gap (Ofran et al. 2005) after *AlphaFold2* has substantially quenched the sequence–structure gap (Porta-Pardo et al. 2022) (i.e., the difference between proteins of known sequence and known structure) (Rost and Sander 1996). Explicitly, pLMs capture sequential patterns in the input by either predicting the next residue (amino acid; earlier pLMs) or by recovering masked-out residues from those surrounding it (later pLMs, typically 15%). Thereby, language models (LMs) from NLP implicitly learn the grammar of written languages. Similarly, pLMs implicitly learn aspects of the language of life as written in proteins (Fig. 2; Step 1). The information learned by such pLMs (e.g., by inputting a protein sequence into the network and construct-

ing vectors from the values representing the last hidden layers of the pLM) yields a representation of protein sequences referred to as embeddings (Fig. 2). This allows to transfer features learned by the pLM to any downstream (prediction) task requiring numerical protein representations (i.e., *transfer learning*) (Fig. 2; Step 2). In fact, the so-called *EvoFormer* in *AlphaFold2* that extracts information from an MSA is a modified version of the pLM-based solution learning directly from MSAs, namely, the *MSA-Transformer* (Rao et al. 2021). However, instead of pretraining on large,

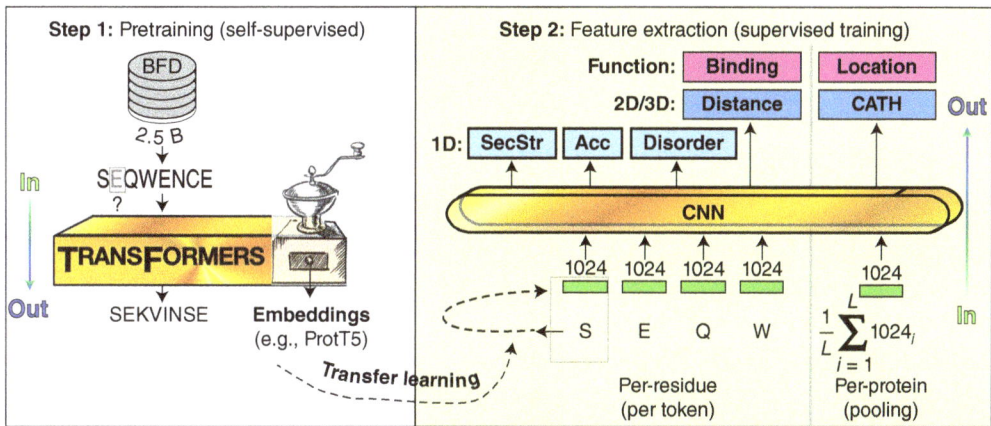

Figure 2. Embeddings from protein language model (pLM) for transfer learning. The sketch illustrates generic pLMs. *Step 1 (left)*: pictures the self-supervised learning of the context of protein sequences from training on very large sequence databases (e.g., *ProtT5* [Elnaggar et al. 2021] required *BFD* [Steinegger et al. 2019] with 2.5 billion sequences). Large transformer models (*ProtT5* has 3 billion connections, *Ankh* 1.7b, and ESM2 15b) inputting and outputting the same sequences, learn the *grammar of life* (i.e., implicitly extract the rules underlying the generation of known protein sequences). This extracted information is contained in the hidden layers of the transformers, which can be extracted in the form of vector representations, so-called embeddings. In *Step 2 (right)*: the embeddings are used as exclusive input for subsequent artificial intelligence (AI) trained in supervised manner on different tasks (e.g., per-residue prediction of secondary structure, accessibility, disorder, binding residues, or per-protein prediction of subcellular location or *CATH* numbers). *Step 2* is considered *transfer learning* because the *grammar* implicitly extracted from sequences in *Step 1* is transferred to increase the efficiency and success of *Step 2*. The number of units of the last hidden layers differs between pLMs (e.g., *ProtT5*) describes each residue (amino acid position) by a vector of 1024 dimensions. Therefore, the subsequent prediction method (here assumed to be a convolutional neural network, CNN) will use 1024 input units for each residue (i.e., L*1024 for a protein with L residues). The simplest way to condense this information for problems that require per-protein rather than per-residue predictions (e.g., per-residue: three-dimensional [3D] structure coordinate, per-protein: structural class such as CATH [Sillitoe et al. 2021; Nallapareddy et al. 2023]) is to average over each dimension). Although this resembles the concept of amino acid composition, which is well known to be informative of protein function, it is a priori not evident that such a crude average carries any meaningful description of protein features. In addition to per-residue (raw embeddings learned by pLM) and per-protein (pooling/averaging) embeddings, transformers also learn so-called *Attention Heads* (e.g., 732 for ProtT5) reflecting the importance (attention) from the entire protein onto a particular residue i. The latter is crucial when using embeddings to predict 2D and 3D structures (Weissenow et al. 2022a,b).

unlabeled data and transferring the resulting knowledge, the *EvoFormer* is only trained on the orders of magnitude smaller but labeled structure-prediction data set. Yet, the task of reconstructing masked amino acids from unmasked context, which is usually used to pretrain pLMs, was also used as part of *AlphaFold2* training.

Generic pLMs benefit from the transformer technology (Vaswani et al. 2017), which uses large neural networks to recover masked-out amino acids from the context of the entire input sequence. Several aspects seem crucial after 3 years of experience with the tool. Firstly, we need to begin with a very large sequence database (Elnaggar et al. 2022). UniProt (The UniProt Consortium 2021), with more than 220 million sequences, clearly outperforms Swiss-Prot (curated fraction of UniProt) and for some aspects the 10-fold larger BFD (Steinegger et al. 2019) might be needed. This implies that directly building generic pLMs from minute subsets such as *Homo sapiens* or eukaryotes, at this point, constitute a rather bad idea unless such specializations are used to refine existing models. Secondly, solving the task of optimally recovering the sequence from the masked version appears another ill-advised objective (Heinzinger et al. 2019; Elnaggar et al. 2022). After all, the number of pretraining tokens outweighs the number of model parameters usually by orders of magnitude, which prevents the model from learning a perfect reconstruction but instead forces the model to learn some compression of the input. This, in fact, implies that training pLMs is a task relatively free of the typical challenge for AI, namely, over-training or overfitting. Instead, pLMs are optimized to detect, combine, and compress reoccurring patterns in the input, which is an optimal starting point for a variety of transfer learning tasks (Fig. 2B), and this objective cannot easily be summarized by any reasonable single number (Heinzinger et al. 2019; Elnaggar et al. 2022, 2023). The emphasis here is on *reasonable*: while there are many nonsense averages, a solution successfully processing different prediction tasks remains wanted. This strategy optimizes the generic usefulness of embeddings from pLMs, thereby also considering the aspect of energy consumption: Training pLMs is extremely resource-inten-

sive (Heinzinger et al. 2019; Elnaggar et al. 2022, 2023; Lin et al. 2023), with the latest versions even needing the new Google TPUs (Elnaggar et al. 2023). This investment appears justified if and only if the resulting pLMs are sufficiently generic to serve for a large diversity of transfer learning solutions.

Successful transfer solutions exist for diverse aspects of protein prediction (Fig. 2) ranging from 3D structure (Rao et al. 2020; Bhattacharya et al. 2021; Chowdhury et al. 2022; Wang et al. 2022; Weissenow et al. 2022a,b; Wu et al. 2022; Lin et al. 2023), transmembrane regions (Bernhofer and Rost 2022; Hallgren et al. 2022), and intrinsically disordered regions (IDR/IDP) (Ilzhöefer et al. 2022) to various aspects of function (Littmann et al. 2021a; Stärk et al. 2021; Villegas-Morcillo et al. 2021; Bileschi et al. 2022; Heinzinger et al. 2022; Nallapareddy et al. 2023). Distance in embedding space correlates more with protein function than with sequence similarity (Littmann et al. 2021a) and can help with clustering proteins into families (Littmann et al. 2021a; Bileschi et al. 2022; Heinzinger et al. 2022).

Embedding-Based Outperform MSA-Based Predictions

Using embeddings from pLMs instead of EI from MSAs simplifies and speeds up protein prediction. For several tasks, including the prediction of secondary structure (Elnaggar et al. 2022, 2023), transmembrane helices and strands (Bernhofer and Rost 2022; Hallgren et al. 2022), signal peptides (Teufel et al. 2022), subcellular location (Stärk et al. 2021), or binding residues (Littmann et al. 2021b). The latter might be the most impressive example for the concept of transfer learning: the reliable experimental data about binding continues to be so sparse that machine learning cannot easily manage the complexity of the problem. The result is that solutions require models with minimal input information (e.g., neural networks with fewer than 20 input units). Using embeddings from pLMs allows for scaling up two orders of magnitude, simply because the embeddings are extremely informative (Littmann et al. 2021b).

Cite this article as *Cold Spring Harb Perspect Biol* doi: 10.1101/cshperspect.a041458

Despite substantial effort, 2 years after the first introduction of *AlphaFold2*, no method outperforms this remarkable solution despite ample reengineering attempts. What about embedding-based methods? In fact, the first pLMs have not even been able to reach the level of performance of the previous state-of-the-art (SOTA) reached by, for example, *AlphaFold1* (Senior et al. 2020) or *Raptor-X* (Wang et al. 2016); instead it required the more advanced transformers to reach parity (Weissenow et al. 2022a). However, even the better pLM-based methods, such as *ESMFold* (Lin et al. 2022), still remain substantially below *AlphaFold2* (and below other solutions reengineering that solution such as *RoseTTAFold*; Baek et al. 2021). For the time being, it remains unclear whether or not future pLMs might leap above the *AlphaFold2* mark.

If you wanted to apply pLMs to your prediction task, how to know whether it will work? Although we have no comprehensive answer, a few empirical observations might help. (1) Assume that you try to predict a feature with too little reliable data for the complexity of the task. The odds are good that embeddings might help (e.g., prediction of binding residues) (Littmann et al. 2021b), unless existing methods are extremely complex and optimized for this particular task. For instance, impressively complex solutions predict intrinsically disordered regions (Del Conte et al. 2023). Similarly, transmembrane helices, signal peptides, and subcellular location are predicted by well-tuned methods. In all of these cases, however, there are pLM-based solutions that reach or outperform the SOTA after some additional adjustments (Stärk et al. 2021; Bernhofer and Rost 2022; Hallgren et al. 2022; Ilzhöefer et al. 2022; Teufel et al. 2022). Predicting the effect of sequence variation is an example for lack of data for which advanced MSA-based methods still tend to have the upper hand. (2) Conversely, for tasks for which ample data exist (e.g., the prediction of protein inter-residue distances [2D structure] or actual 3D structure), solutions using very informative MSAs clearly dominate (although when inputting single sequence rather than MSAs AlphaFold2 is outperformed by pLM-based solutions; Lin et al. 2022; Weissenow et al. 2022b; Wu et al.

2022). However, the same is not true for the much simpler secondary structure prediction for which rich, informative MSAs also yield much better results, but for which pLM-based predictions appear to approach the ultimate possible. The reason why 1D secondary structure and 2D interresidue distances or 3D coordinates behave so differently remains unknown. In lieu of knowledge, we might speculate two possible answers: (1) saturation effect: secondary structure prediction might be closer to the top level of performance possible, namely, the "experimental error" than 3D prediction, or (2) simplicity: possibly pLMs condense less information than MSAs and advanced modules in AlphaFold2 might turn any additional information into better predictions.

Protein-Specific Rather than Family-Averaged Predictions

One technical aspect of pLM-based prediction methods is that they do not need MSAs. This has two advantages. Firstly, with growing databases MSA generation requires resources. Although *MMseqs2* (Mirdita et al. 2019) as used for *ColabFold* (Mirdita et al. 2022) is now so fast that this hardly matters in everyday applications (Bernhofer et al. 2021). Secondly, for some proteins, such as intrinsically disordered proteins (Dunker et al. 2013) or the *dark proteome* (Perdigão et al. 2015) MSA generation becomes problematic. Possibly more substantial, however, is another advantage: pLM-based methods enable protein-specific rather than family-averaged predictions. For instance, this permits to predict the effects of sequence variation upon function (Meier et al. 2021; Marquet et al. 2022; Dunham et al. 2023). Unfortunately, solutions predicting the effects upon structure appear to be largely confined to signals captured by the transformers rather than by the differential structure prediction (Weissenow et al. 2022a, b). Even a feature as intricately linked to EI as the "conservation" within a family can be predicted surprisingly well by rather simplistic convolution neural networks (Marquet et al. 2022). Does this imply that pLMs have captured evolutionary information? Although we have collected ample indirect evidence pointing to an affirmative answer (K Erckert and B Rost, unpubl.), we continue to doubt this to be the case. One reason for this is that pLMs are apparently extremely sensitive to differential data sets: A difference of less than an order of magnitude between the least (cysteine) and most frequent amino acid (leucine) leads to a substantially poorer resolution of cysteine. In contrast, the "dilution" of family relations is at least five orders of magnitude smaller: large families have $<10^4$ members, BFD holds $>10^9$ proteins (i.e., the difference between the number of related (same family) and unrelated (different family) sequences any protein will pick up exceeds 10^5. Although not impossible, this is unlikely to be picked up by today's pLMs. If so, why do they capture information relevant to predict family conservation? Most likely because the grammar of the language of life as written in proteins is coined by the same constraints written into the conservation profiles reflected by MSAs. In other words, both pLMs and MSAs capture aspects of the grammar, one due to biophysical constraints imprinted into sequences, the other due to constraints regulating what is observed —in MSAs—and what is not.

Beginning of the End for Alignment Methods?

Leveraging the sparse experimental annotations available 70 years ago, methods comparing protein sequences (Schwartz and Dayhoff 1978; Smith and Waterman 1981) have arguably been at the center of development for the field of computational biology since its existence. Although protein sequence databases have been growing faster than the speed at which computers can cope with this wealth of data since the mid-1990s, blazingly fast solutions such as *MMseqs2* (Mirdita et al. 2019) have succeeded in staying on top of the seemingly lost challenge through finding shortcuts with acceptable performance loss for most users. In fact, *MMseqs2* is even able to cope with challenges amounting to *Gargantuan* tasks such as the more than 3.1 quintillion ($2.5 \times 1.25 \times 10^{18}$) pair comparisons required for an all-on-all of BFD (Steinegger et al. 2019). However, except for requiring resources, sequence comparisons have been limited by the simple assumption that the alignment at positions i and j are statistically independent of each other. It continues to be stunning how successful an entire field can become even when built upon such a blatantly incorrect assumption. The only approach surmounting this problem, the Genetic Algorithm-based T-Coffee (Notredame et al. 2000), was already too slow for comparing to entire databases 20 years ago. The advent of pLMs offers another stab at this problem: if we could replace sequence comparisons by the generalized sequences generated by embeddings, we might be able to capture correlations between residues i and j. In its simplest implementation of per-protein comparisons, this approach indeed works very well to predict GeneOntology (GO) numbers (Littmann et al. 2021a), Pfam families (Bileschi et al. 2022), and CATH numbers (Heinzinger et al. 2022; Nallapareddy et al. 2023), ultimately, because similarity in embedding space is more informative for inferring similarity in function than the similarity in sequence space (Littmann et al. 2021a). Existing embedding-based protein comparisons also benefit from immense speed: by describing the protein as one average number (e.g., with 1024 dimensions for ProtT5), they project the task of comparison to a simple vector product. To picture the power of embeddings, just imagine you wanted to compare two proteins based on their 20D vectors of amino acid composition: clearly not sufficient to distinguish between two proteins with similar GO numbers from among hundreds of thousands! The devil in the details is that most proteins have more than one domain, and as many have three or more than that (Liu and Rost 2002). As long as we use a per-protein average embedding, we will not succeed in capturing domain similarities unless we base the comparison on domains such as those described by Pfam and CATH (Schütze et al. 2022). When comparing multidomain full-length proteins, we need to refine by actually comparing k-mers of per-residue embeddings between pairs of proteins (Schütze et al. 2022). Although this is both possible and successful, it comes with an overhead in runtime, and for the time being the costs seem not to justify the gain (Schütze et al. 2022). This result brings

Cite this article as *Cold Spring Harb Perspect Biol* doi: 10.1101/cshperspect.a041458

up another issue: so far, due to limitations in resources, generic pLMs have largely been built upon the software tools developed for NLP (i.e., the LMs). Ultimately, LMs will become available for handheld telephones. At that point, pLMs might retire traditional protein alignment methods.

CONCLUSIONS

In the first approximation, *AlphaFold2* solved the protein 3D structure prediction through a combination of advanced AI with advanced EI from MSAs generated from ever-growing protein sequence databases. Combining AI and EI as it peaked in *AlphaFold2* has been the winning card for almost three decades (Fig. 1). The *AlphaFold2* leap that could not have been realized more than 5 years ago has already spawned a flurry of important new developments. One of those, *Foldseek*, enables extremely reliable and fast comparison of millions of predicted or observed protein 3D structures. This tool immensely increases the value of databases with hundreds of millions of accurate 3D predictions from *AlphaFold2* (*AlphaFoldDB*) or *ESMFold*, and it might do more than any other solution to complement (or even replace at some point) the tools for protein sequence comparisons (alignment methods) upon which *Foldseek* is based. Another equally explosive and orthogonal development has been the introduction and rapid improvement of pLMs (Fig. 2). Through successful transfer learning (Fig. 2) that leverages information from unannotated protein sequences to help in learning from (even very little) experimental data, these tools begin to replace MSAs and open the age of making protein-specific trump family-averaged predictions. The combination of those two revolutions from successful 3D prediction and pLMs begins to spawn successful protein design beyond what was possible before and will help fictive Dr. Elena to solve real-world problems that we are already facing today.

ACKNOWLEDGMENTS

Thanks primarily to Chris Dallago (NVIDIA), Martin Steinegger (Seoul National University), and Christine Orengo (UCL) for invaluable collaborations that have strongly supported our work, and to Ahmed Elnaggar (TUM) for his immense panache in helping to push the development of protein language models in our group at TUM. M.H. and B.R. were supported by the Bavarian Ministry of Education through funding to the TUM, by a grant from the Alexander von Humboldt Foundation through the German Ministry for Research and Education (BMBF: Bundesministerium für Bildung und Forschung), and by a grant from Deutsche Forschungsgemeinschaft (DFG-GZ: RO1320/4-1). Last but not least, thanks to all those who maintain public sequence databases, in particular Alex Bateman (UniProt, EBI Hinxton), Johannes Söding (MPI Göttingen), Martin Steinegger (Seoul National University) and their crews, and to all experimentalists who enabled this analysis by making their data publicly available.

REFERENCES

Akdel M, Pires DEV, Pardo EP, Jänes J, Zalevsky AO, Mészáros B, Bryant P, Good LL, Laskowski RA, Pozzati G, et al. 2022. A structural biology community assessment of AlphaFold2 applications. *Nat Struct Mol Biol* **29:** 1056–1067. doi:10.1038/s41594-022-00849-w

Alley EC, Khimulya G, Biswas S, AlQuraishi M, Church GM. 2019. Unified rational protein engineering with sequence-based deep representation learning. *Nat Methods* **16:** 1315–1322. doi:10.1038/s41592-019-0598-1

Altschul SF, Madden TL, Schaeffer AA, Zhang J, Zhang Z, Miller W, Lipman DJ. 1997. Gapped blast and PSI-blast: a new generation of protein database search programs. *Nucleic Acids Res* **25:** 3389–3402. doi:10.1093/nar/25.17.3389

Andersen CAF, Palmer AG, Brunak S, Rost B. 2002. Continuum secondary structure captures protein flexibility. *Structure* **10:** 175–184. doi:10.1016/s0969-2126(02)00700-1

Andreeva A, Kulesha E, Gough J, Murzin AG. 2020. The SCOP database in 2020: expanded classification of representative family and superfamily domains of known protein structures. *Nucleic Acids Res* **48:** D376–D382. doi:10.1093/nar/gkz1064

Baek M, DiMaio F, Anishchenko I, Dauparas J, Ovchinnikov S, Lee GR, Wang J, Cong Q, Kinch LN, Schaeffer RD, et al. 2021. Accurate prediction of protein structures and interactions using a three-track neural network. *Science* **373:** 871–876. doi:10.1126/science.abj8754

Baker D, Sali A. 2001. Protein structure prediction and structural genomics. *Science* **294:** 93–96. doi:10.1126/science.1065659

Balakrishnan S, Kamisetty H, Carbonell JG, Lee SI, Langmead CJ. 2011. Learning generative models for protein fold families. *Proteins* **79:** 1061–1078. doi:10.1002/prot.22934

Bepler T, Berger B. 2019. Learning protein sequence embeddings using information from structure. arXiv doi:10.48550/arXiv.1902.08661

Bepler T, Berger B. 2021. Learning the protein language: evolution, structure, and function. *Cell Syst* **12**: 654–669.e3. doi:10.1016/j.cels.2021.05.017

Bernhofer M, Rost B. 2022. TMbed: transmembrane proteins predicted through Language Model embeddings. *BMC Bioinformatics* **23**: 326. doi:10.1186/s12859-022-04873-x

Bernhofer M, Dallago C, Karl T, Satagopam V, Heinzinger M, Littmann M, Olenyi T, Qiu J, Schoetze K, Yachdav G, et al. 2021. PredictProtein—predicting protein structure and function for 29 years. *Nucleic Acids Res* **49**: W535–W540. doi:10.1093/nar/gkab354

Bhattacharya N, Thomas N, Rao R, Dauparas J, Koo PK, Baker D, Song YS, Ovchinnikov S. 2021. Interpreting potts and transformer protein models through the lens of simplified attention. *Biocomputing* **2022**: 34–45. doi:10.1142/9789811250477_0004

Biasini M, Bienert S, Waterhouse A, Arnold K, Studer G, Schmidt T, Kiefer F, Gallo Cassarino T, Bertoni M, Bordoli L, et al. 2014. SWISS-MODEL: modelling protein tertiary and quaternary structure using evolutionary information. *Nucleic Acids Res* **42**: W252–W258. doi:10.1093/nar/gku340

Bileschi ML, Belanger D, Bryant DH, Sanderson T, Carter B, Sculley D, Bateman A, DePristo MA, Colwell LJ. 2022. Using deep learning to annotate the protein universe. *Nat Biotechnol* **40**: 932–937. doi:10.1038/s41587-021-01179-w

Bohr H, Bohr J, Brunak S, Cotterill RMJ, Lautrup B, Nørskov L, Olsen OH, Petersen SB. 1988. Protein secondary structure and homology by neural networks. The α-helices in rhodopsin. *FEBS Lett* **241**: 223–228. doi:10.1016/0014-5793(88)81066-4

Bonneau R, Tsai J, Ruczinski I, Chivian D, Rohl C, Strauss CE, Baker D. 2001. Rosetta in CASP4: progress in ab initio protein structure prediction. *Proteins* **45**: 119–126. doi:10.1002/prot.1170

Bordin N, Sillitoe I, Nallapareddy V, Rauer C, Lam SD, Waman VP, Sen N, Heinzinger M, Littmann M, Kim S, et al. 2023. Alphafold2 reveals commonalities and novelties in protein structure space for 21 model organisms. *Commun Biol* **6**: 160. doi:10.1038/s42003-023-04488-9

Bryant P, Pozzati G, Zhu W, Shenoy A, Kundrotas P, Elofsson A. 2022. Predicting the structure of large protein complexes using AlphaFold and Monte Carlo tree search. *Nat Commun* **13**: 6028. doi:10.1038/s41467-022-33729-4

Burke DF, Bryant P, Barrio-Hernandez I, Memon D, Pozzati G, Shenoy A, Zhu W, Dunham AS, Albanese P, Keller A, et al. 2023. Towards a structurally resolved human protein interaction network. *Nat Struct Mol Biol* **30**: 216–225. doi:10.1038/s41594-022-00910-8

Burley SK, Bhikadiya C, Bi C, Bittrich S, Chao H, Chen L, Craig PA, Crichlow GV, Dalenberg K, Duarte JM, et al. 2023. RCSB protein Data Bank (RCSB.org): delivery of experimentally-determined PDB structures alongside one million computed structure models of proteins from artificial intelligence/machine learning. *Nucleic Acids Res* **51**: D488–D508. doi:10.1093/nar/gkac1077

Bystroff C, Baker D. 1998. Prediction of local structure in proteins using a library of sequence-structure motifs. *J Mol Biol* **281**: 565–577. doi:10.1006/jmbi.1998.1943

Chowdhury R, Bouatta N, Biswas S, Floristean C, Kharkar A, Roy K, Rochereau C, Ahdritz G, Zhang J, Church GM, et al. 2022. Single-sequence protein structure prediction using a language model and deep learning. *Nat Biotechnol* **40**: 1617–1623. doi:10.1038/s41587-022-01432-w

Del Conte A, Bouhraoua A, Mehdiabadi M, Clementel D, Monzon AM, CAID predictors, Tosatto SCE, Piovesan D. 2023. CAID prediction portal: a comprehensive service for predicting intrinsic disorder and binding regions in proteins. *Nucleic Acids Res* **51**: W62–W69. doi:10.1093/nar/gkad430

Dunham AS, Beltrao P, AlQuraishi M. 2023. High-throughput deep learning variant effect prediction with sequence UNET. *Genome Biol* **24**: 110. doi:10.1101/2022.05.23.493038

Dunker AK, Babu MM, Barbash E, Blackledge M, Bondos SE, Dosztányi Z, Dyson HJ, Forman-Kay JD, Fuxreiter M, Gsponer J, et al. 2013. What's in a name? Why these proteins are intrinsically disordered. *Intrinsically Disord Proteins* **1**: e24157. doi:10.4161/idp.24157

Elnaggar A, Heinzinger M, Dallago C, Rehawi G, Yu W, Jones L, Gibbs T, Feher T, Angerer C, Steinegger M, et al. 2022. Prottrans: toward understanding the language of life through self-supervised learning. *IEEE Trans Pattern Anal Mach Intell* **44**: 7112–7127. doi:10.1109/TPAMI.2021.3095381

Elnaggar A, Essam H, Salah-Eldin W, Mousafa W, Elkerdawy M, Rochereau C, Rost B. 2023. Ankh: optimized protein language model unlocks general-purpose modelling. bioRxiv doi:10.48550/arXiv.2301.06568

Evans R, O'Neill M, Pritzel A, Antropova N, Senior A, Green T, Žídek A, Bates R, Blackwell S, Yim J, et al. 2021. Protein complex prediction with AlphaFold-Multimer. bioRxiv doi:10.1101/2021.10.04.463034

Guo HB, Perminov A, Bekele S, Kedziora G, Farajollahi S, Varaljay V, Hinkle K, Molinero V, Meister K, Hung C, et al. 2022. Alphafold2 models indicate that protein sequence determines both structure and dynamics. *Sci Rep* **12**: 10696. doi:10.1038/s41598-022-14382-9

Hallgren J, Tsirigos KD, Pedersen MD, Almagro Armenteros JJ, Marcatili P, Nielsen H, Krogh A, Winther O. 2022. DeepTMHMM predicts α and β transmembrane proteins using deep neural networks. bioRxiv doi:10.1101/2022.04.08.487609

Hamp T, Rost B. 2015. More challenges for machine-learning protein interactions. *Bioinformatics* **31**: 1521–1525. doi:10.1093/bioinformatics/btu857

Heinzinger M, Elnaggar A, Wang Y, Dallago C, Nechaev D, Matthes F, Rost B. 2019. Modeling aspects of the language of life through transfer-learning protein sequences. *BMC Bioinformatics* **20**: 723. doi:10.1186/s12859-019-3220-8

Heinzinger M, Littmann M, Sillitoe I, Bordin N, Orengo C, Rost B. 2022. Contrastive learning on protein embeddings enlightens midnight zone. *NAR Genom Bioinform* **4**: lqac043. doi:10.1093/nargab/lqac043

Ilzhöefer D, Heinzinger M, Rost B. 2022. SETH predicts nuances of residue disorder from protein embeddings. *Front Bioinform* **2**: 1019597. doi:10.3389/fbinf.2022.1019597

Cite this article as *Cold Spring Harb Perspect Biol* doi: 10.1101/cshperspect.a041458

Johansson-Åkhe I, Wallner B. 2022. Improving peptide-protein docking with AlphaFold-multimer using forced sampling. *Front Bioinform* **2**: 959160. doi:10.3389/fbinf.2022.959160

Jones DT, Buchan DWA, Cozzetto D, Pontil M. 2012. PSI-COV: precise structural contact prediction using sparse inverse covariance estimation on large multiple sequence alignments. *Bioinformatics* **28**: 184–190. doi:10.1093/bioinformatics/btr638

Jumper J, Evans R, Pritzel A, Green T, Figurnov M, Ronneberger O, Tunyasuvunakool K, Bates R, Žídek A, Potapenko A, et al. 2021. Highly accurate protein structure prediction with AlphaFold. *Nature* **596**: 583–589. doi:10.1038/s41586-021-03819-2

Klausen MS, Jespersen MC, Nielsen H, Jensen KK, Jurtz VI, Sønderby CK, Sommer MOA, Winther O, Nielsen M, Petersen B, et al. 2019. NetSurfP-2.0: improved prediction of protein structural features by integrated deep learning. *Proteins* **87**: 520–527. doi:10.1002/prot.25674

Kolodny R, Koehl P, Levitt M. 2005. Comprehensive evaluation of protein structure alignment methods: scoring by geometric measures. *J Mol Biol* **346**: 1173–1188. doi:10.1016/j.jmb.2004.12.032

Kryshtafovych A, Fidelis K, Moult J. 2007. Progress from CASP6 to CASP7. *Proteins* **69** (Suppl 8): 194–207. doi:10.1002/prot.21769

Kryshtafovych A, Schwede T, Topf M, Fidelis K, Moult J. 2021. Critical assessment of methods of protein structure prediction (CASP)—round XIV. *Proteins* **89**: 1607–1617. doi:10.1002/prot.26237

Lapedes AS, Liu L, Stormo GD. 1999. Correlated mutations in models of protein sequences: phylogenetic and structural effects. In *Proceedings of the IMS/AMS International Conference on Statistics in Molecular Biology and Genetics*, pp. 236–256. Institute for Mathematical Statistics, Hayward, CA.

Laurents DV. 2022. Alphafold 2 and NMR spectroscopy: partners to understand protein structure, dynamics and function. *Front Mol Biosci* **9**: 906437. doi:10.3389/fmolb.2022.906437

Lin Z, Akin H, Rao R, Hie BL, Zhu Z, Lu W, dos Santos Costa A, Fazel-Zarandi M, Sercu T, Candido S, et al. 2022. Language models of protein sequences at the scale of evolution enable accurate structure prediction. bioRxiv doi:10.1101/2022.07.20.500902

Lin Z, Akin H, Rao R, Hie B, Zhu Z, Lu W, Smetanin N, Verkuil R, Kabeli O, Shmueli Y, et al. 2023. Evolutionary-scale prediction of atomic-level protein structure with a language model. *Science* **379**: 1123–1130. doi:10.1126/science.ade2574

Littmann M, Heinzinger M, Dallago C, Olenyi T, Rost B. 2021a. Embeddings from deep learning transfer GO annotations beyond homology. *Sci Rep* **11**: 1160. doi:10.1038/s41598-020-80786-0

Littmann M, Heinzinger M, Dallago C, Weissenow K, Rost B. 2021b. Protein embeddings and deep learning predict binding residues for various ligand classes. *Sci Rep* **11**: 23916. doi:10.1038/s41598-021-03431-4

Liu J, Rost B. 2002. Target space for structural genomics revisited. *Bioinformatics* **18**: 922–933. doi:10.1093/bioinformatics/18.7.922

Marks DS, Colwell LJ, Sheridan R, Hopf TA, Pagnani A, Zecchina R, Sander C. 2011. Protein 3D structure computed from evolutionary sequence variation. *PLoS ONE* **6**: e28766. doi:10.1371/journal.pone.0028766

Marquet C, Heinzinger M, Olenyi T, Dallago C, Erckert K, Bernhofer M, Nechaev D, Rost B. 2022. Embeddings from protein language models predict conservation and variant effects. *Hum Genet* **141**: 1629–1647. doi:10.1007/s00439-021-02411-y

Marx V. 2022. Method of the year: protein structure prediction. *Nat Methods* **19**: 5–10. doi:10.1038/s41592-021-01359-1

Meier J, Rao R, Verkuil R, Liu J, Sercu T, Rives A. 2021. Language models enable zero-shot prediction of the effects of mutations on protein function. bioRxiv doi:10.1101/2021.07.09.450648

Mikolov T, Chen K, Corrado G, Dean J. 2013. Efficient estimation of word representations in vector space. arXiv doi:10.48550/arXiv.1301.3781

Millán C, Keegan RM, Pereira J, Sammito MD, Simpkin AJ, McCoy AJ, Lupas AN, Hartmann MD, Rigden DJ, Read RJ. 2021. Assessing the utility of CASP14 models for molecular replacement. *Proteins* **89**: 1752–1769. doi:10.1002/prot.26214

Mirdita M, Steinegger M, Söding J. 2019. MMseqs2 desktop and local web server app for fast, interactive sequence searches. *Bioinformatics* **35**: 2856–2858. doi:10.1093/bioinformatics/bty1057

Mirdita M, Schütze K, Moriwaki Y, Heo L, Ovchinnikov S, Steinegger M. 2022. Colabfold: making protein folding accessible to all. *Nat Methods* **19**: 679–682. doi:10.1038/s41592-022-01488-1

Moult J, Pedersen JT, Judson R, Fidelis K. 1995. A large-scale experiment to assess protein structure prediction methods. *Proteins* **23**: ii–iv. doi:10.1002/prot.340230303

Moult J, Hubbard T, Bryant SH, Fidelis K, Pedersen JT. 1999. Critical assessment of methods of protein structure prediction (CASP): round III. *Proteins* **3**: 2–6. doi:10.1002/(SICI)1097-0134(1999)37:3+<2::AID-PROT2>3.0.CO;2-2

Moult J, Fidelis K, Kryshtafovych A, Rost B, Hubbard T, Tramontano A. 2007. Critical assessment of methods of protein structure prediction-round VII. *Proteins* **69** (Suppl 8): 9–10. doi:10.1002/prot.21767

Nallapareddy V, Bordin N, Sillitoe I, Heinzinger M, Littmann M, Waman VP, Sen N, Rost B, Orengo C. 2023. CATHe: detection of remote homologues for CATH superfamilies using embeddings from protein language models. *Bioinformatics* **39**: btad029. doi:10.1093/bioinformatics/btad029

Notredame C, Higgins DG, Heringa J. 2000. T-Coffee: a novel method for fast and accurate multiple sequence alignment. *J Mol Biol* **302**: 205–217. doi:10.1006/jmbi.2000.4042b

Ofran Y, Punta M, Schneider R, Rost B. 2005. Beyond annotation transfer by homology: novel protein-function prediction methods to assist drug discovery. *Drug Discov Today* **10**: 1475–1482. doi:10.1016/S1359-6446(05)03621-4

Park Y, Marcotte EM. 2012. Flaws in evaluation schemes for pair-input computational predictions. *Nat Methods* **9**: 1134–1136. doi:10.1038/nmeth.2259

Perdigão N, Heinrich J, Stolte C, Sabir KS, Buckley MJ, Tabor B, Signal B, Gloss BS, Hammang CJ, Rost B, et al. 2015.

Unexpected features of the dark proteome. *Proc Natl Acad Sci* 112: 15898–15903. doi:10.1073/pnas.1508380112

Peters ME, Neumann M, Iyyer M, Gardner M, Clark C, Lee K, Zettlemoyer L. 2018. Deep contextualized word representations. arXiv doi:10.48550/arXiv.1802.05365

Pieper U, Webb BM, Barkan DT, Schneidman-Duhovny D, Schlessinger A, Braberg H, Yang Z, Meng EC, Pettersen EF, Huang CC, et al. 2011. Modbase, a database of annotated comparative protein structure models, and associated resources. *Nucleic Acids Res* 39: D465–D474. doi:10.1093/nar/gkq1091

Porta-Pardo E, Ruiz-Serra V, Valentini S, Valencia A. 2022. The structural coverage of the human proteome before and after AlphaFold. *PLoS Comput Biol* 18: e1009818. doi:10.1371/journal.pcbi.1009818

Punta M, Rost B. 2005a. PROFcon: novel prediction of long-range contacts. *Bioinformatics* 21: 2960–2968. doi:10.1093/bioinformatics/bti454

Punta M, Rost B. 2005b. Protein folding rates estimated from contact predictions. *J Mol Biol* 348: 507–512. doi:10.1016/j.jmb.2005.02.068

Qian N, Sejnowski TJ. 1988. Predicting the secondary structure of globular proteins using neural network models. *J Mol Biol* 202: 865–884. doi:10.1016/0022-2836(88)90564-5

Rao R, Meier J, Sercu T, Ovchinnikov S, Rives A. 2020. Transformer protein language models are unsupervised structure learners. bioRxiv doi:10.1101/2020.12.15.422761

Rao RM, Liu J, Verkuil R, Meier J, Canny J, Abbeel P, Sercu T, Rives A. 2021. MSA transformer. In *Proceedings of the 38th International Conference on Machine Learning* (ed. Marina M, Tong Z), pp. 8844–8856. *PMLR* 139: 8844–8856.

Rives A, Meier J, Sercu T, Goyal S, Lin Z, Liu J, Guo D, Ott M, Zitnick CL, Ma J, et al. 2021. Biological structure and function emerge from scaling unsupervised learning to 250 million protein sequences. *Proc Natl Acad Sci* 118: e2016239118. doi:10.1073/pnas.2016239118

Rost B. 1993. *Neural networks and evolution—advanced prediction of protein secondary structure*. Departments of Physics and Astronomy, University of Heidelberg, Germany.

Rost B. 1996. PHD: predicting one-dimensional protein structure by profile based neural networks. *Methods Enzymol* 266: 525–539. doi:10.1016/s0076-6879(96)66033-9

Rost B. 2001. Protein secondary structure prediction continues to rise. *J Struct Biol* 134: 204–218. doi:10.1006/jsbi.2001.4336

Rost B, Sander C. 1992. Jury returns on structure prediction. *Nature* 360: 540. doi:10.1038/360540b0

Rost B, Sander C. 1993. Prediction of protein secondary structure at better than 70% accuracy. *J Mol Biol* 232: 584–599. doi:10.1006/jmbi.1993.1413

Rost B, Sander C. 1996. Bridging the protein sequence-structure gap by structure predictions. *Annu Rev Biophys Biomol Struct* 25: 113–136. doi:10.1146/annurev.bb.25.060196.000553

Rost B, Sander C, Schneider R. 1994. Redefining the goals of protein secondary structure prediction. *J Mol Biol* 235: 13–26. doi:10.1016/s0022-2836(05)80007-5

Rost B, Liu J, Nair R, Wrzeszczynski KO, Ofran Y. 2003. Automatic prediction of protein function. *Cell Mol Life Sci* 60: 2637–2650. doi:10.1007/s00018-003-3114-8

Schlessinger A, Punta M, Rost B. 2007. Natively unstructured regions in proteins identified from contact predictions. *Bioinformatics* 23: 2376–2384. doi:10.1093/bioinformatics/btm349

Schütze K, Heinzinger M, Steinegger M, Rost B. 2022. Nearest neighbor search on embeddings rapidly identifies distant protein relations. *Front Bioinform* 2: 1033775. doi:10.3389/fbinf.2022.1033775

Schwartz RM, Dayhoff MO. 1978. Origins of prokaryotes, eukaryotes, mitochondria, and chloroplasts. *Science* 199: 395–403. doi:10.1126/science.202030

Seemayer S, Gruber M, Söding J. 2014. CCMpred—fast and precise prediction of protein residue–residue contacts from correlated mutations. *Bioinformatics* 30: 3128–3130. doi:10.1093/bioinformatics/btu500

Senior AW, Evans R, Jumper J, Kirkpatrick J, Sifre L, Green T, Qin C, Žídek A, Nelson AWR, Bridgland A, et al. 2020. Improved protein structure prediction using potentials from deep learning. *Nature* 577: 706–710. doi:10.1038/s41586-019-1923-7

Sillitoe I, Bordin N, Dawson N, Waman VP, Ashford P, Scholes HM, Pang CSM, Woodridge L, Rauer C, Sen N, et al. 2021. CATH: increased structural coverage of functional space. *Nucleic Acids Res* 49: D266–D273. doi:10.1093/nar/gkaa1079

Smith TF, Waterman MS. 1981. Identification of common molecular subsequences. *J Mol Biol* 147: 195–197. doi:10.1016/0022-2836(81)90087-5

Stärk H, Dallago C, Heinzinger M, Rost B. 2021. Light attention predicts protein location from the language of life. *Bioinform Adv* 1: vbab035. doi:10.1093/bioadv/vbab035

Steinegger M, Mirdita M, Söding J. 2019. Protein-level assembly increases protein sequence recovery from metagenomic samples manyfold. *Nat Methods* 16: 603–606. doi:10.1038/s41592-019-0437-4

Taylor WR, Orengo CA. 1989. Protein structure alignment. *J Mol Biol* 208: 1–22. doi:10.1016/0022-2836(89)90084-3

Teufel F, Almagro Armenteros JJ, Johansen AR, Gíslason MH, Pihl SI, Tsirigos KD, Winther O, Brunak S, von Heijne G, Nielsen H. 2022. Signalp 6.0 predicts all five types of signal peptides using protein language models. *Nat Biotechnol* 40: 1023–1025. doi:10.1038/s41587-021-01156-3

The UniProt Consortium. 2021. Uniprot: the universal protein knowledgebase in 2021. *Nucleic Acids Res* 49: D480–D489. doi:10.1093/nar/gkaa1100

Thorn A. 2022. Artificial intelligence in the experimental determination and prediction of macromolecular structures. *Curr Opin Struct Biol* 74: 102368. doi:10.1016/j.sbi.2022.102368

Tsaban T, Varga JK, Avraham O, Ben-Aharon Z, Khramushin A, Schueler-Furman O. 2022. Harnessing protein folding neural networks for peptide-protein docking. *Nat Commun* 13: 176. doi:10.1038/s41467-021-27838-9

Tunyasuvunakool K, Adler J, Wu Z, Green T, Zielinski M, Žídek A, Bridgland A, Cowie A, Meyer C, Laydon A, et al. 2021. Highly accurate protein structure prediction for the human proteome. *Nature* 596: 590–596. doi:10.1038/s41586-021-03828-1 van den Oord A, Vinyals O, Kavukcuoglu K. 2017. Neural discrete representation learning. arXiv doi:10.48550/arXiv.1711.00937

van Kempen M, Kim SS, Tumescheit C, Mirdita M, Lee J, Gilchrist CLM, Söding J, Steinegger M. 2024. Fast and accurate protein structure search with Foldseek. *Nat Biotechnol* **42:** 243–246. doi:10.1038/s41587-023-01773-0

Vaswani A, Shazeer N, Parmar N, Uszkoreit J, Jones L, Gomez AN, Kaiser Ł, Polosukhin I. 2017. Attention is all you need. In *Proceedings of the 31st International Conference on Neural Information Processing Systems*, pp. 6000–6010. Curran, Long Beach, CA.

Villegas-Morcillo A, Makrodimitris S, van Ham R, Gomez AM, Sanchez V, Reinders MJT. 2021. Unsupervised protein embeddings outperform hand-crafted sequence and structure features at predicting molecular function. *Bioinformatics* **37:** 162–170. doi:10.1093/bioinformatics/btaa701

Wang S, Li W, Liu S, Xu J. 2016. RaptorX-Property: a web server for protein structure property prediction. *Nucleic Acids Res* **44:** W430–W435. doi:10.1093/nar/gkw306

Wang S, Sun S, Li Z, Zhang R, Xu J. 2017. Accurate de novo prediction of protein contact map by ultra-deep learning model. *PLOS Comput Biol* **13:** e1005324. doi:10.1371/journal.pcbi.1005324

Wang G, Fang X, Wu Z, Liu Y, Xue Y, Xiang Y, Yu D, Wang F, Ma Y. 2022. Helixfold: an efficient implementation of AlphaFold2 using PaddlePaddle. arXiv doi:10.48550/arXiv.2207.05477

Weigt M, White RA, Szurmant H, Hoch JA, Hwa T. 2009. Identification of direct residue contacts in protein-protein interaction by message passing. *Proc Natl Acad Sci* **106:** 67–72. doi:10.1073/pnas.0805923106

Weissenow K, Heinzinger M, Rost B. 2022a. Protein language model embeddings for fast, accurate, and alignment-free protein structure prediction. *Structure* **30:** 1169–1177.e4. doi:10.1016/j.str.2022.05.001

Weissenow K, Heinzinger M, Steinegger M, Rost B. 2022b. Ultra-fast protein structure prediction to capture effects of sequence variation in mutation movies. bioRxiv doi:10.1101/2022.11.14.516473

Wu R, Ding F, Wang R, Shen R, Zhang X, Luo S, Su C, Wu Z, Xie Q, Berger B, et al. 2022. High-resolution de novo structure prediction from primary sequence. bioRxiv doi:10.1101/2022.07.21.500999

Yang J, Anishchenko I, Park H, Peng Z, Ovchinnikov S, Baker D. 2020. Improved protein structure prediction using predicted interresidue orientations. *Proc Natl Acad Sci* **117:** 1496–1503. doi:10.1073/pnas.1914677117

Zvelebil MJ, Barton GJ, Taylor WR, Sternberg MJE. 1987. Prediction of protein secondary structure and active sites using the alignment of homologous sequences. *J Mol Biol* **195:** 957–961. doi:10.1016/0022-2836(87)90501-8

Building Representation Learning Models for Antibody Comprehension

Justin Barton, Aretas Gaspariunas, Jacob D. Galson, and Jinwoo Leem

Alchemab Therapeutics Ltd, London N1C 4AX, United Kingdom

Correspondence: jin@alchemab.com; jake@alchemab.com

Antibodies are versatile proteins with both the capacity to bind a broad range of targets and a proven track record as some of the most successful therapeutics. However, the development of novel antibody therapeutics is a lengthy and costly process. It is challenging to predict the functional and biophysical properties of antibodies from their amino acid sequence alone, requiring numerous experiments for full characterization. Machine learning, specifically deep representation learning, has emerged as a family of methods that can complement wet lab approaches and accelerate the overall discovery and engineering process. Here, we review advances in antibody sequence representation learning, and how this has improved antibody structure prediction and facilitated antibody optimization. We discuss challenges in the development and implementation of such models, such as the lack of publicly available, well-curated antibody function data and highlight opportunities for improvement. These and future advances in machine learning for antibody sequences have the potential to increase the success rate in developing new therapeutics, resulting in broader access to transformative medicines and improved patient outcomes.

B cells provide immune protection through the production of antibodies, which are soluble proteins that can bind almost any target molecule with high specificity. These proteins begin as B-cell receptors (BCRs) on the surfaces of B cells. Antigen recognition then activates the B cell to produce and secrete its BCR as antibodies (Fig. 1A). In humans, each BCR is made of two heavy-light chain pairs. At the distal protrusions of each chain, there are the variable domains, VH and VL, which recognize the antigen. The remainder of the antibody, or the constant region, is relatively conserved. Each variable domain has three polymorphic complementarity determining regions (CDRs) that largely form the BCR's binding site, the paratope (Fig. 1B). Much of a BCR's diversity is focused in the third CDR of the heavy chain (CDRH3). Figure 1C shows the sequence and structural alignment of certolizumab (pink), atezolizumab (blue), and ipilumab (orange), all of which bind different antigens. The structures of the three antibodies have good alignment, apart from the CDRH3; this alludes to the pivotal role of CDRH3 in antigen recognition.

To cover the huge breadth of antigens, B cells rely on several diversification mechanisms (Georgiou et al. 2014). First, B cells undergo V(D)J recombination to produce its BCR. From a vast pool of somatically encoded gene

Cite this article as *Cold Spring Harb Perspect Biol* doi: 10.1101/cshperspect.a041462

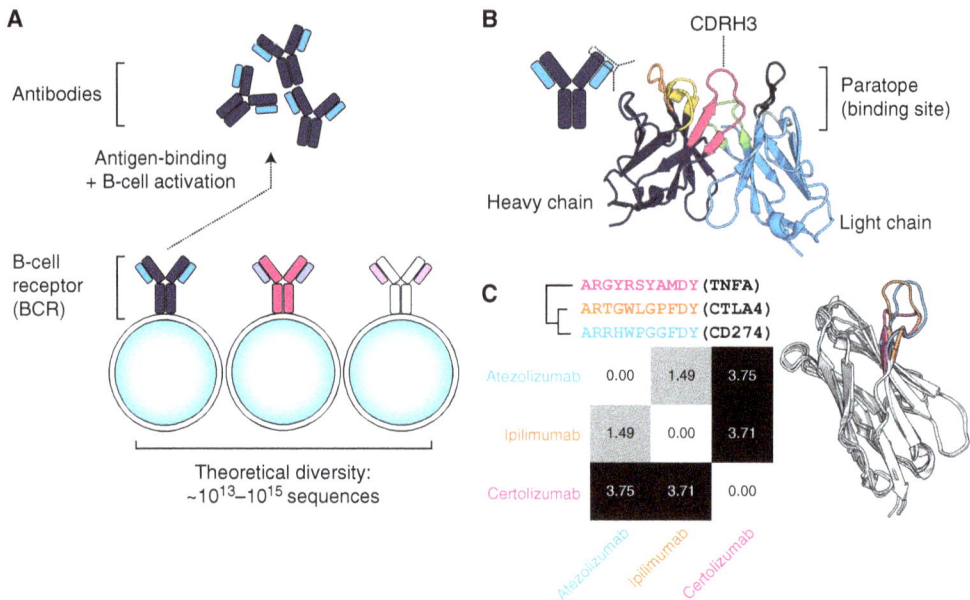

Figure 1. Overview of B-cell receptor ([BCR] antibody) structure. (*A*) BCRs on the surfaces of B cells are secreted as antibodies upon antigen recognition and subsequent B-cell activation. (*B*) BCRs and antibodies are composed of two heavy-light chain dimers. Within each chain are the CDRs: CDRH1 (yellow), CDRH2 (orange), CDRH3 (pink), CDRL1 (black), CDRL2 (gray), and CDRL3 (green). In most antibodies, the ensemble of the six CDRs largely form the binding site, or "paratope." (*C*) Despite the conserved fold, certolizumab (pink), atezolizumab (blue), and ipilumab (orange) bind different antigens, and this is reflected by variations in their CDRH3 sequence and structure. The heatmap shows the pairwise root-mean-square deviations (measured in angstroms) between CDRH3 loop structures of the three antibodies.

segments, one variable (V), one joining (J), and one diversity (D) gene segment are randomly shuffled and joined. Heavy chains recombine V, D, and J segments, while light chains only recombine V and J. The assembly of heavy and light chains introduces a further level of diversity, with some heavy chains being capable of forming multiple different heavy-light chain pairings (Jaffe et al. 2022). Once a B cell is activated following antigen recognition, it triggers somatic hypermutation, which introduces a series of mutations to the BCR sequence to improve selectivity and affinity.

Together, this leads to a comprehensive BCR repertoire with an estimated diversity of $\sim 10^{15}$ variants (Rees 2020). Understanding an individual's BCR repertoire has proven to be highly valuable for gaining new insights into disease biology (Bashford-Rogers et al. 2019; Galson et al. 2020; Park et al. 2022; Wang et al. 2022a; Yu et al. 2022),

building diagnostic tools (Konishi et al. 2019; Zaslavsky et al. 2022), and discovering novel therapeutics (Krawczyk et al. 2019, 2021). Collecting data at the scale of the BCR repertoire has been facilitated by the evolution of modern next-generation sequencing (NGS) platforms and software, which has increased throughput and quality, while reducing the costs of sequencing (Chaudhary and Wesemann 2018).

The most common approach of sequencing the BCR repertoire is "bulk sequencing" of the RNA encoding the BCR heavy chain; it is now possible to recover tens to hundreds of thousands of BCR heavy chain sequences per sample. A related approach is to computationally reconstruct BCR reads from transcriptomics data sets, using methods such as TRUST4 (Song et al. 2021). Both approaches come at the expense of losing the heavy-light chain pairing information for constructing the full BCR. Single-cell sequencing can provide the sequences of the paired BCR heavy and light chains, but it comes at the cost of throughput. For a more extensive overview on sequencing approaches, we refer the reader to Zheng et al. (2022). There are now several community efforts to collect the growing resource of publicly available BCR repertoire data from bulk and single-cell methods (Corrie et al. 2018; Kovaltsuk et al. 2018). For example, the Observed Antibody Space (OAS) has a catalog of 2.4 billion unpaired BCR sequences and 1.5 million paired BCRs, while the iReceptor database contains 232 million unpaired BCRs.

Acquiring function data at the scale of the BCR repertoire is difficult due to the costly nature of in vitro experimentation. However, the large amount of publicly available data and the underlying biological complexity suggest that antibody sequences are a model system for predictive analyses using machine learning techniques. Indeed,

machine learning has been applied to a wide range of tasks related to modeling antibody structure and function (Fig. 2), such as humanization (Prihoda et al. 2022), thermostability (Harmalkar et al. 2023; Nijkamp et al. 2023), and binding site (paratope) prediction (Liberis et al. 2018; Ambrosetti et al. 2020; Del Vecchio et al. 2021; Leem et al. 2022). For all these studies, the key ingredient behind their success is being able to represent antibody sequences in a numeric format that can act as input "features" for machine learning models. However, learning representations of antibody sequences is itself a major challenge; recent advancements in deep learning for natural language processing (NLP) have inspired a new generation of models.

In this work, we focus on the development and application of antibody sequence representation models, and comment on their impact in therapeutic antibody discovery. First, we provide

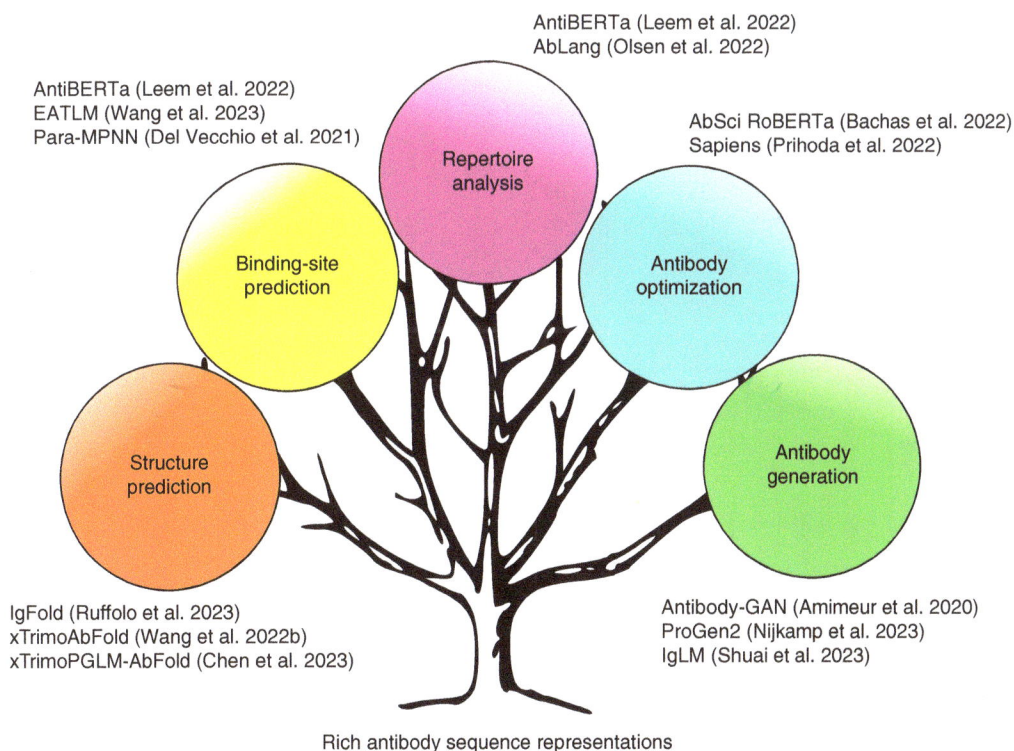

AntiBERTa (Leem et al. 2022)
AbLang (Olsen et al. 2022)

AntiBERTa (Leem et al. 2022)
EATLM (Wang et al. 2023)
Para-MPNN (Del Vecchio et al. 2021)

AbSci RoBERTa (Bachas et al. 2022)
Sapiens (Prihoda et al. 2022)

Repertoire analysis

Binding-site prediction

Antibody optimization

Structure prediction

Antibody generation

IgFold (Ruffolo et al. 2023)
xTrimoAbFold (Wang et al. 2022b)
xTrimoPGLM-AbFold (Chen et al. 2023)

Antibody-GAN (Amimeur et al. 2020)
ProGen2 (Nijkamp et al. 2023)
IgLM (Shuai et al. 2023)

Rich antibody sequence representations

Figure 2. Applications enabled by machine learning (ML)-enriched antibody sequence representations. ML has been deployed across many parts of the antibody discovery and engineering process. Each of these areas benefit from having a strong foundational model that can represent antibody sequences in a numeric format.

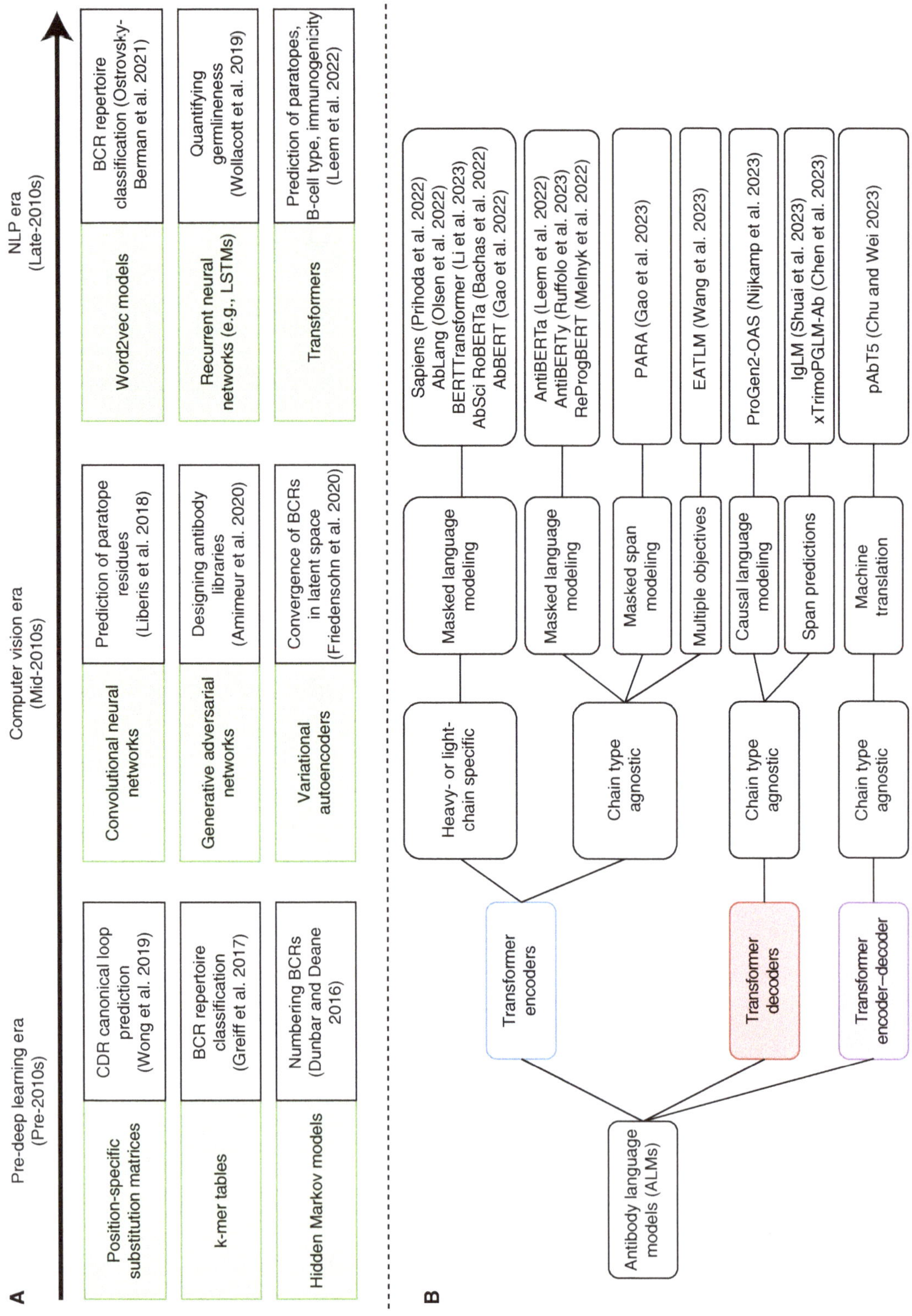

Figure 3. (*See following page for legend.*)

an overview of historical work on learning antibody sequence representations before the widespread adoption of NLP methods. We then describe the cross-pollination of NLP onto general protein representation learning, and how that has inspired training antibody-specific models. Next, we discuss the application of antibody-specific models in two areas: antibody structure prediction and antibody engineering. Finally, we provide a perspective on some of the challenges in the development of such models and provide our thoughts on the future use of antibody language models (ALMs) for antibody discovery.

EVOLUTION OF ANTIBODY SEQUENCE REPRESENTATION MODELS

Antibodies are an atypical subset of sequences in the general protein space. Owing to somatic hypermutation, antibodies can exhibit a great deal of length and sequence variation. This diversity is especially present within the CDRs at the antigen-binding interface. This is the converse of the general protein case, where protein–protein interaction interfaces are generally well conserved and other regions are not as strictly conserved (Esmaielbeiki et al. 2016). These fundamental differences suggest that specialized models may be required; indeed, there is evidence in both bioinformatics (Sippl 1990; Imrie et al. 2018) and NLP (Chalkidis et al. 2020; Xue et al. 2020; Gu et al. 2021) to suggest that domain-specific models can outperform general models in domain-specific tasks.

There is an extensive history of antibody sequence modeling approaches, and we notionally divide this into three "eras" (Fig. 3A). The first era we call the "pre-deep learning era," which comprises invariant, context-free representations such as Hidden Markov Models (HMMs) (Dunbar and Deane 2016), position-specific substitu-

tion matrix (PSSM) probabilities (Wong et al. 2019), physicochemical vectors such as Atchley factors (Townsend et al. 2016), and vectors of k-mer frequencies (Greiff et al. 2017; Weber et al. 2022). Many of these methods were introduced very early in the history of bioinformatics tools; for example, HMMs were used since the 1990s to predict membrane protein topology (Bystroff and Krogh 2008). However, they were only deployed for antibody-specific tasks in the latter 2000s with a growing volume of antibody sequence and structural data.

The second era of approaches is hallmarked by machine learning models from computer vision, such as convolutional neural networks (Liberis et al. 2018; Konishi et al. 2019; Mason et al. 2021), variational autoencoders (Friedensohn et al. 2020), and generative adversarial networks (Amimeur et al. 2020; Lim et al. 2022). In the present era, inspiration has been drawn from NLP, applying skip-gram models such as word2-vec (Chen et al. 2020; Ostrovsky-Berman et al. 2021), recurrent neural networks such as long short-term memory networks (Wollacott et al. 2019; Saka et al. 2021), or gated recurrent unit networks (Akbar et al. 2021). Remarkably, within the last 2 years alone, there has been an explosion of transformer neural networks for antibody sequence representation learning. We show a breakdown of these approaches according to the class of model architecture and the pretraining mechanism (Fig. 3B; Ruffolo et al. 2021; Bachas et al. 2022; Gao et al. 2022, 2023; Leem et al. 2022; Melnyk et al. 2022; Olsen et al. 2022; Prihoda et al. 2022; Shuai et al. 2023; Chen et al. 2023; Chu and Wei 2023; Nijkamp et al. 2023).

The transformer model (Vaswani et al. 2017) and its variants, such as bidirectional encoder representations from transformers (BERTs) (Devlin et al. 2018) and generative pretrained transformer 2 (GPT-2) (Radford et al. 2019),

---

Figure 3. Evolution of antibody sequence representation learning models. (*A*) We separate the evolution of representation learning approaches into three eras. Some approaches, despite being introduced in an "earlier" era, may still have been used in a later time point. For example, the position-specific substitution matrix (PSSM) is a classic representation learning model from bioinformatics techniques that were still being used in 2020, despite the use of convolutional neural networks in 2018. (*B*) Taxonomy of transformer architectures for building antibody language models (ALMs). (NLP) Natural language processing, (BCR) B-cell receptor.

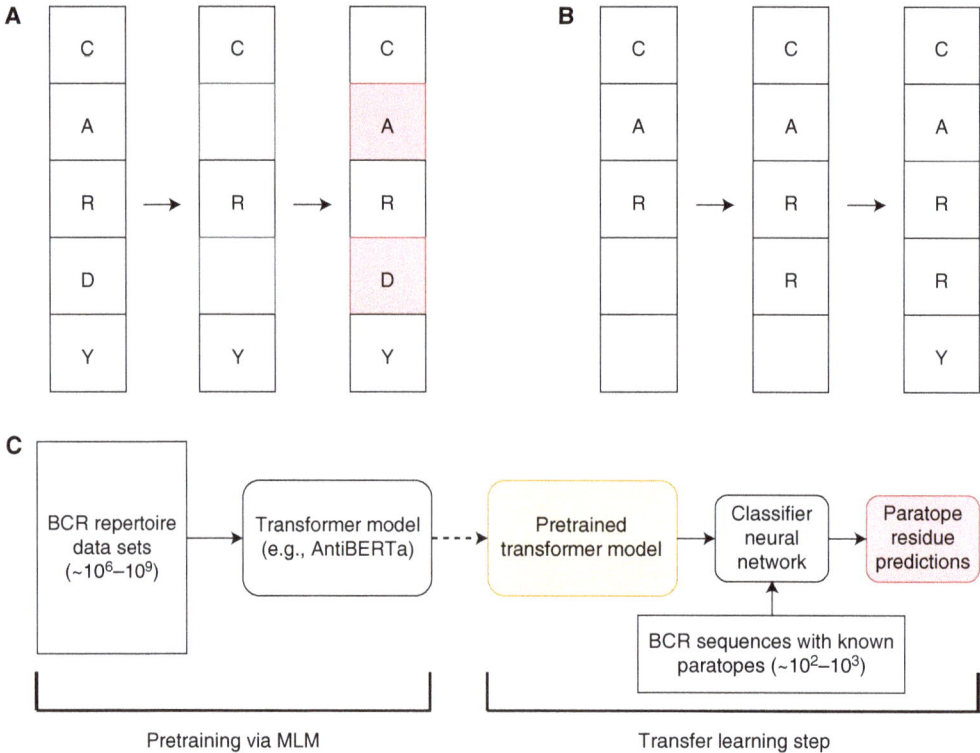

Figure 4. Pretraining and fine-tuning regime for language models. (*A*) Masked language modeling for proteins and antibodies involves randomly perturbing a subset of residues and reconstructing the perturbed positions. (*B*) Causal language modeling is a next residue prediction task where the model is given information only up to the position it needs to predict. (*C*) Two-step training regime for paratope prediction using transformer-based antibody language models (ALMs). First, a transformer model is pretrained on a self-supervised task, such as masked language modeling (MLM), with a very large antibody sequence data set. Next, in the transfer learning step, the pretrained transformer's outputs act as the input for a second neural network with a lower volume of labeled data. (BCR) B-cell receptor.

have radically shifted the paradigm in NLP. In brief, transformers use an attention mechanism to learn a contextualized representation, or embedding, for each word in an input sentence. These word embeddings can then be used for a wide range of applications, such as text classification and text generation.

Transformers are conceived in three different forms: an encoder–decoder model (e.g., for text summarization), an encoder-only form (e.g., for sentiment analysis), or as a decoder-only model (e.g., for sequence generation). These transformer-based "language models" (LMs) are first pretrained on a self-supervised task, such as masked language modeling (MLM) for BERT variants (Fig. 4A), or causal language modeling (CLM)

for GPT-2 variants (Fig. 4B). In a second step, the pretrained model is specialized to predict an outcome of interest (Devlin et al. 2018; Raffel et al. 2019). This can be done by using, for example, a single neural network layer that uses the pretrained model's embeddings as its features (Fig. 4C).

The main advantage of LMs and the two-step transfer learning approach is that they harness big volumes of unlabeled text data to develop the initial comprehension of the language. The pretrained LM is also generalizable and can be adapted for classification or regression tasks. To learn latent patterns in huge text corpora, transformer models are often set up with millions of learnable parameters; for instance, BERT-base contains 110 million parameters

(Devlin et al. 2018). By scaling up the number of parameters, in some cases to many billions of parameters, a transformer model can capture highly complex, nuanced patterns in a language. These large models can also thrive in "few-shot" learning scenarios (i.e., cases where there are shallow amounts of "labeled" data). All these factors have contributed to the huge popularity of transformers and their variants in building LMs (Lin et al. 2022).

The rapid evolution of transformer-based LMs prompted adaptations of transformers for building protein LMs (PLMs), starting with evolutionary scale modeling (ESM) (Rives et al. 2021), ProtTrans (Elnaggar et al. 2022), and a multitude of PLMs since. Much like the LMs developed for NLP applications, PLMs learn protein sequence representations from a large corpora of unlabeled protein sequences. Furthermore, PLMs have been designed with the aim to have a single model that can generalize across a mixture of downstream tasks for any protein family.

Sequence representations from PLMs have been applied to various types of antibody property prediction. Embeddings from ESM models have been used as input to regression and classification models predicting antibody solubility (Feng et al. 2022) and thermostability (Zainchkovskyy et al. 2022; Harmalkar et al. 2023). Estimated sequence likelihoods from ESM, UniRep, and ProGen2 models were tested for rank correlations with measures of thermostability and expression quality with mixed results (Harmalkar et al. 2023; Nijkamp et al. 2023).

Motivated by the differences between antibodies and other proteins, various ALMs have been developed in parallel using transformers. Since the first ALM described by Prihoda et al. (2022), 14 different transformer-based ALMs have been described in the literature in total, which is by far the most for a single protein family (Fig. 3B). The key difference between PLMs and ALMs is their underlying pretraining corpus. PLMs are typically pretrained on nonredundant splits of UniRef, covering a broad spectrum of protein families. In contrast, ALMs are exclusively pretrained on nonredundant collections of BCR repertoires, such as a clustered, nonredundant subset of the OAS database.

The most common strategy for constructing ALMs is using encoder-only models that are derivatives of BERT (Devlin et al. 2018); 10 of the 14 ALMs in the literature are based on this architecture (Fig. 3B). The underlying bidirectional attention mechanism of these encoder-only architectures make them ideal for learning representations of sequences. For example, models like AntiBERTa are pretrained on millions of BCR sequences from the OAS database using MLM. In this task, 15% of the amino acids in the BCR heavy or light chain are randomly perturbed, and the model is tasked with reconstructing the correct amino acids in their places (Fig. 4A). AntiBERTa then leverages its self-attention mechanism to understand the sequence context before and after the masked positions to predict the correct amino acids (Leem et al. 2022). Through pretraining, AntiBERTa learns to extract latent information from antibody sequences, and express this information in its output embeddings.

Similar to how LMs provide embeddings for each word in a sentence, ALMs compute embeddings for each amino acid in the broader amino acid sequence. In contrast to previous representation learning approaches for antibody sequences, ALM embeddings leverage sequence context. In other words, the representation for a single amino acid at a specific position will vary depending on the sequence upstream or downstream of that position.

Ultimately, these contextualized embeddings provide a valuable starting point for downstream modeling tasks and can yield highly accurate models (Fig. 4C). For instance, AntiBERTa has been used for predicting the paratopes of antibody sequences with high accuracy. Although there are only hundreds of antibody sequences with known paratope information, the bulk of the pattern recognition is handled during pretraining on millions of sequences. Effectively, pretraining makes paratope prediction a much more amenable task.

The remaining four ALMs are generative. ProGen2-OAS, IgLM, pAbT5, and xTrimoPGLM contain decoder layers, making them more suited for synthesizing novel antibody sequences in silico. For example, ProGen2-OAS is pretrained using a CLM objective, where amino acids are autoregressively predicted on a position-by-position basis, from the amino to carboxyl terminus (Fig. 4B). Like the encoder-only architectures described above, generative models also learn internal representations during CLM pretraining. However, due to their autoregressive nature, decoders use a unidirectional attention mechanism attending only to the portion of the sequence preceding each residue. As such, their representations can be less performant than bidirectional encoders in downstream prediction tasks (Devlin et al. 2018; Elnaggar et al. 2022). To circumvent these limitations, xTrimoPGLM-Ab has recently been proposed to leverage bidirectional attention while using a decoder architecture. This is done by first pretraining with MLM, then using an alternation between MLM and CLM (Chen et al. 2023). The representation capacities of xTrimoPGLM-Ab have only been tested on a limited number of tasks, although it has shown promising results.

MLM is the dominant pretraining strategy among ALMs, although there can be variations. For instance, Gao et al. (2023) employ higher masking rates than the traditional 15%, while Gao et al. (2022) only mask the CDR residues as opposed to the entire heavy chain. Generative models such as IgLM and xTrimoPGLM-Ab rely on span predictions (Fig. 4B), while pAbT5 uses machine translation (Chu and Wei 2023). Currently, it is unclear which pretraining objective is supreme, as each of these ALMs have been evaluated on separate bespoke tasks. For instance, AntiBERTy has been trained using MLM and used for antibody structure prediction and engineering. On the other hand, IgLM has only been tested for its capability of generating new variant antibodies. This makes it challenging to speculate how IgLM will perform on structure prediction.

STRUCTURE PREDICTION DRIVEN BY ALMs

Antibody structure prediction is a subproblem within the field of protein structure prediction. Accurate structure predictions are a precursor to many subsequent analyses, such as antibody–antigen docking and optimization (Hummer et al. 2022). Accurate antibody structure predictions can shed light on the residues that are important for binding the target (paratope). Structure predictions can also highlight where the antibody may bind on the target (epitope) to deconvolute the antibody's mechanism of action. Finally, structural models can pinpoint liabilities to the development and manufacturing process. Most structure prediction tools can predict the overall fold of the antibody structure with high accuracy, although the CDRs can pose a significant obstacle for current models. In particular, the CDRH3 is often poorly modeled. This is expected, as the CDRH3 is the most polymorphic in terms of sequence length, sequence diversity, and conformational plasticity (Fig. 1C; Fernández-Quintero et al. 2023a).

A fundamental ingredient of modern protein structure prediction tools such as Alpha-Fold2 and RoseTTAFold is the input multiple sequence alignment (MSA) (Baek et al. 2021; Jumper et al. 2021). Sites of sequence covariation in the MSA encode contacts in three-dimensional space, and thus act as constraints for structure prediction (Kuhlman and Bradley 2019). However, as antibody sequences are hypervariable, especially in the CDRs, it is impractical to build a sufficiently deep MSA for high-resolution antibody models. This has prompted the development of antibody structure prediction tools that leverage components in AlphaFold2 but avoid MSA inputs. ALMs offer a unique solution for this problem as they can represent salient features of the antibody sequence without necessitating homologous antibody sequences in their training sets. Furthermore, self-attention matrices from ALMs can allude to potential structural contacts within the antibody (Leem et al. 2022; Prihoda et al. 2022).

To our knowledge, four antibody structure prediction tools use ALMs: AbBERT-HMPN (Gao et al. 2022), IgFold (Ruffolo et al. 2023), xTrimoAbFold (Wang et al. 2022b), and xTrimoPGLM-AbFold (Chen et al. 2023). AbBERT-HMPN uses the AbBERT ALM to generate embeddings that act as features for a structure prediction graph neural network. IgFold extracts antibody sequence embeddings and attention matrices from the AntiBERTy ALM as node

Cite this article as *Cold Spring Harb Perspect Biol* doi: 10.1101/cshperspect.a041462

and edge features for a graph transformer. This effectively replaces the MSA of AlphaFold2. In a more explicit substitution of the MSA, xTrimoAbFold uses the embeddings and attention matrices from an ALM as direct inputs to AlphaFold2's Evoformer and structure modules. xTrimoPGLM-AbFold is conceptually similar to xTrimoAbFold, although it uses fewer Evoformer and structure module blocks. All four approaches report excellent accuracies, although IgFold, xTrimoAbFold, and xTrimoPGLM-AbFold have the added advantage of being able to predict structures in seconds. This makes it feasible to use these tools for B-cell repertoire data sets, allowing users to structurally cluster leads and identify functionally convergent clones (Raybould et al. 2021; Robinson et al. 2021).

Ignoring ALMs and PLMs altogether are ImmuneBuilder and EquiFold (Lee et al. 2022; Abanades et al. 2023). Like xTrimoAbFold, ImmuneBuilder depends on AlphaFold2's structure module for coordinate generation. Instead of using an ALM input, ImmuneBuilder uses a one-hot encoding of the antibody sequence. On the other hand, EquiFold uses a bespoke antibody sequence representation for predicting the structure using the Equiformer model (Liao and Smidt 2022). Despite having simpler input features, both models can generate exceptional predictions across all antibody regions. However, structure prediction tools that leverage ALMs produce the most accurate results, with xTrimoPGLM-AbFold reporting the lowest root-mean-square deviations (RMSDs).

One of the challenges in evaluating the state of antibody structure prediction is the lack of a blinded, community-accepted benchmark, along with an accepted comparison metric, akin to critical assessment of protein structure prediction (CASP) (Moult et al. 1995). As a result, it is difficult to determine the precise level of value that ALMs add to structure prediction accuracy. It is also worth noting that publicly available antibody structures cover a biased proportion of the total antibody sequence space. For example, there is an overwhelming coverage of SARS-CoV-2-binding antibodies, but a paucity of antibodies binding targets related to other diseases (Dunbar et al. 2014). Training on structural data

alone, as in the case of ImmuneBuilder and EquiFold, may yield models that only perform well on a subset of antibodies (e.g., antibodies with shorter CDRH3 lengths) or antibodies that are more rigid and crystallizable. ALM-based models may be more robust to these scenarios as they would have been exposed to a broader spectrum of antibody sequences through their pretraining procedure.

ANTIBODY ENGINEERING

The "fitness" of an antibody is a highly nuanced term, encompassing a wide range of properties, such as thermostability and biochemical activity (Dallago et al. 2021). An antibody engineer must consider not only antibody thermostability and developability but also the target that the antibody binds, the affinity at which the target is bound, and the specificity of binding to a particular region within a target. Further still, function can also refer to downstream effects of the antibody, such as the antibody's agonistic or antagonistic properties (Schardt et al. 2022), its capacity to produce an immune response (referred to as "immunogenicity"), and off-target toxicity (Fernández-Quintero et al. 2023b). Successful modulation of these properties, and others, contributes to making a safe and efficacious therapeutic antibody. To date, ALMs have only been used to engineer the variable domain sequence; antibody function can also be affected by the constant region, which has been outside the remit of ALM-based techniques.

As with general proteins, the landscape for each of these properties is hilly, and combining mutations may have nonadditive epistatic effects. Moreover, optimizing for a single property in isolation may be detrimental to another. Computational methods can reduce the risks of antibody engineering by helping protein engineers distil a set of advantageous mutations from the wider combinatoric space. Typical pipelines feature a model that predicts a single aspect of antibody function, such as thermostability, from which mutations are then proposed. For therapeutic applications, ALM embeddings have mostly been used to predict binding affinity, safety, or developability.

Binding Affinity Optimization Using ALMs

Affinity maturation models typically harness sequence representations as the input for a regression model on a quantitative measure of affinity, such as K_D. Both PLMs and ALMs have been shown to be adaptable to this purpose (Hie et al. 2022a; Li et al. 2023). These models have reported success in training target-specific regression models when provided with training data of 10^3–10^5 binders and nonbinders to a target antigen from high-throughput assays. For instance, the AlphaSeq assay uses a yeast mating system to obtain hundreds of thousands of antibody–antigen interactions for fine-tuning ALMs (Engelhart et al. 2022; Li et al. 2023). The ACE assay from Bachas et al. (2022) is another route to obtaining high-throughput data sets that can fuel ALMs.

One study that demonstrates how ALMs can tie in with computational optimization algorithms is the work by Li et al. (2023). Using the AlphaSeq data set from Engelhart et al. (2022), a pretrained BERT transformer is fine-tuned to predict affinity and carve a mutational landscape to prioritize mutations. The proposed design variants achieve nearly 30-fold higher affinity than a baseline approach using PSSMs. Another striking aspect of the model is that despite being only trained on sequences with at most three mutations in the CDRs, the optimization model can generate variants with over 20 mutations. However, the implications of such a mutated antibody sequence with respect to immunogenicity and safety were not assessed. While this is a promising technique, it may be impractical for many, as the model required $\sim 10^4$ antibodies with known binding affinities. Furthermore, the target is a peptide; antigen-binding dynamics will likely be very different for larger proteins with conformational epitopes.

Although a generalized regression model that can predict binding to any arbitrary target is of great interest, at this time it is beyond reach, given the publicly available training data. To date, most antibody-binding data sets are either narrow and deep (i.e., many antibody variants for a single antigen) (Mason et al. 2021; Engelhart et al. 2022) or wide and shallow (i.e., anti-

bodies against many different targets, but with very few examples per target) (Dunbar et al. 2014; Hie et al. 2022a).

An alternative approach to predicting affinity is to rank sequences as opposed to regressing a specific value. For instance, Ruffolo et al. (2023) ordered ALM embeddings along an affinity maturation pathway using the evolutionary velocity technique (Hie et al. 2022b). In brief, an ALM is first used to calculate the likelihoods of various antibody sequence mutants, and sequences are sorted by difference in likelihood from a parent sequence.

Another possible strategy for affinity maturation that has not yet been extensively explored is leveraging paratope predictions. ALMs can predict paratopes with good accuracy (Leem et al. 2022; Wang et al. 2023), and are comparable in performance to structure-based paratope prediction tools. By identifying sites that are most likely to affect antigen binding, this can help prioritize positions to manipulate for affinity maturation. While this would not necessarily provide guidance on the impact of mutations (i.e., does a provisional change increase or decrease affinity), it could be coupled with methods such as phage display to design new variants.

Safety

In terms of safety, ALMs have been promising for antibody "humanization." This involves engineering antibodies extracted from immunized animals to resemble a human antibody sequence and reduce immunogenic risk (Prihoda et al. 2022). The ability of ALMs to make residue-level humanness predictions, alongside a conditional probability distribution for each amino acid at a given position, makes them useful for guiding humanization campaigns.

The Sapiens ALM from Prihoda et al. (2022) can propose mutations that match decisions that would have been undertaken by experimental scientists. On the other hand, IgLM takes a slightly different approach; instead of point mutations, it proposes a contiguous "span" of mutations to facilitate humanization (Shuai et al. 2023). As IgLM is a generative model, it is possible to tune the model to explore a larger set of

 Cite this article as *Cold Spring Harb Perspect Biol* doi: 10.1101/cshperspect.a041462

possible mutations, as opposed to taking the most probable mutation. While both models can increase humanness scores, the scores themselves are weakly correlated with experimentally determined immunogenicity (Prihoda et al. 2022); moreover, there are only ~200 antibody sequences with experimental immunogenicity data in the public domain (Marks et al. 2021). Thus, while these two approaches demonstrate the proof-of-principle that ALMs can help create safer therapeutics, more predictive safety metrics, standardized assays for robust validation, and the data to support creating such metrics, are critical (Ducret et al. 2022).

Developability

When manufacturing an antibody molecule for therapeutic use, it is imperative that it retains favorable biophysical characteristics, such as high thermostability and low aggregation propensity. One solution is to predict sequence liabilities (e.g., solvent-accessible deamidation motifs) within the variable domains using structural models (Leem et al. 2016; Raybould et al. 2019). Many of the factors that influence the shelf-life of a molecule can be outside of the antibody sequence itself, such as formulation (Fernández-Quintero et al. 2023b). Nevertheless, modifications of the amino acid sequence can influence developability, and several studies have attempted using ALMs for enhancing developability.

Harmalkar et al. (2023) used ESM-1b and AntiBERTy to predict thermostability. The authors found that AntiBERTy had poorer zero-shot performance than ESM-1b, although in fine-tuning, the ALM was superior and had stronger out-of-distribution performance. When taken forward for optimization, ESM-1b had better agreement with experimental data than structure-based methods or AntiBERTy. However, the caveat of this work is that it only features 20 experimental data points. It is worth noting that large-scale developability data sets are rare in the public domain. To our knowledge, the largest such set only has ~400 antibody sequences' worth of data (Shehata et al. 2019), while the next largest contains ~140 antibodies (Jain et al. 2017). Neither provide a sufficient

volume for deep learning models to learn the patterns underpinning thermostability. We expect with more data, there will be a more optimistic outlook for ALMs in supporting antibody developability.

Multiparameter Optimization

Optimizing an antibody sequence should ideally be done within a multiparameter optimization (MPO) framework to directly model the trade-offs of optimizing one property over another. Recently, a Bayesian approach has been published for antibody MPO (Khan et al. 2023), although it does not use an ALM and is outside the scope of this review.

Two studies have attempted to harmonize the various facets of antibody fitness into one global metric for optimization. Bachas et al. (2022) calculated the pseudo–log likelihood of an antibody sequence from their ALM, which they referred to as "naturalness." They showed that naturalness is broadly correlated with immunogenicity, developability, and expression titer, but not binding affinity. Nijkamp et al. (2023) used a similar concept of calculating the perplexity of an antibody sequence from the Pro-Gen2-OAS ALM. Perplexity was tested for its correlation with thermostability and binding affinity but was found to be weakly correlated with both. This may be explained by the fact that representations from decoders are generally less performant for sequence classification tasks than those from encoder-based models (Elnaggar et al. 2022). Computing a single metric that can capture the multitude of antibody properties remains an unsolved challenge. However, distilling such a metric will be a boon for antibody engineering as it will simplify optimization routines and lead to more interpretable outcomes.

CHALLENGES AND OPPORTUNITIES

While ALMs have been successful in many applications, their predictive power can be variable; in some applications, there is a clear gap in the volume of labeled data, which could give a "false negative" view on the utility of ALMs. Two key challenges remain in the pathway toward a general-purpose ALM that can truly revolutionize antibody drug discovery: creating large-scale data sets for machine learning, and technical standardization.

Creating Large-Scale Data Sets for Machine Learning

Data is the bedrock for successful machine learning. Ideal data sets for testing ALMs should have as many of the following characteristics as possible: low measurement noise, broad antibody sequence and broad target coverage, high-throughput, and translational relevance. To our knowledge, no publicly available data set satisfies all these requirements.

Antibody data sets are typically "narrow and deep" in nature. For example, the recently published data set of antibody-binding data to SARS-CoV-2 peptides contains over 100,000 antibody sequences (Engelhart et al. 2022). It can be useful for screening and designing new antibodies against the SARS-CoV-2 virus (Li et al. 2023). However, it is challenging to extrapolate how an ALM fine-tuned on this data set will generalize to other antigens, especially nonviral targets. Another narrow and deep data set is the publicly available set of more than 30,000 trastuzumab variants that were screened using yeast display (Mason et al. 2021). While this data set is valuable for training and evaluating ALMs that predict HER2 binding, all the antibody sequence variation is focused in the CDRH3 and CDRL3. This would raise challenges on whether an ALM would be generalizable in predicting variants outside these regions. Here, we believe that assays like ACE, AlphaSeq, and MIPSA are particularly promising, as they can sample thousands of antibody–antigen interactions (Younger et al. 2017; Bachas et al. 2022; Credle et al. 2022). Data sets can also be "wide and shallow"; SAbDab is a primary example, which contains antibody-binding affinity data for a diverse set of antibodies and antigens. However, for most antigens, there are only one or two cognate antibodies.

We believe that benchmark data sets in deep learning research, such as GLUE for NLP, or ImageNet for computer vision, provide an ideal template for future work in developing ALMs (Russakovsky et al. 2015; Wang et al. 2018). Both GLUE and ImageNet have thousands of examples for various subtasks, giving researchers access to large data sets for training and evaluating models, and have many different tasks to obtain a more holistic view on the power of their models.

SAbDab follows this premise, as it can act as a benchmark for binding affinity prediction, paratope prediction, or structure prediction but at low volumes for each separate task (Dunbar et al. 2014). In addition, SAbDab is highly skewed with antiviral antibodies, particularly SARS-CoV-2 binders, which may hamper the translational impact of models. The recently published ATUE benchmark is another excellent attempt at creating a community benchmark for gauging model performance (Wang et al. 2023). So far in ATUE, there are four different tasks: classifying trastuzumab variants for antigen-binding, paratope prediction, B-cell-type prediction, and ranking SARS-CoV-2 binders from collections of BCR repertoires. Apart from paratope prediction, each data set contains over 10,000 training examples, making it a good initial platform for evaluating models. The most impactful data sets will only come about through closer partnerships between experimental and dry-lab scientists, as well as interorganizational collaborations, especially on core questions in drug discovery.

Technical Standardization

Given the relatively recent development of ALMs, it is expected that there is not as extensive literature on best practices for building these models. Indeed, many of the hyperparameter choices for setting up an ALM (e.g., type of transformer architecture, number of layers) are based on successful recipes in NLP. For instance, large LMs in NLP can feature many self-attention heads, but the justification for having so many self-attention heads for antibody sequences has yet to be established. In fact, correct design choices for many aspects of ALMs are unresolved. For example, some ALMs are heavy or light-chain-specific (e.g., AbLang-VH and AbLang-VL; Olsen et al. 2022), while others accept either chain (e.g., AntiBERTa; Leem et al., 2022). Only xTrimoPGLM-Ab, to our knowledge, can accept both chains simultaneously as its input (Chen et al. 2023).

The closest attempt at investigating "good standards" for building ALMs has been the work by Gao et al. (2023). Here, the authors ran several ablations to determine how certain hyperparameter choices can affect reconstruction of masked sequences. Their work provides a template for some of the considerations that should be made in building ALMs. However, this study did not consider the impact of pre-training data size, nor how the changes in these pretraining regimes affect downstream fine-tuning performance. This is where studies in PLM design, primarily those investigating the impact of scaling up models, can provide a blueprint for how we should consider designing new ALMs in the future (Elnaggar et al. 2023; Lin et al. 2023; Nijkamp et al. 2023).

Another initiative that can facilitate standardization is agreeing on a set of software libraries and packages for development. This should enable more robust comparisons between models and reduce engineering burden. Most ALMs use PyTorch and the HuggingFace transformers library (Leem et al. 2022; Nijkamp et al. 2023; Ruffolo et al. 2023; Shuai et al. 2023), although some use FAIRSeq (Olsen et al. 2022; Prihoda et al. 2022).

CONCLUDING REMARKS

In this work, we describe the evolution and application of ALMs for representation learning of antibody sequences. Following a long history of using context-free methods, transformer-based ALMs have changed the paradigm for antibody sequence representation learning in a span of 2 years. They are now a "Swiss-army knife" that can be used across a suite of bioinformatics challenges in therapeutic antibody discovery, including antibody structure prediction and antibody engineering. A striking feature of ALMs is their sheer generalizability: a prime example is the AntiBERTy model, which has been used for structure prediction and engineering several different antibody properties. Considering that

AntiBERTy is one of many encoder-only models, other ALMs such as AntiBERTa, AbLang, and Sapiens can equally be customized for these applications as well.

We anticipate further adoption of ALMs, and an increasing volume of experimental evidence that validates the use of these models. The field is still rapidly evolving; xTrimoPGLM-Ab has recently broken the one billion parameter barrier, and we expect further scaling up of ALMs. Furthermore, we anticipate tighter integration between ALMs and other types of deep learning architectures, such as diffusion models (Luo et al. 2022). Taking inspiration from the broader NLP domain, we also believe that ALMs will become multimodal: whether that is integration with structural data, imaging data, or other types of phenotypic readouts that can map the relationship between an antibody's sequence with respect to function.

ACKNOWLEDGMENTS

We would like to thank the wider team at Alchemab Therapeutics for their thoughts and discussion points to prepare the manuscript.

REFERENCES

Abanades B, Wong WK, Boyles F, Georges G, Bujotzek A, Deane CM. 2023. Immunebuilder: deep-learning models for predicting the structures of immune proteins. *Commun Biol* **6**: 575. doi:10.1038/s42003-023-04927-7

Akbar R, Robert PA, Pavlović M, Jeliazkov JR, Snapkov I, Slabodkin A, Weber CR, Scheffer L, Miho E, Haff IH, et al. 2021. A compact vocabulary of paratope-epitope interactions enables predictability of antibody-antigen binding. *Cell Rep* **34**: 108856. doi:10.1016/j.celrep.2021.108856

Ambrosetti F, Olsen TH, Olimpieri PP, Jiménez-García B, Milanetti E, Marcatilli P, Bonvin AMJJ. 2020. proABC-2: prediction of antibody contacts v2 and its application to information-driven docking. *Bioinformatics* **36**: 5107–5108. doi:10.1093/bioinformatics/btaa644

Amimeur T, Shaver JM, Ketchem RR, Taylor JA, Clark RH, Smith J, Van Citters D, Siska CC, Smidt P, Sprague M, et al. 2020. Designing feature-controlled humanoid antibody discovery libraries using generative adversarial networks. bioRxiv doi:10.1101/2020.04.12.024844

Bachas S, Rakocevic G, Spencer D, Sastry AV, Haile R, Sutton JM, Kasun G, Stachyra A, Gutierrez JM, Yassine E, et al. 2022. Antibody optimization enabled by artificial intelligence predictions of binding affinity and naturalness. bioRxiv doi:10.1101/2022.08.16.504181

Baek M, DiMaio F, Anishchenko I, Dauparas J, Ovchinnikov S, Lee GR, Wang J, Cong Q, Kinch LN, Schaeffer RD, et al. 2021. Accurate prediction of protein structures and interactions using a three-track neural network. *Science* **373:** 871–876. doi:10.1126/science.abj8754

Bashford-Rogers RJM, Bergamaschi L, McKinney EF, Pombal DC, Mescia F, Lee JC, Thomas DC, Flint SM, Kellam P, Jayne DRW, et al. 2019. Analysis of the B cell receptor repertoire in six immune-mediated diseases. *Nature* **574:** 122–126. doi:10.1038/s41586-019-1595-3

Bystroff C, Krogh A. 2008. Hidden Markov models for protein structure prediction. *Methods Mol Biol* **413:** 173–198. doi:10.1007/978-1-59745-574-9_7

Chalkidis I, Fergadiotis M, Malakasiotis P, Aletras N, Androutsopoulos I. 2020. LEGAL-BERT: the muppets straight out of law school. arXiv doi:10.48550/arXiv.2010.02559

Chaudhary N, Wesemann DR. 2018. Analyzing immunoglobulin repertoires. *Front Immunol* **9:** 462. doi:10.3389/fimmu.2018.00462

Chen X, Dougherty T, Hong C, Schibler R, Zhao YC, Sadeghi R, Matasci N, Wu YC, Kerman I. 2020. Predicting antibody developability from sequence using machine learning. bioRxiv doi:10.1101/2020.06.18.159798

Chen B, Cheng X, Geng Y, Li S, Zeng X, Wang B, Gong J, Liu C, Zeng A, Dong Y, et al. 2023. xTrimoPGLM: unified 100B-scale pre-trained transformer for deciphering the language of protein. bioRxiv doi:10.1101/2023.07.05.547496

Chu SKS, Wei KY. 2023. Conditional generation of paired antibody chain sequences through encoder-decoder language model. arXiv doi:10.458550/arXiv.2301.02748

Corrie BD, Marthandan N, Zimonja B, Jaglale J, Zhou Y, Barr E, Knoetze N, Breden FMW, Christley S, Scott JK, et al. 2018. Ireceptor: a platform for querying and analyzing antibody/B-cell and T-cell receptor repertoire data across federated repositories. *Immunol Rev* **284:** 24–41. doi:10.1111/imr.12666

Credle JJ, Gunn J, Sangkhapreecha P, Monaco DR, Zheng XA, Tsai HJ, Wilbon A, Morgenlander WR, Rastegar A, Dong Y, et al. 2022. Unbiased discovery of autoantibodies associated with severe COVID-19 via genome-scale self-assembled DNA-barcoded protein libraries. *Nat Biomed Eng* **6:** 992–1003. doi:10.1038/s41551-022-00925-y

Dallago C, Mou J, Johnston KE, Wittmann BJ, Bhattacharya N, Goldman S, Madani A, Yang KK. 2021. FLIP: benchmark tasks in fitness landscape inference for proteins. bioRxiv doi:10.1101/2021.11.09.467890

Del Vecchio A, Deac A, Liò P, Veličković P. 2021. Neural message passing for joint paratope-epitope prediction. arXiv doi:10.458550/arXiv.2106.00757

Devlin J, Chang MW, Lee K, Toutanova K. 2018. BERT: pre-training of deep bidirectional transformers for language understanding. arXiv doi:10.48550/arXiv.1810.04805

Ducret A, Ackaert C, Bessa J, Bunce C, Hickling T, Jawa V, Kroenke MA, Lamberth K, Manin A, Penny HL, et al. 2022. Assay format diversity in pre-clinical immunogenicity risk assessment: toward a possible harmonization of antigenicity assays. *MAbs* **14:** 1993522. doi:10.1080/19420862.2021.1993522

Dunbar J, Deane CM. 2016. ANARCI: antigen receptor numbering and receptor classification. *Bioinformatics* **32:** 298 300. doi:10.1093/bioinformatics/btv552

Dunbar J, Krawczyk K, Leem J, Baker T, Fuchs A, Georges G, Shi J, Deane CM. 2014. SAbdab: the structural antibody database. *Nucleic Acids Res* **42:** D1140–D1146. doi:10.1093/nar/gkt1043

Elnaggar A, Heinzinger M, Dallago C, Rehawi G, Wang Y, Jones L, Gibbs T, Feher T, Angerer C, Steinegger M, et al. 2022. Prottrans: toward understanding the language of life through self-supervised learning. *IEEE Trans Pattern Anal Mach Intell* **44:** 7112–7127. doi:10.1109/TPAMI.2021.3095381

Elnaggar A, Essam H, Salah-Eldin W, Moustafa W, Elkerdawy M, Rochereau C, Rost B. 2023. Ankh: optimized protein language model unlocks general-purpose modelling. arXiv doi:10.48550/arXiv.2301.06568

Engelhart E, Emerson R, Shing L, Lennartz C, Guion D, Kelley M, Lin C, Lopez R, Younger D, Walsh ME. 2022. A dataset comprised of binding interactions for 104,972 antibodies against a SARS-CoV-2 peptide. *Sci Data* **9:** 653. doi:10.1038/s41597-022-01779-4

Esmaielbeiki R, Krawczyk K, Knapp B, Nebel JC, Deane CM. 2016. Progress and challenges in predicting protein interfaces. *Brief Bioinform* **17:** 117–131. doi:10.1093/bib/bbv027

Feng J, Jiang M, Shih J, Chai Q. 2022. Antibody apparent solubility prediction from sequence by transfer learning. *iScience* **25:** 105173. doi:10.1016/j.isci.2022.105173

Fernández-Quintero ML, Kokot J, Waibl F, Fischer ALM, Quoika PK, Deane CM, Liedl KR. 2023a. Challenges in antibody structure prediction. *MAbs* **15:** 2175319. doi:10.1080/19420862.2023.2175319

Fernández-Quintero ML, Ljungars A, Waibl F, Greiff V, Andersen JT, Gjølberg TT, Jenkins TP, Voldborg BG, Grav LM, Kumar S, et al. 2023b. Assessing developability early in the discovery process for novel biologics. *MAbs* **15:** 2171248. doi:10.1080/19420862.2023.2171248

Friedensohn S, Neumeier D, Khan TA, Csepregi L, Parola C, de Vries ARG, Erlach L, Mason DM, Reddy ST. 2020. Convergent selection in antibody repertoires is revealed by deep learning. bioRxiv doi:10.1101/2020.02.25.965673

Galson JD, Schaetzle S, Bashford-Rogers RJM, Raybould MIJ, Kovaltsuk A, Kilpatrick GJ, Minter R, Finch DK, Dias J, James LK, et al. 2020. Deep sequencing of B cell receptor repertoires from COVID-19 patients reveals strong convergent immune signatures. *Front Immunol* **11:** 605170. doi:10.3389/fimmu.2020.605170

Gao K, Wu L, Zhu J, Peng T, Xia Y, He L, Xie S, Qin T, Liu H, He K, et al. 2022. Incorporating pre-training paradigm for antibody sequence-structure co-design. arXiv doi:10.48550/arXiv.2211.0840

Gao X, Cao C, Lai L. 2023. Pre-training with a rational approach for antibody. bioRxiv doi:10.1101/2023.01.19.524683

Georgiou G, Ippolito GC, Beausang J, Busse CE, Wardemann H, Quake SR. 2014. The promise and challenge of high-throughput sequencing of the antibody repertoire. *Nat Biotechnol* **32:** 158–168. doi:10.1038/nbt.2782

Greiff V, Weber CR, Palme J, Bodenhofer U, Miho E, Menzel U, Reddy ST. 2017. Learning the high-dimensional immunogenomic features that predict public and private

Cite this article as *Cold Spring Harb Perspect Biol* doi: 10.1101/cshperspect.a041462

antibody repertoires. *J Immunol* **199**: 2985–2997. doi:10 .4049/jimmunol.1700594

Gu Y, Tinn R, Cheng H, Lucas M, Usuyama N, Liu X, Naumann T, Gao J, Poon H. 2021. Domain-specific language model pretraining for biomedical natural language processing. *Acm Trans Comput Healthc* **3**: 1–23. doi:10.1145/ 3458754

Harmalkar A, Rao R, Xie YR, Honer J, Deisting W, Anlahr J, Hoenig A, Czwikla J, Sienz-Widmann E, Rau D, et al. 2023. Toward generalizable prediction of antibody thermostability using machine learning on sequence and structure features. *MAbs* **15**: 2163584. doi:10.1080/ 19420862.2022.2163584

Hie BL, Shanker VR, Xu D, Bruun TUJ, Weidenbacher PA, Tang S, Kim PS. 2022a. Efficient evolution of human antibodies from general protein language models. *Nat Biotechnol* doi:10.1038/s41587-023-01763-2

Hie BL, Yang KK, Kim PS. 2022b. Evolutionary velocity with protein language models predicts evolutionary dynamics of diverse proteins. *Cell Syst* **13**: 274–285.e6. doi:10.1016/j .cels.2022.01.003

Hummer AM, Abanades B, Deane CM. 2022. Advances in computational structure-based antibody design. *Curr Opin Struc Biol* **74**: 102379. doi:10.1016/j.sbi.2022 .102379

Imrie F, Bradley AR, van der Schaar M, Deane CM. 2018. Protein family-specific models using deep neural networks and transfer learning improve virtual screening and highlight the need for more data. *J Chem Inf Model* **58**: 2319–2330. doi:10.1021/acs.jcim.8b00350

Jaffe DB, Shahi P, Adams BA, Chrisman AM, Finnegan PM, Raman N, Royall AE, Tsai F, Vollbrecht T, Reyes DS, et al. 2022. Functional antibodies exhibit light chain coherence. *Nature* **611**: 352–357. doi:10.1038/s41586-022-05371-z

Jain T, Sun T, Durand S, Hall A, Houston NR, Nett JH, Sharkey B, Bobrowicz B, Caffry I, Yu Y, et al. 2017. Biophysical properties of the clinical-stage antibody landscape. *Proc Natl Acad Sci* **114**: 944–949. doi:10.1073/ pnas.1616408114

Jumper J, Evans R, Pritzel A, Green T, Figurnov M, Ronneberger O, Tunyasuvunakool K, Bates R, Žídek A, Potapenko A, et al. 2021. Highly accurate protein structure prediction with AlphaFold. *Nature* **596**: 583–589. doi:10 .1038/s41586-021-03819-2

Khan A, Cowen-Rivers AI, Grosnit A, Deik DGX, Robert PA, Greiff V, Smorodina E, Rawat P, Akbar R, Dreczkowski K, et al. 2023. Toward real-world automated antibody design with combinatorial Bayesian optimization. *Cell Rep Methods* **3**: 100374. doi:10.1016/j.crmeth.2022.100374

Konishi H, Komura D, Katoh H, Atsumi S, Koda H, Yamamoto A, Seto Y, Fukayama M, Yamaguchi R, Imoto S, et al. 2019. Capturing the differences between humoral immunity in the normal and tumor environments from repertoire-seq of B-cell receptors using supervised machine learning. *BMC Bioinformatics* **20**: 267. doi:10.1186/ s12859-019-2853-y

Kovaltsuk A, Leem J, Kelm S, Snowden J, Deane CM, Krawczyk K. 2018. Observed antibody space: a resource for data mining next-generation sequencing of antibody repertoires. *J Immunol* **201**: 2502–2509. doi:10.4049/jimmu nol.1800708

Krawczyk K, Raybould MIJ, Kovaltsuk A, Deane CM. 2019. Looking for therapeutic antibodies in next-generation sequencing repositories. *MAbs* **11**: 1197–1205. doi:10.1080/ 19420862.2019.1633884

Krawczyk K, Buchanan A, Marcatili P. 2021. Data mining patented antibody sequences. *MAbs* **13**: 1892366. doi:10 .1080/19420862.2021.1892366

Kuhlman B, Bradley P. 2019. Advances in protein structure prediction and design. *Nat Rev Mol Cell Bio* **20**: 681–697. doi:10.1038/s41580-019-0163-x

Lee JH, Yadollahpour P, Watkins A, Frey NC, Leaver-Fay A, Ra S, Cho K, Gligorijević V, Regev A, Bonneau R. 2022. Equifold: protein structure prediction with a novel coarse-grained structure representation. bioRxiv doi:10 .1101/2022.10.07.511322

Leem J, Dunbar J, Georges G, Shi J, Deane CM. 2016. A Bodybuilder: automated antibody structure prediction with data–driven accuracy estimation. *MAbs* **8**: 1259– 1268. doi:10.1080/19420862.2016.1205773

Leem J, Mitchell LS, Farmery JHR, Barton J, Galson JD. 2022. Deciphering the language of antibodies using self-supervised learning. *Patterns* **3**: 100513. doi:10.1016/j .patter.2022.100513

Li L, Gupta E, Spaeth J, Shing L, Jaimes R, Engelhart E, Lopez R, Caceres RS, Bepler T, Walsh ME. 2023. Machine learning optimization of candidate antibodies yields highly diverse sub-nanomolar affinity antibody libraries. *Nat Commun* **14**: 3454. doi:10.1038/s41467-023-39022-2

Liao YL, Smidt T. 2022. Equiformer: equivariant graph attention transformer for 3D atomistic graphs. arXiv doi:10 .48550/arXiv.2206.11990

Liberis E, Veličković P, Sormanni P, Vendruscolo M, Liò P. 2018. Parapred: antibody paratope prediction using convolutional and recurrent neural networks. *Bioinformatics* **34**: 2944–2950. doi:10.1093/bioinformatics/bty305

Lim YW, Adler AS, Johnson DS. 2022. Predicting antibody binders and generating synthetic antibodies using deep learning. *MAbs* **14**: 2069075. doi:10.1080/19420862.2022 .2069075

Lin T, Wang Y, Liu X, Qiu X. 2022. A survey of transformers. *AI Open* **3**: 111–132. doi:10.1016/j.aiopen.2022.10.001

Lin Z, Akin H, Rao R, Hie B, Zhu Z, Lu W, Smetanin N, Verkuil R, Kabeli O, Shmueli Y, et al. 2023. Evolutionary-scale prediction of atomic-level protein structure with a language model. *Science* **379**: 1123–1130. doi:10.1126/sci ence.ade2574

Luo S, Su Y, Peng X, Wang S, Peng J, Ma J. 2022. Antigen-specific antibody design and optimization with diffusion-based generative models for protein structures. bioRxiv doi:10.1101/2022.07.10.499510

Marks C, Hummer AM, Chin M, Deane CM. 2021. Humanization of antibodies using a machine learning approach on large-scale repertoire data. *Bioinformatics* **37**: 4041– 4047. doi:10.1093/bioinformatics/btab434

Mason DM, Friedensohn S, Weber CR, Jordi C, Wagner B, Meng SM, Ehling RA, Bonati L, Dahinden J, Gainza P, et al. 2021. Optimization of therapeutic antibodies by predicting antigen specificity from antibody sequence via deep learning. *Nat Biomed Eng* **5**: 600–612. doi:10.1038/ s41551-021-00699-9

Melnyk I, Chenthamarakshan V, Chen PY, Das P, Dhurand-har A, Padhi I, Das D. 2022. Reprogramming large pre-trained language models for antibody sequence infilling. arXiv doi:10.48550/arXiv.2210.07144

Moult J, Pedersen JT, Judson R, Fidelis K. 1995. A large-scale experiment to assess protein structure prediction meth-ods. *Proteins* **23**: ii–iv. doi:10.1002/prot.340230303

Nijkamp E, Ruffolo J, Weinstein EN, Naik N, Madani A. 2023. ProGen: exploring the boundaries of protein lan-guage models. *Cell Syst* **14**: 968–978.e3. doi:10.1016/j.cels.2023.10.002

Olsen TH, Moal IH, Deane CM. 2022. Ablang: an antibody language model for completing antibody sequences. *Bio-inform Adv* **2**: vbac046. doi:10.1093/bioadv/vbac046

Ostrovsky-Berman M, Frankel B, Polak P, Yaari G. 2021. Immune2vec: embedding B/T cell receptor sequences in ℝN using natural language processing. *Front Immunol* **12**: 680687. doi:10.3389/fimmu.2021.680687

Park JC, Noh J, Jang S, Kim KH, Choi H, Lee D, Kim J, Chung J, Lee DY, Lee Y, et al. 2022. Association of B cell profile and receptor repertoire with the progression of Alzheimer's disease. *Cell Rep* **40**: 111391. doi:10.1016/j.celrep.2022.111391

Prihoda D, Maamary J, Waight A, Juan V, Fayadat-Dilman L, Svozil D, Bitton DA. 2022. Biophi: a platform for anti-body design, humanization, and humanness evaluation based on natural antibody repertoires and deep learning. *MAbs* **14**: 2020203. doi:10.1080/19420862.2021.2020203

Radford A, Wu J, Child R, Luan D, Amodei D, Sutskever I. 2019. Language models are unsupervised multitask learn-ers. *Open AI* **1**: 9. https://d4mucfpksywv.cloudfront.net/better-language-models/language_models_are_unsupervised_multitask_learners.pdf

Raffel C, Shazeer N, Roberts A, Lee K, Narang S, Matena M, Zhou Y, Li W, Liu PJ. 2019. Exploring the limits of trans-fer learning with a unified text-to-text transformer. arXiv doi:10.48550/arXiv.1910.10683

Raybould MIJ, Marks C, Krawczyk K, Taddese B, Nowak J, Lewis AP, Bujotzek A, Shi J, Deane CM. 2019. Five com-putational developability guidelines for therapeutic anti-body profiling. *Proc Natl Acad Sci* **116**: 4025–4030. doi:10.1073/pnas.1810576116

Raybould MIJ, Marks C, Kovaltsuk A, Lewis AP, Shi J, Deane CM. 2021. Public baseline and shared response structures support the theory of antibody repertoire functional com-monality. *PLoS Comput Biol* **17**: e1008781. doi:10.1371/journal.pcbi.1008781

Rees AR. 2020. Understanding the human antibody reper-toire. *MAbs* **12**: 1729683. doi:10.1080/19420862.2020.1729683

Rives A, Meier J, Sercu T, Goyal S, Lin Z, Liu J, Guo D, Ott M, Zitnick CL, Ma J, et al. 2021. Biological structure and function emerge from scaling unsupervised learning to 250 million protein sequences. *Proc Natl Acad Sci* **118**: e2016239118. doi:10.1073/pnas.2016239118

Robinson SA, Raybould MIJ, Schneider C, Wong WK, Marks C, Deane CM. 2021. Epitope profiling using com-putational structural modelling demonstrated on coro-navirus-binding antibodies. *PLoS Comput Biol* **17**: e1009675. doi:10.1371/journal.pcbi.1009675

Ruffolo JA, Gray JJ, Sulam J. 2021. Deciphering antibody affinity maturation with language models and weakly su-pervised learning. arXiv doi:10.48550/arXiv.2112.07782

Ruffolo JA, Chu LS, Mahajan SP, Gray JJ. 2023. Fast, accurate antibody structure prediction from deep learning on mas-sive set of natural antibodies. *Nat Commun* **14**: 2389. doi:10.1038/s41467-023-38063-x

Russakovsky O, Deng J, Su H, Krause J, Satheesh S, Ma S, Huang Z, Karpathy A, Khosla A, Bernstein M, et al. 2015. Imagenet large scale visual recognition challenge. *Int J Comput Vision* **115**: 211–252. doi:10.1007/s11263-015-0816-y

Saka K, Kakuzaki T, Metsugi S, Kashiwagi D, Yoshida K, Wada M, Tsunoda H, Teramoto R. 2021. Antibody design using LSTM based deep generative model from phage display library for affinity maturation. *Sci Rep* **11**: 5852. doi:10.1038/s41598-021-85274-7

Schardt JS, Jhajj HS, O'Meara RL, Lwo TS, Smith MD, Tes-sier PM. 2022. Agonist antibody discovery: experimental, computational, and rational engineering approaches. *Drug Discov Today* **27**: 31–48. doi:10.1016/j.drudis.2021.09.008

Shehata L, Maurer DP, Wec AZ, Lilov A, Champney E, Sun T, Archambault K, Burnina I, Lynaugh H, Zhi X, et al. 2019. Affinity maturation enhances antibody specificity but compromises conformational stability. *Cell Rep* **28**: 3300–3308.e4. doi:10.1016/j.celrep.2019.08.056

Shuai RW, Ruffolo JA, Gray JJ. 2023. IgLM: infilling lan-guage modelling for antibody sequence design. *Cell Syst* **14**: 979–989.e4. doi:10.1016/j.cels.2023.10.001

Sippl MJ. 1990. Calculation of conformational ensembles from potentials of mena force. *J Mol Biol* **213**: 859–883. doi:10.1016/S0022-2836(05)80269-4

Song L, Cohen D, Ouyang Z, Cao Y, Hu X, Liu XS. 2021. TRUST4: immune repertoire reconstruction from bulk and single-cell RNA-seq data. *Nat Methods* **18**: 627–630. doi:10.1038/s41592-021-01142-2

Townsend CL, Laffy JMJ, Wu YCB, O'Hare JS, Martin V, Kipling D, Fraternali F, Dunn-Walters DK. 2016. Signifi-cant differences in physicochemical properties of human immunoglobulin kappa and lambda CDR3 regions. *Front Immunol* **7**: 388. doi:10.3389/fimmu.2016.00388

Vaswani A, Shazeer N, Parmar N, Uszkoreit J, Jones L, Go-mez AN, Kaiser L, Polosukhin I. 2017. Attention is all you need. arXiv doi:10.48550/arXiv.1706.03762

Wang A, Singh A, Michael J, Hill F, Levy O, Bowman SR. 2018. GLUE: a multi-task benchmark and analysis plat-form for natural language understanding. arXiv doi:10.48550/arXiv.1804.07461

Wang P, Luo M, Zhou W, Jin X, Xu Z, Yan S, Li Y, Xu C, Cheng R, Huang Y, et al. 2022a. Global characterization of peripheral b cells in Parkinson's disease by single-cell RNA and BCR sequencing. *Front Immunol* **13**: 814239. doi:10.3389/fimmu.2022.814239

Wang Y, Gong X, Li S, Yang B, Sun Y, Shi C, Wang Y, Yang C, Li H, Song L. 2022b. xTrimoABFold: de novo antibody structure prediction without MSA. arXiv doi:10.48550/arXiv.2212.00735

Wang D, Ye F, Zhou H. 2023. On pre-trained language mod-els for antibody. bioRxiv doi:10.1101/2023.01.29.525793

Weber CR, Rubio T, Wang L, Zhang W, Robert PA, Akbar R, Snapkov I, Wu J, Kuijjer ML, Tarazona S, et al. 2022. Reference-based comparison of adaptive immune receptor repertoires. *Cell Rep Methods* **2:** 100269. doi:10.1016/j.crmeth.2022.100269

Wollacott AM, Xue C, Qin Q, Hua J, Bohnuud T, Viswanathan K, Kolachalama VB. 2019. Quantifying the nativeness of antibody sequences using long short-term memory networks. *Protein Eng Des Sel* **32:** 347–354. doi:10.1093/protein/gzz031

Wong WK, Georges G, Ros F, Kelm S, Lewis AP, Taddese B, Leem J, Deane CM. 2019. SCALOP: sequence-based antibody canonical loop structure annotation. *Bioinformatics* **35:** 1774–1776. doi:10.1093/bioinformatics/bty877

Xue L, Constant N, Roberts A, Kale M, Al-Rfou R, Siddhant A, Barua A, Raffel C. 2020. Mt5: a massively multilingual pre-trained text-to-text transformer. arXiv doi:10.48550/arXiv.2010.11934

Younger D, Berger S, Baker D, Klavins E. 2017. High-throughput characterization of protein–protein interactions by reprogramming yeast mating. *Proc Natl Acad Sci* **114:** 12166–12171. doi:10.1073/pnas.1705867114

Yu K, Ravoor A, Malats N, Pineda S, Sirota M. 2022. A pancancer analysis of tumor-infiltrating B cell repertoires. *Front Immunol* **12:** 790119. doi:10.3389/fimmu.2021.790119

Zainchkovskyy Y, Ferkinghoff-Borg J, Bennett A, Egebjerg T, Lorenzen N, Greisen PJ, Hauberg S, Stahlhut C. 2022. Probabilistic thermal stability prediction through sparsity promoting transformer representation. arXiv doi:10.48550/arXiv.2211.05698

Zaslavsky ME, Craig E, Michuda JK, Ram-Mohan N, Lee JY, Nguyen KD, Hoh RA, Pham TD, Parsons ES, Macwana SR, et al. 2022. Disease diagnostics using machine learning of immune receptors. bioRxiv doi:10.1101/2022.04.26.489314

Zheng B, Yang Y, Chen L, Wu M, Zhou S. 2022. B-cell receptor repertoire sequencing: deeper digging into the mechanisms and clinical aspects of immune-mediated diseases. *iScience* **25:** 105002. doi:10.1016/j.isci.2022.105002

Engineering Proteins Using Statistical Models of Coevolutionary Sequence Information

Jerry C. Dinan,[1,2,3] James W. McCormick,[1,2,3] and Kimberly A. Reynolds[1,2,3]

[1]The Green Center for Systems Biology; [2]The Lyda Hill Department of Bioinformatics; [3]The Department of Biophysics, University of Texas Southwestern Medical Center, Dallas, Texas 75390, USA

Correspondence: kimberly.reynolds@utsouthwestern.edu

Homologous protein sequences are wonderfully diverse, indicating many possible evolutionary "solutions" to the encoding of function. Consequently, one can construct statistical models of protein sequence by analyzing amino acid frequency across a large multiple sequence alignment. A central premise is that covariance between amino acid positions reflects coevolution due to a shared functional or biophysical constraint. In this review, we describe the implementation and discuss the advantages, limitations, and recent progress on two coevolution-based modeling approaches: (1) Potts models of protein sequence (direct coupling analysis [DCA]-like), and (2) the statistical coupling analysis (SCA). Each approach detects interesting features of protein sequence and structure—the former emphasizes local physical contacts throughout the structure, while the latter identifies larger evolutionarily coupled networks of residues. Recent advances in large-scale gene synthesis and high-throughput functional selection now motivate additional work to benchmark model performance across quantitative function prediction and de novo design tasks.

More than 60 years have passed since Anfinsen's elegant experiments demonstrated the sufficiency of protein sequence to specify fold and function, yet a quantitative model relating sequence, structure, and biochemical activity remains elusive (Anfinsen 1973; Anfinsen and Scheraga 1975). Understanding how amino acid sequence influences function would transform our ability to create bespoke enzymatic catalysts, design synthetic cellular systems, engineer biosensors, and unravel the molecular basis of phenotype. A tsunami of sequencing data paired with advances in machine learning has now opened new avenues for creating statistical models of protein sequence and function (The UniProt Consortium 2023).

In contrast to structure-based design, which uses atomistic physics-based potentials to identify residues that stabilize functionally competent conformation(s), statistical sequence models are not directly informed by tertiary structure and physical mechanism. Instead, the idea is to determine the statistical constraints on a protein sequence by looking at many diverse examples of solutions produced through natural selection (Baker 2014; Ferguson and Ranganathan 2021). Structurally and functionally homologous proteins often have sequence identities as low as

15%–20%, suggesting there are numerous degenerate solutions to the encoding of function (Rost 1999). Similarly, deep mutational scanning experiments revealed that many proteins are fairly tolerant of substitutions (McLaughlin et al. 2012; Thompson et al. 2020; Ding et al. 2022). Together, these data indicate that the core encoding of protein function is achieved with a relatively sparse number of sequence constraints, and tuned by additional modifications in individual family members. Coevolutionary models attempt to glean these family-wide constraints by following two guiding principles: (1) that evolutionary conservation is an indicator of functional importance, and (2) that coevolution between pairs of positions reflects a shared functional or thermodynamic constraint. The resulting model captures constraints on amino acid sequence imposed by evolutionary pressures experienced across the protein family and is neither biased nor informed by knowledge of biochemical or biophysical mechanism.

Our review centers on pairwise statistical potentials for protein sequence, which stem from amino acid frequencies in a multiple sequence alignment (MSA) and have their basis in statistical physics. While these relatively "simple" models overlook the potential for higher-order interactions (i.e., among groups of three or more residues), they still require the inference of roughly 10^5–10^7 parameters for an average-sized protein (Ferguson and Ranganathan 2021). When properly parameterized, such models capture the higher-order statistics of protein sequences on par with some complex deep-learning models, and suffice to design small folded, functional protein domains (Russ et al. 2005, 2020; Socolich et al. 2005; Tian et al. 2018; McGee et al. 2021). Deep-learning approaches have displayed an incredible ability to directly capture higher-order couplings throughout a sequence, and recent work indicates they may improve our capacity to design clade-specific functional features (Ferguson and Ranganathan 2021; McGee et al. 2021; Lian et al. 2022). Yet pairwise potentials shine in their interpretability and capacity to generate insight into the physical architecture of proteins. For example, coevolution-based models have led to the hypothesis that proteins contain cooperative groups of collectively evolving residues (termed sectors) that are modularly associated to particular protein functions (Halabi et al. 2009). This has led to new ideas for the design of allostery and the evolution of protein specificity and activity (Lee et al. 2008; Raman et al. 2016; Pincus et al. 2017, 2018). Thus, by creating pairwise models alongside high-dimensional, deep-learning architectures, one can explore what information is both necessary and sufficient to encode function and extract general principles for the design of novel proteins.

STRATEGIES FOR STATISTICAL ANALYSIS OF MULTIPLE SEQUENCE ALIGNMENTS

The two primary approaches for creating pairwise statistical models of protein sequence are: (1) the statistical coupling analysis (SCA), and (2) a growing collection of algorithms for inferring Potts models of protein sequence (we group these under the header of direct coupling analysis [DCA]-like methods) (Fig. 1; Rivoire et al. 2016; Cocco et al. 2018). SCA was originally developed by Ranganathan and colleagues with the goal of mapping the thermodynamic couplings underpinning information transmission in proteins (Lockless and Ranganathan 1999). This method identified evolutionarily correlated networks of residues that formed physically contiguous paths in the protein tertiary structure. These distributed networks typically encompass 15%–30% of amino acid positions and were observed to link allosteric regulatory sites to active sites in several proteins (Süel et al. 2003; Shulman et al. 2004; Ferguson et al. 2007; Halabi et al. 2009; Rivoire et al. 2016). In contrast, DCA was originally reported as a method for local contact identification with the goal of facilitating structure prediction (Morcos et al. 2011). Early work on DCA expressly aimed to disentangle direct interactions (physical contacts), from indirect interactions emerging from transitive chains (if A interacts with B, and B interacts with C, then A will appear to interact with C) (Stein et al. 2015; Anishchenko et al. 2017). Drawing from earlier work on protein covariation analysis, isolation of direct contacts was accomplished by treating protein sequence modeling as an inverse problem, with the goal of discovering the maxi-

Cite this article as *Cold Spring Harb Perspect Biol* doi: 10.1101/cshperspect.a041463

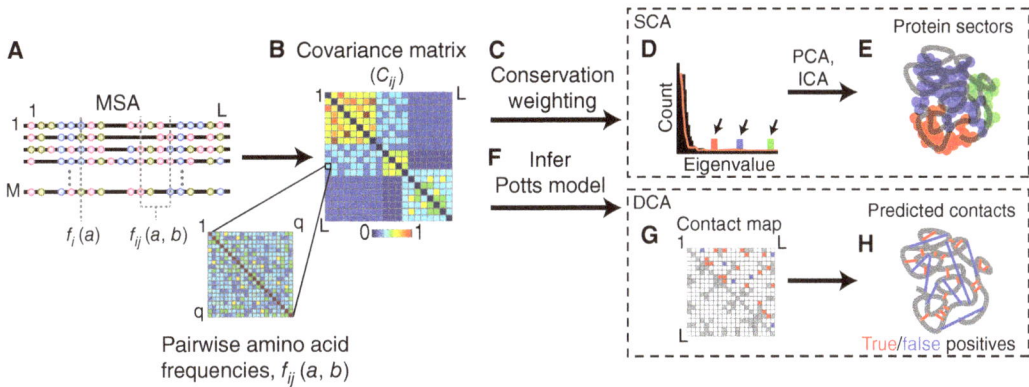

Figure 1. Overview of statistical coupling analysis (SCA) and direct coupling analysis (DCA). (*A*) Single-site and pairwise amino acid frequencies are calculated from an alignment of evolutionarily related sequences. (*B*) A four-dimensional covariance tensor is computed from the pairwise amino acid frequencies between each site. (*C*) In SCA, the covariance matrix is weighted based on conservation. This emphasizes correlations in amino acid frequency that may have functional relevance. (*D*) The eigenspectrum of the covariance matrix is shown as black bars. The eigenspectrum expected from randomly shuffled alignments is shown as a red curve. The top eigenmodes above random noise are selected. (*E*) Clustering from principal component analysis (PCA) is optimized by independent component analysis (ICA). Protein sectors are then defined from the resulting clusters. When mapped onto protein structures, sectors tend to form networks of physical contacts. (*F*) In DCA, maximum entropy methods are used to infer a Potts Hamiltonian model from the covariance matrix. (*G–H*) Residue pairs with a high direct information (DI) are predicted to be in contact in the folded protein.

mum entropy set of interactions underlying an observed pattern of amino acid covariances (Korber et al. 1993; Lapedes et al. 1999). DCA thus belongs to a larger family of pairwise maximum-entropy models, in which the probability of a given system configuration (here a specific protein sequence) is described by a Boltzmann distribution, and the system Hamiltonian is given by a Potts model (Stein et al. 2015). While individual methods vary in their strategy for inferring the Potts model parameters, DCA and DCA-related methods excel at identifying contacts and provide a probabilistic score of the compatibility of specific sequences with the statistical constraints of a given alignment (Zhang et al. 2021). These scores have some ability to predict the effects of mutation in sequence, but sometimes miss nonlocal interactions critical to function and evolvability (Hopf et al. 2017; Ding et al. 2022). Further work is necessary to better understand the mathematical relationship between SCA and DCA and their respective capacity to design protein fold and function; here we highlight recent progress and tools for assessing statistical sequence models.

A Model Is Only as Good as the Data: Evaluating Diversity and Phylogeny in Alignment Construction

Both SCA and DCA-like approaches begin by construction of a large and diverse MSA (Fig. 1A). This alignment is of dimensions M sequences by L positions, where each position (i), or column, contains one of 21 states (q, spanning the 20 natural amino acids and a gap). Well-made alignments are available through protein family servers, including ECOD, EGGNOG, UniProt, and InterPro (Cheng et al. 2014; Huerta-Cepas et al. 2019; Paysan-Lafosse et al. 2023; The UniProt Consortium 2023). However, alignment content will deeply influence the derived model. Custom alignments may be best depending on the application. Three (interrelated) factors to consider are alignment: (1) size, (2) diversity, and (3) phylogenetic structure.

Generally, larger alignments provide more statistical power. This is especially important for DCA-like models since they use information from less conserved positions and lower eigenvectors to capture contacts while SCA focuses on top eigenvectors (Cocco et al. 2013). In either case, typical alignment sizes are on the order of 10^3–10^5 sequences, while the number of parameters to be inferred is on the order of 10^5–10^7. Thus, even the largest alignments are undersampled. "Pseudocounting" of alignment frequencies is employed to avoid overfitting; this acknowledges that amino acid frequencies of zero likely reflect undersampling rather than a true impossibility of an amino acid at a site (Morcos et al. 2011). A related method is standard L2 regularization of the inferred model parameters (Kleeorin et al. 2023). Recent work using toy and real alignments has shown that alignment size and regularization strength profoundly impact detected sequence features. Lower regularization strengths improved prediction of pairwise interactions while higher regularization strengths favored detection of sector-like collections of residues (Kleeorin et al. 2023).

Alignment diversity can be assessed by examining the distribution of pairwise sequence identities across an MSA (Fig. 2A). SCA is typically applied to alignments with a mean pairwise sequence identity near 20%–40%; performance on more conserved alignments has not been thoroughly assessed (Rivoire et al. 2016). Because DCA uses information at less conserved sites, sequence diversity is also essential. Both approaches use sequence weights to downweight high-identity sequences. Given sequence weights, one may calculate the number of effective sequences (M_{eff}): $M_{eff} = \sum_s w_s$, where $w_s = 1/N_s$ is the sequence weight, and N_s is the number of sequences with an identity to sequence s above a threshold of 80% by default. Alignments for SCA and DCA should generally have more than 100 effective sequences (Rivoire et al. 2016).

Finally, alignment phylogenetic structure is key to identifying residues associated with clade-specific or class-specific features. For example, SCA of the mitogen activating protein kinase (MAPK) family identified a sector connecting the MAPK-binding groove to the active site. However, SCA of a eukaryotic protein kinome-wide alignment misses or "averages out" this clade-specific feature (Pincus et al. 2018). Inclusion of the noncatalytic haptoglobins in a broader alignment of S1A serine proteases permits identification of residues associated with catalytic function (Fig. 2B). Likewise, inclusion of the non-allosteric Hsp110 sequences in a broader alignment of Hsp70 molecular chaperones permits identification of residues associated with allosteric regulation with SCA (Halabi et al. 2009; Smock et al. 2010). Given this, SCA makes use of singular value decomposition (SVD) to relate coevolution among residues to functional and evolutionary divergences of sequences (Fig. 2B–D). SVD can generate new hypotheses about the functional role and evolutionary origin of sectors (Rivoire et al. 2016). However, for other tasks (like contact prediction) it may be desirable to remove phylogenetic structure. Entropy-based methods to correct for the coevolutionary signal include APC (average product correction), LRS (low-rank and sparse decomposition), and BND (balanced network deconvolution) are commonly used in DCA (Dunn et al. 2008; Sun et al. 2015; Zhang et al. 2016). Recent work from Wang and colleagues has proposed a spectral regularization strategy that allows one to partly remove phylogenetic signal from the covariance matrix prior to downstream statistical model creation or deep-learning applications (Wang et al. 2022).

Given that alignment size, regularization, and composition all influence the resulting statistical model, it is imperative to consider the robustness of any predictions to variations in these factors.

A Mathematical Overview of SCA and DCA

Once the alignment is assembled and appropriately weighted, SCA and DCA operate on the regularized single-site frequencies ($f_i(a)$) and pairwise covariances ($f_{ij}(a, b)$) computed across positions. These data are organized into an $L \times q \times L \times q$ tensor (Fig. 3, Equation 1) or alternately compressed into an $Lq \times Lq$ matrix, describing the covariance between each pair of amino acids at every pair of positions (Figs. 1B and 3A). From there, SCA and DCA take different approaches to analyze the pattern of covariance and distill out interacting residue groups (Figs. 1 and 3B,C). Below, we provide an overview of both methods and highlight three key differences: (1) DCA operates directly on the frequency-based covariance tensor (Fig. 1B), whereas SCA uses a conservation-weighted version that emphasizes

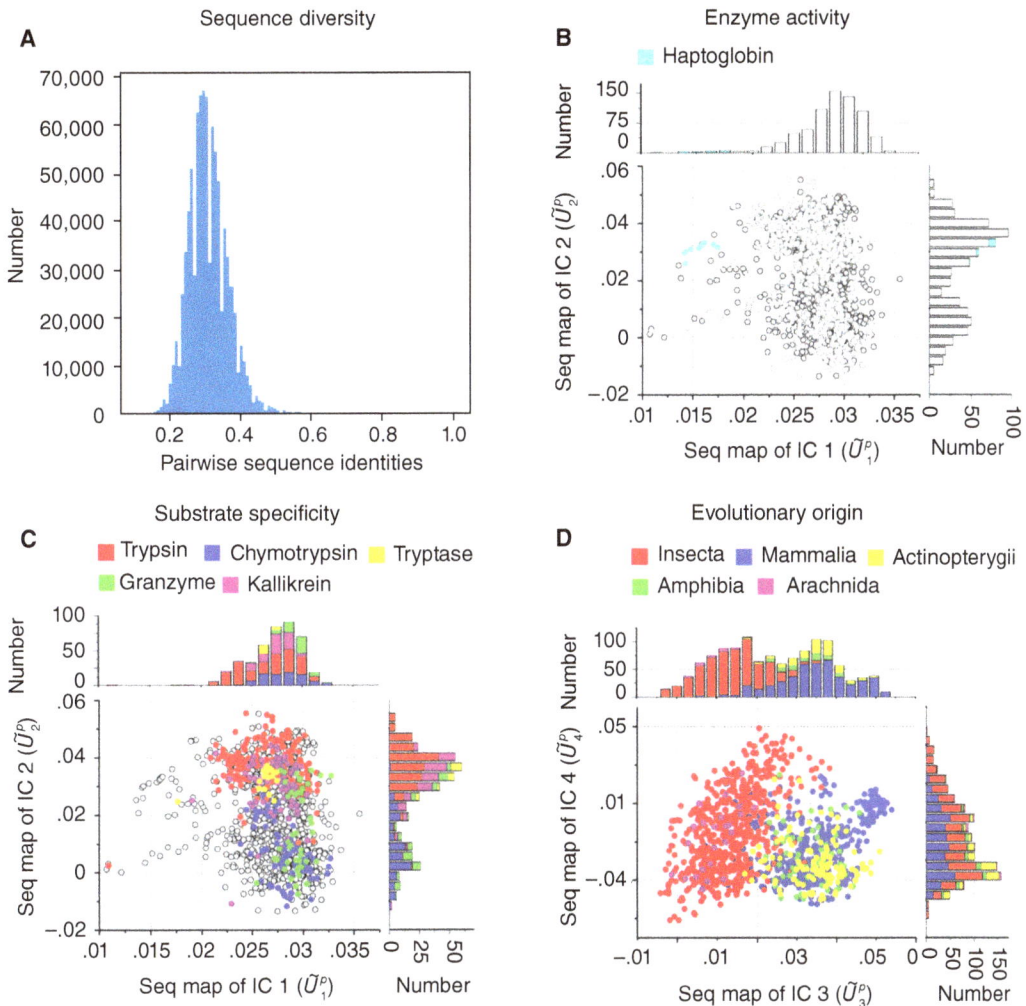

Figure 2. Alignment composition is important for statistical coupling analysis (SCA). (*A*) A histogram of pairwise sequence identity for an alignment of the S1A serine protease family. The sequences are normally distributed with a mean identity of ∼30%. (*B–D*) Using spectral value decomposition to interpret sectors. In the following plots, the contribution of individual sequences to the top singular vectors of the sequence correlation matrix are plotted as circles and color coded. The goal is to investigate the pattern of sequence divergences associated with positional correlations in the green, red, and blue sectors of the S1A serine protease. (*B*) IC1 of the S1A serine protease family separates functional sequences from the nonfunctional haptoglobins, consistent with the interpretation that the green sector (associated to the top singular vector of the positional correlation matrix) is associated with catalytic function. (*C*) IC2 of the S1A serine protease family separates sequences based on substrate specificity, consistent with the interpretation that the red sector (associated to the second singular vector of the positional correlation matrix) is associated with specificity. (*D*) IC3 of the S1A serine protease family separates sequences based on invertebrate/vertebrate origin. In this case, the blue sector (associated to the third singular vector of the positional correlation matrix) is associated to stability. (Panels *B–D* reprinted from Rivoire et al. 2016 under the terms of the Creative Commons Attribution License.)

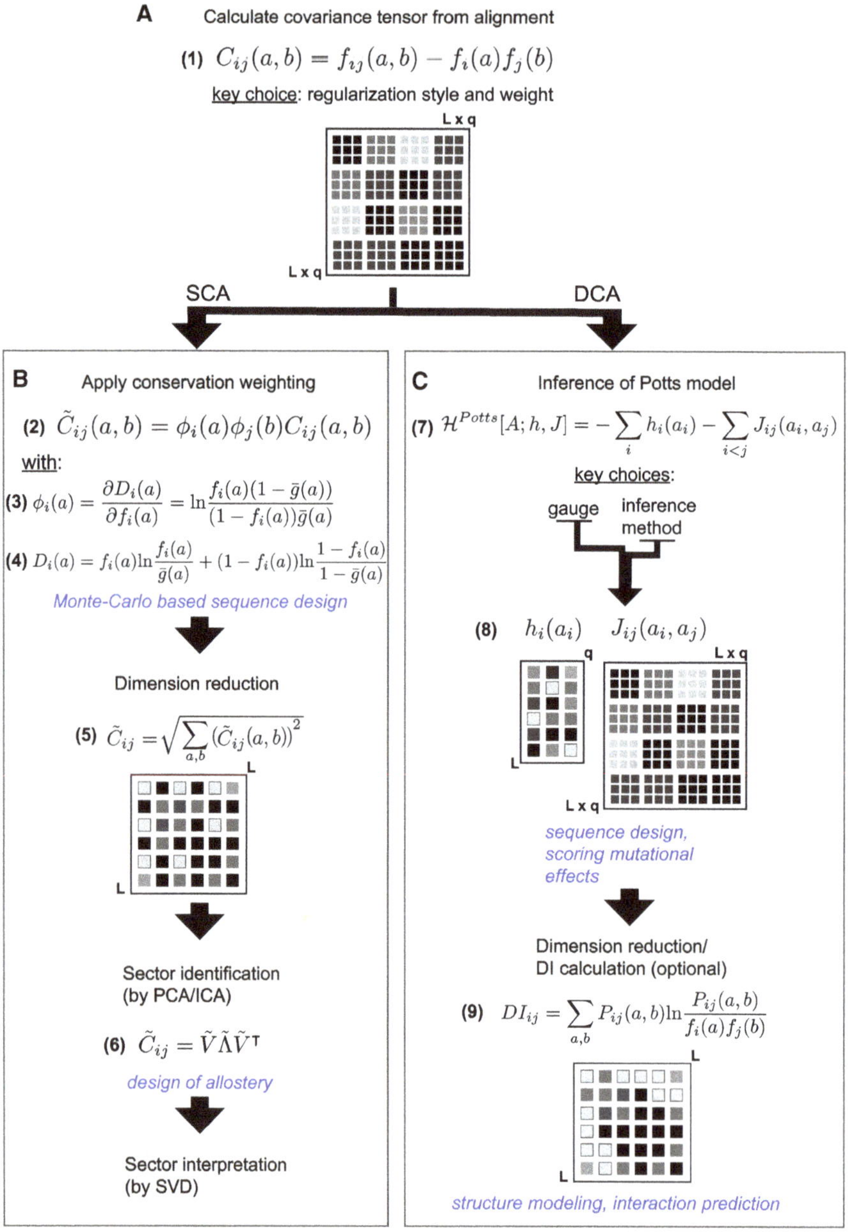

A   Calculate covariance tensor from alignment

(1) $C_{ij}(a,b) = f_{ij}(a,b) - f_i(a)f_j(b)$

<u>key choice</u>: regularization style and weight

SCA   DCA

B   Apply conservation weighting

(2) $\tilde{C}_{ij}(a,b) = \phi_i(a)\phi_j(b)C_{ij}(a,b)$

<u>with</u>:

(3) $\phi_i(a) = \dfrac{\partial D_i(a)}{\partial f_i(a)} = \ln\dfrac{f_i(a)(1-\bar{g}(a))}{(1-f_i(a))\bar{g}(a)}$

(4) $D_i(a) = f_i(a)\ln\dfrac{f_i(a)}{\bar{g}(a)} + (1-f_i(a))\ln\dfrac{1-f_i(a)}{1-\bar{g}(a)}$

Monte-Carlo based sequence design

Dimension reduction

(5) $\tilde{C}_{ij} = \sqrt{\sum_{a,b}(\tilde{C}_{ij}(a,b))^2}$

Sector identification
(by PCA/ICA)

(6) $\tilde{C}_{ij} = \tilde{V}\tilde{\Lambda}\tilde{V}^{\mathsf{T}}$

design of allostery

Sector interpretation
(by SVD)

C   Inference of Potts model

(7) $\mathcal{H}^{Potts}[A;h,J] = -\sum_i h_i(a_i) - \sum_{i<j} J_{ij}(a_i,a_j)$

<u>key choices</u>:

gauge   inference
          method

(8) $h_i(a_i)$ $J_{ij}(a_i,a_j)$

*sequence design,
scoring mutational
effects*

Dimension reduction/
DI calculation (optional)

(9) $DI_{ij} = \sum_{a,b} P_{ij}(a,b)\ln\dfrac{P_{ij}(a,b)}{f_i(a)f_j(b)}$

structure modeling, interaction prediction

Figure 3. The mathematical strategy of statistical coupling analysis (SCA) and direct coupling analysis (DCA). (*A*) Both methods begin by calculating a $L \times q \times L \times q$ tensor, which can be compressed into an $Lq \times Lq$ matrix. (*B*) SCA proceeds by applying conservation weights ϕ_i^a to the matrix $C_{ij}(a,b)$ to obtain the conservation weighted matrix $\tilde{C}_{ij}(a,b)$ (Equations 2–4). Next, the matrix is further compressed to obtain a measure of coevolution between positions (Equation 5). From there, principal component analysis (PCA) and independent component analysis (ICA) are applied to identify correlated residue groups (Equation 6); SVD can additionally be applied to relate positional correlations to sequence variation. (*C*) DCA proceeds by inferring a Hamiltonian describing the statistical energy $\mathcal{H}^{Potts}$ of an amino acid sequence A given the fields ($h_i(a)$) and couplings ($J_{ij}(a,b)$) (Equations 7 and 8). Key choices prior to model inference include gauge fixing, regularization strategy, and inference method. Once the Potts model is inferred, it can be used directly to score mutational effects and design new sequences; or further dimension reduced to identify direct residue contacts (Equation 9).

correlations at evolutionarily conserved positions (Fig. 1C), (2) DCA is an inverse method, whereas SCA proceeds by direct analysis of the covariance matrix, and (3) DCA provides a single-sequence score describing the likelihood that a given sequence is part of the modeled family, whereas SCA has no well-defined notion of a single sequence score. (All equations referenced in the text appear in Fig. 3.)

SCA

The original implementation of SCA took inspiration from double-mutant cycle analysis, an experimental approach to identify thermodynamic couplings between two residues in a protein (Horovitz and Fersht 1990; Lockless and Ranganathan 1999). The basic idea was to start by subsampling an MSA such that all sequences have amino acid a at site i, creating a first "mutation." Then, the effect of this perturbation on the conservation of amino acid frequencies at other sites, j, was quantified as a statistical free energy. The resulting measure captured correlations in the evolutionary conservation between sites, rather than correlations in frequency, a point which is important to newer SCA implementations. This perturbation-based approach allowed one to compute couplings between pairs of moderately conserved amino acids. However, the subsampling approach was difficult to implement for all pairs of positions due to severe undersampling.

The SCA approach was updated by Halabi et al. (2009), following the recognition that correlations in evolutionary conservation $\tilde{C}_{ij}(a, b) = \langle D_{i,A}^a D_{j,A}^b \rangle_A - \langle D_{i,A}^a \rangle_A \langle D_{j,A}^b \rangle_A$ observed across jackknife resampling of the alignment are equivalent to $\tilde{C}_{ij}(a, b) = \phi_i^a \phi_j^b C_{ij}(a, b)$ (Figure 3, Equation 2), a conservation weighted covariance matrix (Halabi et al. 2009). To explain, D_i^a is the Kullback–Leibler (KL) relative entropy of amino acid a at position i. The KL relative entropy provides a measure how "surprising" it is to observe the frequency of amino acid a at site i, $f_i(a)$, given the background expectation, $\bar{g}(a)$, of observing amino acid a in the National Center for Biotechnology Information (NCBI) nonredundant database (Figure 3, Equation 4). During jackknife

resampling, single sequences are iteratively removed from the MSA, creating an ensemble of M alignments, A. This process mimics the perturbation-based spirit of the original SCA implementation. For each alignment, one then computes the resulting pattern of conservation $D_{i,A}^a$. SCA then calculates the correlation in conservation across this ensemble of alignments. Conveniently, this jackknife resampling is not actually necessary because perturbative expansion of $D_i(a)$ as a function of $f_i(a)$ gives an equivalent analytical solution, that is, $\tilde{C}_{ij}(a, b) = \phi_i^a \phi_j^b C_{ij}(a, b)$ (Figure 3, Equation 2). In this equation, ϕ_i^a and ϕ_j^b, are calculated as the partial derivative of the KL divergence with respect to frequency (Figure 3, Equation 3). More conserved positions have greater values of ϕ; conservation weighting thus emphasizes correlations between evolutionarily conserved positions and de-emphasizes correlations at less conserved positions. This promotes the identification of pairwise correlations important for core aspects of protein function.

The resulting tensor after conservation weighting, $\tilde{C}_{ij}(a, b)$, is compressed to an $L \times L$ matrix by taking the Frobenius norm across amino acids (Figure 3, Equation 5). Principal component analysis (PCA) is then used as a dimension reduction technique to identify maximally independent coevolving groups of amino acids from this matrix (Figure 3, Equation 6). However, most of the correlations contained in the matrix are spurious, appearing due to statistical noise from undersampling of sequences in the MSA. To correct for this, SCA uses random matrix theory: eigendecomposition is performed on both $\tilde{C}_{ij}$ and many randomized versions of $\tilde{C}_{ij}$ computed from alignments in which the amino acid identities at each site are shuffled across a position (column) to remove correlations between sites. The eigenspectrum of these shuffled matrices will contain eigenmodes that occur only due to undersampling; eigenvalues greater than those in the control eigenspectrum are considered to represent an independently coevolving group of residues. These top k^* eigenvectors are then transformed with independent component analysis (ICA) to maximize their independence with minimal data loss. The resulting independent components (ICs) are groups of amino acids that represent independent functional modules within a protein family that are termed protein sectors. SVD can then be used to map variation within particular sectors back to specific patterns of sequence divergence (Fig. 2B–D). Based on observations in several protein families, it was proposed that sectors represent quasi-independent units of function within proteins (Halabi et al. 2009).

DCA

DCA-like models describe the probability of a sequence $P(a_1, a_2, \ldots a_L)$ in terms of a Boltzmann distribution, where the Hamiltonian is a generalized q-state Potts model as shown in Figure 3, Equation 7 (Stein et al. 2015). Calculating the Potts model of the MSA requires the inference of Lq fields ($h_i(a_i)$) and $L(L-1)/2 * q^2$ couplings ($J_{ij}(a_i, a_j)$) (Figure 3, Equations 8). These parameters are fit such that the calculated probabilities of individual sequences $P(a_1, a_2, \ldots a_L)$ recapitulate the observed $f_i(a)$ and $f_{ij}(a, b)$ terms from the alignment. However, because there are fewer independent frequency terms than inferred parameters, the model is overparameterized, resulting in a problem called gauge invariance. To infer the full Potts model, one must arbitrarily fix the gauge so that the number of parameters reduces to the number of degrees of freedom that must be satisfied. There are several strategies to fix the gauge, including the lattice gas gauge and the zero-sum gauge, and it is worth considering which is appropriate to a given problem. For example, the lattice gas gauge will arbitrarily remove symmetry in the Potts variables, unlike the zero-sum gauge (Cocco et al. 2018).

The next step is to decide the inference method. One strategy is mean-field approximation, in which the direct couplings are estimated by inverting the covariance matrix, $C_{ij}(a, b):J_{ij}(a, b) = -(C^{-1})_{ij}(a, b)$. The fields $h_i(a)$ and $h_j(b)$ are then selected such that the Potts probability model recapitulates the observed $f_i(a)$ and $f_{ij}(a, b)$ terms given the couplings. While the mean field approach is computationally efficient, and relatively straightforward, it is less accurate in recapitulating pairwise alignment frequencies than other more sophisticated computational methods. Thus, one

may consider other inference approaches including Boltzmann machine learning (bmDCA), gaussian approximation, pseudolikelihood maximization (plmDCA), adaptive cluster expansion (ACE), or autoregressive models (arDCA) (Jones et al. 2012; Baldassi et al. 2014; Ekeberg et al. 2014; Barton et al. 2016; Figliuzzi et al. 2018; Trinquier et al. 2021). Many of these are freely available as open-source code, and are suited to different tasks. For contact prediction, mfDCA is a fast and simple approach, whereas plmDCA is slower, yet more accurate (Ekeberg et al. 2014). For de novo sequence design, methods such as bmDCA, ACE, or arDCA are better choices (Jacquin et al. 2016; Cocco et al. 2018; Trinquier et al. 2021).

Once the Potts model has been inferred, it can be used to sample artificial sequences from the Boltzmann distribution. Additionally, the inferred Potts model enables one to predict contacts within a protein structure by estimating the direct information (DI), a position-level measure of coupling (Figure 3, Equation 9; see Morcos et al. 2011 for additional explanation). Top-ranking DI pairs were shown to be highly enriched for structural contacts (Morcos et al. 2011). Though these contacts represent local pairwise interactions, recent work has examined the connection between these local interactions and larger sector-like networks. Shiau et al. (2022) used a DCA-like model to extract protein sectors through a spectral clustering approach. The resulting sectors could not be identified from contact maps or structures alone. Intriguingly, they found that proteins with highly similar structures did not necessarily share the same sectors.

EVALUATING STATISTICAL MODELS OF PROTEIN SEQUENCE

The statistical models created by SCA and DCA are readily used to generate new synthetic sequences. In the case of DCA, new sequences can be created by Monte Carlo sampling the Boltzmann distribution at a specified temperature, or directly drawing sequences from the probabilistic sequence model (Russ et al. 2020; Trinquier et al. 2021). In the case of SCA, one can use Monte Carlo simulations to iteratively shuffle and select the positions (columns) of a synthetic alignment until it recovers the SCA matrix of a target protein

Cite this article as *Cold Spring Harb Perspect Biol* doi: 10.1101/cshperspect.a041463

family (Socolich et al. 2005; Reynolds et al. 2013). Given the resulting collections of synthetic sequences, we need clear assessment tools to help guide and refine algorithmic choices. Here we discuss quantitative techniques to computationally benchmark design strategies, experimental methods that test the function of designed sequences, and give some sense of the state of the field. Methods for sequence design are diverse, but we refer to them broadly as generative protein sequence models (GPSMs). While we focus on pairwise statistical models, these assessment methods readily generalize to deep-learning models.

Computational Approaches for Assessing Model Fit and Generative Capacity

GPSMs can be evaluated computationally either in the context of "toy models" or with actual sequences. Because toy models are fully controlled, they have proven particularly informative for comparing the effects of varied algorithmic choices on model inference and designed sequence outcome. For example, Kleeorin et al. (2023) used a toy model that specified evolutionary couplings in a 20-amino-acid "protein" to evaluate the impact of alignment size and regularization strength on the detection of interactions. Lattice proteins provide another toy model that additionally captures three-dimensional structure: Jacquin et al. (2016) used these to compare different Potts model inference strategies, and to better understand the interpretation of the fit couplings. Moving beyond toy models, a quality GPSM should (1) replicate the patterns of covariance observed in the natural alignment, (2) produce sequences equally or more diverse than the training set, and (3) replicate phylogenetic subpopulations within natural protein families.

Higher-Order Covariances

An important first step is to evaluate how well the model fits the data. This is typically established by creating correlation plots of pairwise covariation in the natural and synthetic alignments and evaluating the R^2. However, a good generative model should recover features beyond the training set, extending to three-point or higher-order covariances. McGee

et al. (2021) developed a method to systematically evaluate higher-order covariances based on the comparison of "word" frequencies within an MSA. Each word consists of L randomly selected sets of positions in the MSA; increasing the word length amounts to increasing the order of covariance being examined (Fig. 4A,B). The frequency of each word identity in the synthetic MSA is plotted against that of the natural "target" MSA and the Pearson correlation coefficient (r) is calculated. The "r_{20} score" is the average r for the top 20 most frequent word identities over many randomly sampled words. The r_{20} score showed that a Potts model outperformed two deep-learning models in replicating higher-order covariances (McGee et al. 2021). The r_{20} score will prove particularly useful if it is found to be true that replicating covariances at higher than pairwise orders result in improved functionality of artificial sequences.

Sequence Diversity

A straightforward computational method to judge the quality of a GPSM is to assess the diversity of the artificial sequences it generates. This is accomplished by calculating the distribution of Hamming distances of each artificial sequence to the consensus sequence (Jacquin et al. 2016; McGee et al. 2021). A quality GPSM will result in a similar Hamming distance distribution to natural MSAs, and potentially identify lower-identity novel sequences with similar functions (Fig. 4C–E). A more sophisticated approach is to analyze phylogenetic structure. This is accomplished using PCA to visualize natural and artificial sequence distributions in dimensionally reduced principal-component space (Fig. 4F–H; Russ et al. 2020; Trinquier et al. 2021). The clustering observed in this analysis depicts subpopulations within the natural protein family and artificial MSAs from different GPSMs. This method is noteworthy to determine whether a GPSM is capturing subtle diversities within a natural MSA.

Experimental Testing of Computational Protein Designs

While computational approaches toward refinement, benchmarking, and validation are impor-

Recovering high-order covariances

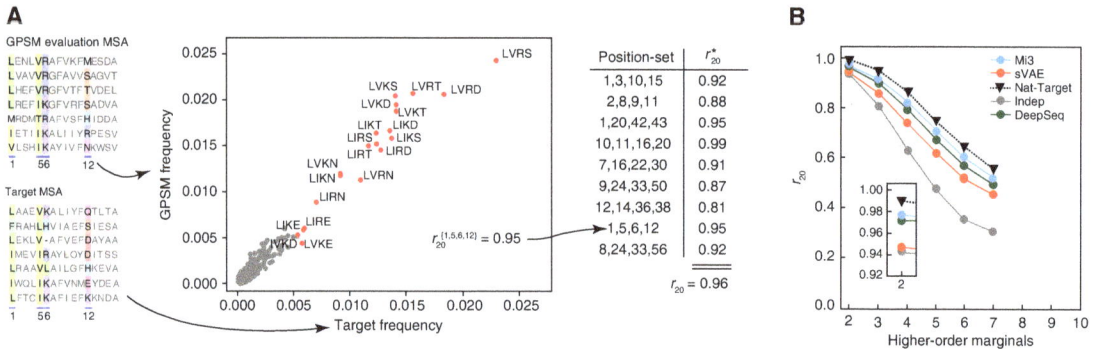

Diversity of designed sequences

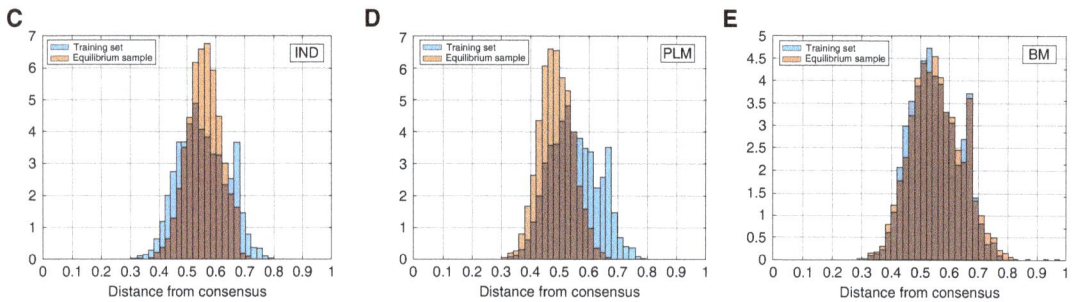

Replicating phylogenetic structure

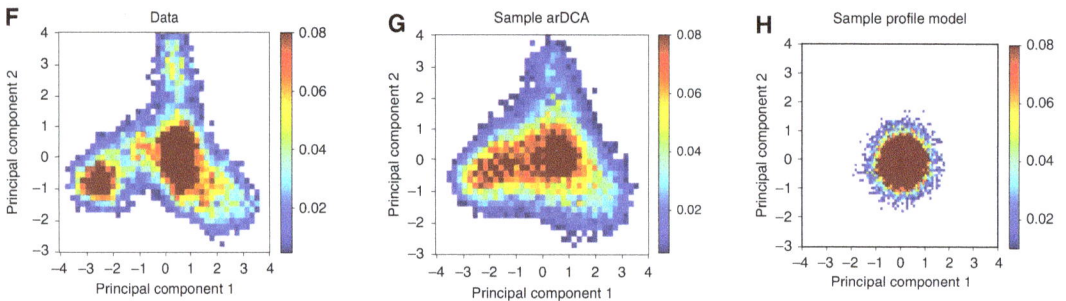

Figure 4. Computational methods for evaluating model performance. (*A*) Explanation of the r_{20} metric for assessing higher-order covariance in designed sequences. The frequency of short collections of residues called "words" in designed sequences are plotted against their frequency in a natural alignment. The r_{20} is the average correlation coefficient for the 20 most frequent words. (*B*) The r_{20} metric can be used to assess the ability of different generative models to recover emergent higher-order covariances from pairwise information. (*C–E*) Distributions of sequence diversity of natural proteins (blue) and designed sequences (orange) for three algorithms that infer Potts model parameters. (*C*) The site-independent model ignores pairwise correlations and serves as a control. (*D*) Pseudolikelihood maximization failed to generate sequences with natural-like diversity. (*E*) The Boltzmann machine algorithm successfully recapitulates the distribution of diversity in natural sequences. (*F–H*) Principal component analysis (PCA) can be used to visualize the phylogenetic structure of natural and artificial sequence alignments. (*F*) An alignment of sequences in the PF00072 protein family visualized in principle component space. Multiple evolutionary subpopulations can be observed in the natural sequences. (*G*) The autoregressive direct coupling analysis (arDCA) generative model can replicate the phylogenetic structure of the natural protein family. (*H*) A control generative model trained on single site frequencies alone cannot replicate the phylogenetic structure. (Panels *A–B* reproduced from McGee et al. 2021 under the terms of Creative Commons Attribution 4.0 International License. Panels *C–E* reproduced from Cocco et al. 2018 with permission from IOP Publishing, Ltd. © 1934. Panels *F–H* reprinted from Trinquier et al. 2021 under a Creative Commons Attribution 4.0 International License.)

tant, assessing the predictive capacity of SCA and DCA with experimental data is essential. Here we discuss three broad areas where SCA and DCA have been applied to make predictions and engineer function: (1) comparisons with existing structural and functional data, (2) design and optimization of allostery, and (3) de novo design of protein function.

Comparisons between Contact Prediction and Structural Data

Abundant high-resolution structural data facilitates testing DCA as a method of contact prediction (Fig. 5A; Berman 2000). DCA is robustly able to predict contacts for use as distance restraints in structure determination; indeed AlphaFold is largely predicated upon the idea of using sequence information to distill out distance restraints for structure prediction (Morcos et al. 2011; Ekeberg et al. 2013, 2014; Jumper et al. 2021). DCA has also been applied to cases beyond soluble single domain proteins including membrane protein structure prediction, identification of contacts underlying alternate conformational states, predicting the structure of non-coding RNA complexes, and interaction prediction for both bacterial as well as human protein complexes (Ovchinnikov et al. 2014; Palovcak et al. 2015; Bitbol et al. 2016; Weinreb et al. 2016; Humphreys et al. 2021). Particularly for protein complexes, contact prediction is quite sensitive to the quality and structure of the MSA (Ekeberg et al. 2013, 2014; Baldassi et al. 2014; dos Santos et al. 2015). Furthermore, subdividing within protein families according to their domain architecture has been shown to strongly improve the predictive power of DCA, illuminating differences between the domains that are hidden when all are taken together (Uguzzoni et al. 2017).

Prediction of Mutational Fitness Effects

Going beyond using coevolutionary information to predict structural information requires quantitative data on protein function both in vitro and in vivo (Fig. 5B). In some cases, a small number of focused experiments have been used to assess coevolutionary predictions. For example, SCA predictions have been compared to double mutant cycles over ligand-binding free energy in the PDZ domain (Lockless and Ranganathan 1999). Additionally, double-mutant cycle analysis between sectors in the S1A serine protease supported the idea that each sector contributes independently to function (Halabi et al. 2009). Increasingly, more comprehensive testing is now possible thanks to recent advances in next-generation sequencing–based measurements of growth rate (Chubiz et al. 2012; Boucher et al. 2014; Fowler and Fields 2014). In this approach, assay conditions are devised such that growth rate is coupled with the activity of the protein of interest. Because next-generation sequencing enables the measurement of thousands of growth rates in a single experiment, it becomes possible to quantify the effects of all possible single mutants, and in some cases double mutants, on function (Olson et al. 2014; Diss and Lehner 2018). Both DCA and SCA have been assessed in comparison to deep mutational scanning data; DCA more extensively so (McLaughlin et al. 2012; Figliuzzi et al. 2016; Hopf et al. 2017; Laine et al. 2019; Trinquier et al. 2021). Perhaps most significantly of these, the DCA-based EVmutation algorithm clearly outperforms coupling-free models, highlighting the importance of epistasis in mutation prediction (Hopf et al. 2017). However, because growth-based outputs often report on several aspects of protein function simultaneously (stability, abundance, and activity), it remains unclear which aspects of function are captured by these coevolutionary models. In this case, in vitro measurements can help distill the predictive capacity for biochemical and biophysical parameters (Xie et al. 2022). Further, recent advances in both the high-throughput microfluidic determination of biochemical parameters and mathematical strategies for inferring quantitative biophysical parameters from growth rate measurements suggest powerful approaches to disambiguating the impact of mutations on protein function at scale (Otwinowski 2018; Markin et al. 2021; Faure et al. 2022; Weng et al. 2022).

Engineering Allosteric Regulation

Because SCA sectors are hypothesized to represent cooperative mechanisms for information transmission in proteins, it is essential to test the

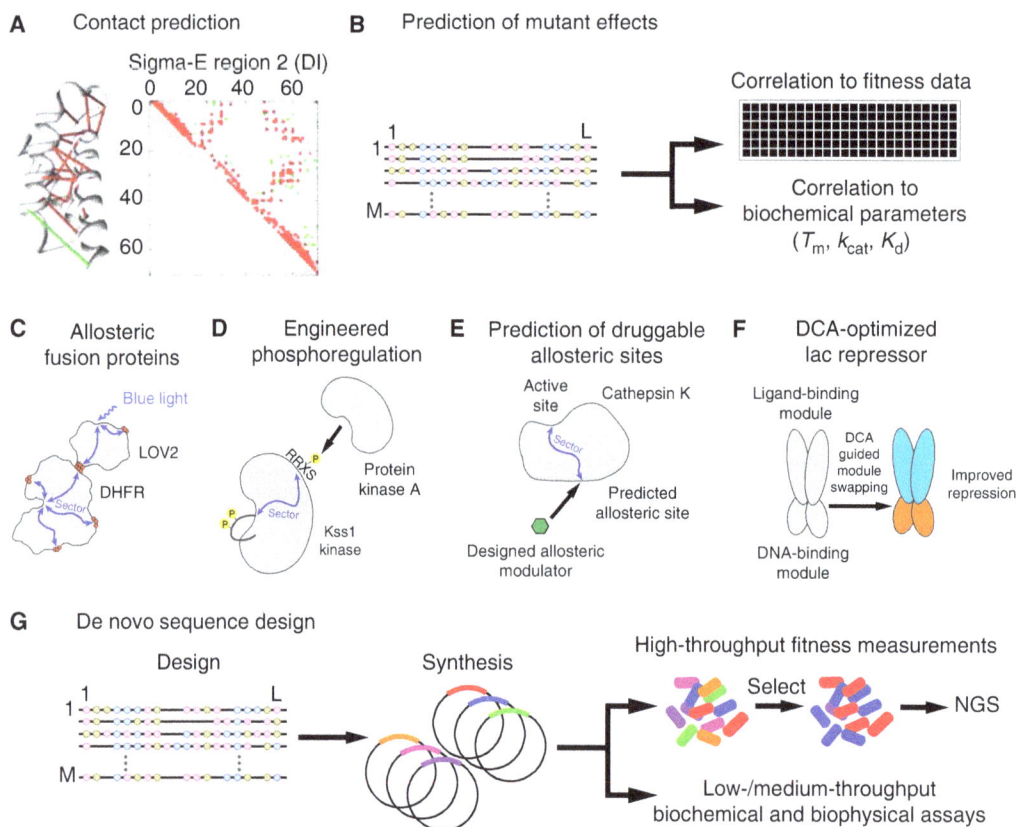

Figure 5. Notable applications of direct coupling analysis (DCA) and statistical coupling analysis (SCA). (*A*) DCA can be used to predict contacts in natively folded proteins. (Panel *A* is reproduced from Morcos et al. 2011 under the terms of Creative Commons Fair Use License.) (*B*) DCA scores can be correlated with high-throughput fitness data or biochemical parameters to predict mutational effects based on DCA score. (*C*–*F*) Examples of SCA and DCA being used to engineer allosteric regulation. (*C*) The light-sensitive LOV2 domain can be fused to sector-contacting surface sites on dihydrofolate reductase (DHFR) to produce light-regulated allosteric regulation of DHFR activity. (*D*) The protein kinase A (PKA) consensus motif RRXS can be engineered into sector-contacting surfaces of Kss1 kinase to create a new mechanism of Kss1 phosphoregulation. (*E*) Small molecules can be designed to bind to sector-connected surface sites to allosterically modulate protein function. (*F*) DCA has been used to predict transcriptional repressor module pairs with optimized gene repression activity. (*G*) High-throughput fitness or biochemical assays can be used to screen designed sequences for functionality.

role of sectors in allostery. Prior work has found that sector-connected edges are "hotspots" for the introduction of new allostery by domain insertion (Fig. 5C), can be used to introduce new phospho-regulation (Fig. 5D), and have been used to identify druggable allosteric sites (Fig. 5E; Lee et al. 2008; Reynolds et al. 2011; Novinec et al. 2014; Pincus et al. 2018). Interestingly, more recent work found that allostery-tuning mutations are depleted within the sector, indicating that the strategies for introducing and optimizing allostery

may differ (McCormick et al. 2021). DCA has also been used to optimize allosteric communication in synthetic transcriptional repressors created by hybridizing DNA-binding domains and ligand-binding domains to rewire genetic networks (Fig. 5F; Dimas et al. 2019; Jiang et al. 2021).

De Novo Sequence Design and Testing

Generated hypothetical sequences are often produced in relatively large numbers to comprehensively test the design capacity of GPSMs and examine statistical "success rates." Moreover, prior work has often compared synthetic sequences to libraries of natural sequences (as a positive control), and libraries created by simpler models (single-site conservation, as a test of necessity of pairwise information) (Socolich et al. 2005; Russ et al. 2020). In any case, testing designed sequences begins with large scale gene synthesis and protein expression (Fig. 5G). The cost of gene synthesis has led many to pursue smaller protein domains as test targets. Once expressed, these sequences can be evaluated either in vivo or in vitro (following purification). While in vitro characterizations provide quantitative structural and biochemical information, they lack the throughput of most in vivo screens.

Prior work using both SCA and DCA led to the design of small ligand binding domains that fold and bind native ligand with good affinity. These synthetic proteins could be designed to new roles or even generate "bridge" sequences that are able to switch between different folds (Alexander et al. 2009; Tian et al. 2018). In one of the earliest examples of statistical sequence design, Ranganathan and colleagues used SCA to design folded, functional WW domains (Russ et al. 2005; Socolich et al. 2005). More recently, DCA was used to design a series of domains that folded into the correct structure and sometimes showed high-affinity ligand binding (SH3, and GA and GB domains of streptococcal protein G, Tian et al. 2018). Going beyond binding proteins, Russ et al. have produced synthetic chorismate mutase enzymes with catalytic activities on par with their natural counterparts (Russ et al. 2020). Together these studies are encouraging for the statistics-based design of sequences and motivate testing on larger and more complex targets.

OPEN PROBLEMS AND FUTURE DIRECTIONS

We identify two major challenges for the de novo design of sequences using statistical methods. First is the need for tools that properly account for variable cellular context. Years of work on orthologous gene complementation provides numerous examples where a functional gene from one organism does not rescue growth or phenotype in the context of another (Michener et al. 2014; Bershtein et al. 2015; Kachroo et al. 2015). Presumably, this occasional lack of fungibility reflects additional cell-context-specific constraints on activity, stability, specificity, or interaction partners. We need new approaches to design sequences with increased compatibility in a given cell environment or context. In prior work, labeled assay data was combined with statistical sequence models to create refined context-specific models (Russ et al. 2020; Hsu et al. 2022). Additionally, recent work indicates that some machine learning algorithms can capture sequence features important to paralog-specific function, enabling a designed protein to function in a particular signaling network without a need for post hoc model refinement on experimental training data (Lian et al. 2022).

The second challenge is creation of truly novel sequences with non-natural function. Current GPSMs create synthetic sequences that mimic their natural counterparts. To some extent, these designs are "chimeras" of evolved proteins that piece together fragments of natural sequences. Encouragingly, recent deep-learning models exhibited some capacity to make predictions for regions of sequence space outside the original training set (Biswas et al. 2018). Additionally, a new class of artificial neural network—called essence neural networks (ENNs)—shows promise in generalizing beyond training data while maintaining more interpretability than standard gradient-based deep-learning approaches (Blazek and Lin 2021).

The need for direct comparison of pairwise statistical models such as SCA or DCA and various machine learning–based models is increasingly apparent. While these approaches can all be used to generate novel protein sequences, their performances may differ for various design tasks. Head-to-head comparisons of these approaches may reveal appropriate use cases for each type of model. For example, EVmutation, a pairwise statistical model, and DeepSequence, a deep-learning model, were both unable to predict mutations to a toxin that rescued binding to a mutated antitoxin, whereas a simple nonlinear machine learning model successfully predicted mutational effects on the toxin–antitoxin interaction (Ding et al. 2022).

Additionally, the incorporation of coevolutionary sequence information into machine learning models encourages further research toward using these approaches synergistically (Hsu et al. 2022). It has been noted that while machine-learning applications such as AlphaFold apply the full coevolution tensor as an input, they do not apply methods to correct for entropic "noise" from phylogeny as in DCA (Wang et al. 2022). Large language models (LLMs) such as ProGen are trained on millions of unaligned protein sequences and lack coevolutionary assumptions (Madani et al. 2023). While LLMs show an impressive ability to design functional protein sequences, incorporating additional coevolutionary constraints may allow them to capture subtle evolutionary features of some protein families.

Regardless, it seems clear that using statistical sequence models to guide directed evolution experiments will prove a powerful approach to create non-natural functions with applications in agriculture, biotechnology, biofuels, and biomedicine (Yang et al. 2019).

CONCLUDING REMARKS

The past two decades have resulted in great strides toward extracting and understanding the evolutionary information contained within protein families. These models enable the accurate prediction of contacts within and between proteins, as well as the identification of coevolved sectors of residues that modularly regulate protein functions. SCA and DCA allow for the identification of allosteric sites to introduce regulation, as well as domain swaps to engineer new function. Solving the protein design problem has been a longstanding challenge in computational biophysics and the application of these statistical approaches to the problem have made impressive advances, especially in the design of enzymes with natural-like functions (Tian et al. 2018; Russ et al. 2020). Efficient experimental validation of GPSMs as well as high-throughput characterization of mutational effects continues to present a major hurdle for the field. Currently, an understanding of the factors that cause non-contacting residues to be epis-

tatically linked, and thus detected, by statistical models of coevolutionary sequence information can only be determined by onerous experimental characterization. Additionally, GPSMs that generate protein sequences with non-natural functions have not yet been introduced. These challenges, if solved, would mark an astonishing breakthrough in the field of protein design, dramatically increasing our ability to engineer new and useful proteins.

ACKNOWLEDGMENTS

The authors thank Marielle A.X. Russo for discussion, and members of the Reynolds laboratory for their discussion and feedback. Our work on this review was supported by NSF CAREER award 1942354 to K.A.R.; J.C.D. was supported in part by NIH training grant T32GM131963.

REFERENCES

Alexander PA, He Y, Chen Y, Orban J, Bryan PN. 2009. A minimal sequence code for switching protein structure and function. *Proc Natl Acad Sci* **106:** 21149–21154. doi:10.1073/pnas.0906408106

Anfinsen CB. 1973. Principles that govern the folding of protein chains. *Science* **181:** 223–230. doi:10.1126/science.181.4096.223

Anfinsen CB, Scheraga HA. 1975. Experimental and theoretical aspects of protein folding. In *Advances in protein chemistry* (ed. Anfinsen CB, et al.), Vol. 29, pp. 205–300. Academic Press, New York.

Anishchenko I, Ovchinnikov S, Kamisetty H, Baker D. 2017. Origins of coevolution between residues distant in protein 3D structures. *Proc Natl Acad Sci* **114:** 9122–9127. doi:10.1073/pnas.1702664114

Baker D. 2014. Centenary award and Sir Frederick Gowland Hopkins memorial lecture. Protein folding, structure prediction and design. *Biochem Soc Trans* **42:** 225–229. doi:10.1042/BST20130055

Baldassi C, Zamparo M, Feinauer C, Procaccini A, Zecchina R, Weigt M, Pagnani A. 2014. Fast and accurate multivariate Gaussian modeling of protein families: predicting residue contacts and protein-interaction partners. *PLoS ONE* **9:** e92721. doi:10.1371/journal.pone.0092721

Barton JP, De Leonardis E, Coucke A, Cocco S. 2016. ACE: adaptive cluster expansion for maximum entropy graphical model inference. *Bioinformatics* **32:** 3089–3097. doi:10.1093/bioinformatics/btw328

Berman HM. 2000. The protein data bank. *Nucleic Acids Res* **28:** 235–242. doi:10.1093/nar/28.1.235

Bershtein S, Serohijos AWR, Bhattacharyya S, Manhart M, Choi JM, Mu W, Zhou J, Shakhnovich EI. 2015. Protein homeostasis imposes a barrier on functional integration of horizontally transferred genes in bacteria. *PLoS Genet* **11:** e1005612. doi:10.1371/journal.pgen.1005612

Biswas S, Kuznetsov G, Ogden PJ, Conway NJ, Adams RP, Church GM. 2018. Toward machine-guided design of proteins. bioRxiv doi:10.1101/337154

Bitbol AF, Dwyer RS, Colwell LJ, Wingreen NS. 2016. Inferring interaction partners from protein sequences. *Proc Natl Acad Sci* **113:** 12180–12185. doi:10.1073/pnas.1606762113

Blazek PJ, Lin MM. 2021. Explainable neural networks that simulate reasoning. *Nat Comput Sci* **1:** 607–618. doi:10.1038/s43588-021-00132-w

Boucher JI, Cote P, Flynn J, Jiang L, Laban A, Mishra P, Roscoe BP, Bolon DNA. 2014. Viewing protein fitness landscapes through a next-gen lens. *Genetics* **198:** 461–471. doi:10.1534/genetics.114.168351

Cheng H, Schaeffer RD, Liao Y, Kinch LN, Pei J, Shi S, Kim BH, Grishin NV. 2014. ECOD: an evolutionary classification of protein domains. *PLoS Comput Biol* **10:** e1003926. doi:10.1371/journal.pcbi.1003926

Chubiz LM, Lee MC, Delaney NF, Marx CJ. 2012. FREQ-Seq: a rapid, cost-effective, sequencing-based method to determine allele frequencies directly from mixed populations. *PLoS ONE* **7:** e47959. doi:10.1371/journal.pone.0047959

Cocco S, Monasson R, Weigt M. 2013. From principal component to direct coupling analysis of coevolution in proteins: low-eigenvalue modes are needed for structure prediction. *PLoS Comput Biol* **9:** e1003176. doi:10.1371/journal.pcbi.1003176

Cocco S, Feinauer C, Figliuzzi M, Monasson R, Weigt M. 2018. Inverse statistical physics of protein sequences: a key issues review. *Rep Prog Phys* **81:** 032601. doi:10.1088/1361-6633/aa9965

Dimas RP, Jiang XL, Alberto de la Paz J, Morcos F, Chan CTY. 2019. Engineering repressors with coevolutionary cues facilitates toggle switches with a master reset. *Nucleic Acids Res* **47:** 5449–5463. doi:10.1093/nar/gkz280

Ding D, Green AG, Wang B, Lite TLV, Weinstein EN, Marks DS, Laub MT. 2022. Co-evolution of interacting proteins through non-contacting and non-specific mutations. *Nat Ecol Evol* **6:** 590–603. doi:10.1038/s41559-022-01688-0

Diss G, Lehner B. 2018. The genetic landscape of a physical interaction. *eLife* **7:** e32472. doi:10.7554/eLife.32472

dos Santos RN, Morcos F, Jana B, Andricopulo AD, Onuchic JN. 2015. Dimeric interactions and complex formation using direct coevolutionary couplings. *Sci Rep* **5:** 13652. doi:10.1038/srep13652

Dunn SD, Wahl LM, Gloor GB. 2008. Mutual information without the influence of phylogeny or entropy dramatically improves residue contact prediction. *Bioinformatics* **24:** 333–340. doi:10.1093/bioinformatics/btm604

Ekeberg M, Lövkvist C, Lan Y, Weigt M, Aurell E. 2013. Improved contact prediction in proteins: using pseudolikelihoods to infer Potts models. *Phys Rev E* **87:** 012707. doi:10.1103/PhysRevE.87.012707

Ekeberg M, Hartonen T, Aurell E. 2014. Fast pseudolikelihood maximization for direct-coupling analysis of protein structure from many homologous amino-acid sequences. *J Comput Phys* **276:** 341–356. doi:10.1016/j.jcp.2014.07.024

Faure AJ, Domingo J, Schmiedel JM, Hidalgo-Carcedo C, Diss G, Lehner B. 2022. Mapping the energetic and allosteric landscapes of protein binding domains. *Nature* **604:** 175–183. doi:10.1038/s41586-022-04586-4

Ferguson AL, Ranganathan R. 2021. 100th anniversary of macromolecular science viewpoint: data-driven protein design. *ACS Macro Lett* **10:** 327–340. doi:10.1021/acsmacrolett.0c00885

Ferguson AD, Amezcua CA, Halabi NM, Chelliah Y, Rosen MK, Ranganathan R, Deisenhofer J. 2007. Signal transduction pathway of TonB-dependent transporters. *Proc Natl Acad Sci* **104:** 513–518. doi:10.1073/pnas.0609887104

Figliuzzi M, Jacquier H, Schug A, Tenaillon O, Weigt M. 2016. Coevolutionary landscape inference and the context-dependence of mutations in beta-lactamase TEM-1. *Mol Biol Evol* **33:** 268–280. doi:10.1093/molbev/msv211

Figliuzzi M, Barrat-Charlaix P, Weigt M. 2018. How pairwise coevolutionary models capture the collective residue variability in proteins? *Mol Biol Evol* **35:** 1018–1027. doi:10.1093/molbev/msy007

Fowler DM, Fields S. 2014. Deep mutational scanning: a new style of protein science. *Nat Methods* **11:** 801–807. doi:10.1038/nmeth.3027

Halabi N, Rivoire O, Leibler S, Ranganathan R. 2009. Protein sectors: evolutionary units of three-dimensional structure. *Cell* **138:** 774–786. doi:10.1016/j.cell.2009.07.038

Hopf TA, Ingraham JB, Poelwijk FJ, Schärfe CPI, Springer M, Sander C, Marks DS. 2017. Mutation effects predicted from sequence co-variation. *Nat Biotechnol* **35:** 128–135. doi:10.1038/nbt.3769

Horovitz A, Fersht AR. 1990. Strategy for analysing the co-operativity of intramolecular interactions in peptides and proteins. *J Mol Biol* **214:** 613–617. doi:10.1016/0022-2836(90)90275-Q

Hsu C, Nisonoff H, Fannjiang C, Listgarten J. 2022. Learning protein fitness models from evolutionary and assay-labeled data. *Nat Biotechnol* **40:** 1114–1122. doi:10.1038/s41587-021-01146-5

Huerta-Cepas J, Szklarczyk D, Heller D, Hernández-Plaza A, Forslund SK, Cook H, Mende DR, Letunic I, Rattei T, Jensen LJ, et al. 2019. eggNOG 5.0: a hierarchical, functionally and phylogenetically annotated orthology resource based on 5090 organisms and 2502 viruses. *Nucleic Acids Res* **47:** D309–D314. doi:10.1093/nar/gky1085

Humphreys IR, Pei J, Baek M, Krishnakumar A, Anishchenko I, Ovchinnikov S, Zhang J, Ness TJ, Banjade S, Bagde SR, et al. 2021. Computed structures of core eukaryotic protein complexes. *Science* **374:** eabm4805. doi:10.1126/science.abm4805

Jacquin H, Gilson A, Shakhnovich E, Cocco S, Monasson R. 2016. Benchmarking inverse statistical approaches for protein structure and design with exactly solvable models. *PLoS Comput Biol* **12:** e1004889. doi:10.1371/journal.pcbi.1004889

Jiang XL, Dimas RP, Chan CTY, Morcos F. 2021. Coevolutionary methods enable robust design of modular repressors by reestablishing intra-protein interactions. *Nat Commun* **12:** 5592. doi:10.1038/s41467-021-25851-6

Jones DT, Buchan DWA, Cozzetto D, Pontil M. 2012. PSICOV: precise structural contact prediction using sparse inverse covariance estimation on large multiple sequence alignments. *Bioinformatics* **28:** 184–190. doi:10.1093/bioinformatics/btr638

Jumper J, Evans R, Pritzel A, Green T, Figurnov M, Ronneberger O, Tunyasuvunakool K, Bates R, Žídek A, Potapenko A, et al. 2021. Highly accurate protein structure prediction with AlphaFold. *Nature* **596**: 583–589. doi:10 .1038/s41586-021-03819-2

Kachroo AH, Laurent JM, Yellman CM, Meyer AG, Wilke CO, Marcotte EM. 2015. Systematic humanization of yeast genes reveals conserved functions and genetic modularity. *Science* **348**: 921–925. doi:10.1126/science.aaa0769

Kleeorin Y, Russ WP, Rivoire O, Ranganathan R. 2023. Undersampling and the inference of coevolution in proteins. *Cell Syst* **14**: 210–219.e7. doi:10.1016/j.cels.2022.12 .013

Korber BT, Farber RM, Wolpert DH, Lapedes AS. 1993. Covariation of mutations in the V3 loop of human immunodeficiency virus type 1 envelope protein: an information theoretic analysis. *Proc Natl Acad Sci* **90**: 7176–7180. doi:10.1073/pnas.90.15.7176

Laine E, Karami Y, Carbone A. 2019. GEMME: a simple and fast global epistatic model predicting mutational effects. *Mol Biol Evol* **36**: 2604–2619. doi:10.1093/molbev/msz179

Lapedes AS, Giraud B, Liu L, Stormo GD. 1999. Correlated mutations in models of protein sequences: phylogenetic and structural effects. In *Institute of mathematical statistics lecture notes—monograph series*, pp. 236–256. Institute of Mathematical Statistics, Hayward, CA.

Lee J, Natarajan M, Nashine VC, Socolich M, Vo T, Russ WP, Benkovic SJ, Ranganathan R. 2008. Surface sites for engineering allosteric control in proteins. *Science* **322**: 438–442. doi:10.1126/science.1159052

Lian X, Praljak N, Subramanian SK, Wasinger S, Ranganathan R, Ferguson AL. 2022. Deep learning-enabled design of synthetic orthologs of a signaling protein. *Mol Biol* doi:10.1101/2022.12.21.521443

Lockless SW, Ranganathan R. 1999. Evolutionarily conserved pathways of energetic connectivity in protein families. *Science* **286**: 295–299. doi:10.1126/science.286.5438.295

Madani A, Krause B, Greene ER, Subramanian S, Mohr BP, Holton JM, Olmos JL, Xiong C, Sun ZZ, Socher R, et al. 2023. Large language models generate functional protein sequences across diverse families. *Nat Biotechnol* **41**: 1099–1106. doi:10.1038/s41587-022-01618-2

Markin CJ, Mokhtari DA, Sunden F, Appel MJ, Akiva E, Longwell SA, Sabatti C, Herschlag D, Fordyce PM. 2021. Revealing enzyme functional architecture via high-throughput microfluidic enzyme kinetics. *Science* **373**: eabf8761. doi:10.1126/science.abf8761

McCormick JW, Russo MA, Thompson S, Blevins A, Reynolds KA. 2021. Structurally distributed surface sites tune allosteric regulation. *eLife* **10**: e68346. doi:10.7554/eLife .68346

McGee F, Hauri S, Novinger Q, Vucetic S, Levy RM, Carnevale V, Haldane A. 2021. The generative capacity of probabilistic protein sequence models. *Nat Commun* **12**: 6302. doi:10.1038/s41467-021-26529-9

McLaughlin RN Jr, Poelwijk FJ, Raman A, Gosal WS, Ranganathan R. 2012. The spatial architecture of protein function and adaptation. *Nature* **491**: 138–142. doi:10.1038/ nature11500

Michener JK, Camargo Neves AA, Vuilleumier S, Bringel F, Marx CJ. 2014. Effective use of a horizontally-transferred pathway for dichloromethane catabolism requires post-transfer refinement. *eLife* **3**: e04279. doi:10.7554/eLife .04279

Morcos F, Pagnani A, Lunt B, Bertolino A, Marks DS, Sander C, Zecchina R, Onuchic JN, Hwa T, Weigt M. 2011. Direct-coupling analysis of residue coevolution captures native contacts across many protein families. *Proc Natl Acad Sci* **108**: E1293–E1301. doi:10.1073/pnas.1111471108

Novinec M, Korenč M, Caflisch A, Ranganathan R, Lenarčič B, Baici A. 2014. A novel allosteric mechanism in the cysteine peptidase cathepsin K discovered by computational methods. *Nat Commun* **5**: 3287. doi:10.1038/ncomms 4287

Olson CA, Wu NC, Sun R. 2014. A comprehensive biophysical description of pairwise epistasis throughout an entire protein domain. *Curr Biol* **24**: 2643–2651. doi:10.1016/j .cub.2014.09.072

Otwinowski J. 2018. Biophysical inference of epistasis and the effects of mutations on protein stability and function. *Mol Biol Evol* **35**: 2345–2354. doi:10.1093/molbev/msy141

Ovchinnikov S, Kamisetty H, Baker D. 2014. Robust and accurate prediction of residue-residue interactions across protein interfaces using evolutionary information. *eLife* **3**: e02030. doi:10.7554/eLife.02030

Palovcak E, Delemotte L, Klein ML, Carnevale V. 2015. Comparative sequence analysis suggests a conserved gating mechanism for TRP channels. *J Gen Physiol* **146**: 37–50. doi:10.1085/jgp.201411329

Paysan-Lafosse T, Blum M, Chuguransky S, Grego T, Pinto BL, Salazar GA, Bileschi ML, Bork P, Bridge A, Colwell L, et al. 2023. Interpro in 2022. *Nucleic Acids Res* **51**: D418–D427. doi:10.1093/nar/gkac993

Pincus D, Resnekov O, Reynolds KA. 2017. An evolution-based strategy for engineering allosteric regulation. *Phys Biol* **14**: 025002. doi:10.1088/1478-3975/aa64a4

Pincus D, Pandey JP, Feder ZA, Creixell P, Resnekov O, Reynolds KA. 2018. Engineering allosteric regulation in protein kinases. *Sci Signal* **11**: eaar3250. doi:10.1126/scisig nal.aar3250

Raman AS, White KI, Ranganathan R. 2016. Origins of allostery and evolvability in proteins: a case study. *Cell* **166**: 468–480. doi:10.1016/j.cell.2016.05.047

Reynolds KA, McLaughlin RN, Ranganathan R. 2011. Hot spots for allosteric regulation on protein surfaces. *Cell* **147**: 1564–1575. doi:10.1016/j.cell.2011.10.049

Reynolds KA, Russ WP, Socolich M, Ranganathan R. 2013. Evolution-based design of proteins. In *Methods in enzymology*, Vol. 523, pp. 213–235. Elsevier, New York.

Rivoire O, Reynolds KA, Ranganathan R. 2016. Evolution-based functional decomposition of proteins. *PLoS Comput Biol* **12**: e1004817. doi:10.1371/journal.pcbi.1004817

Rost B. 1999. Twilight zone of protein sequence alignments. *Protein Eng* **12**: 85–94. doi:10.1093/protein/12.2.85

Russ WP, Lowery DM, Mishra P, Yaffe MB, Ranganathan R. 2005. Natural-like function in artificial WW domains. *Nature* **437**: 579–583. doi:10.1038/nature03990

Russ WP, Figliuzzi M, Stocker C, Barrat-Charlaix P, Socolich M, Kast P, Hilvert D, Monasson R, Cocco S, Weigt M, et al. 2020. An evolution-based model for designing chorismate mutase enzymes. *Science* **369**: 440–445. doi:10.1126/sci ence.aba3304

Shiau C, Wang H, Lee Y, Ovchinnikov S. 2022. Global statistical models of protein coevolution reveal higher-order sectors beyond those obtained from structure alone. bioRxiv doi:10.1101/2022.05.27.493723v1

Shulman AI, Larson C, Mangelsdorf DJ, Ranganathan R. 2004. Structural determinants of allosteric ligand activation in RXR heterodimers. *Cell* **116:** 417–429. doi:10.1016/S0092-8674(04)00119-9

Smock RG, Rivoire O, Russ WP, Swain JF, Leibler S, Ranganathan R, Gierasch LM. 2010. An interdomain sector mediating allostery in Hsp70 molecular chaperones. *Mol Syst Biol* **6:** 414. doi:10.1038/msb.2010.65

Socolich M, Lockless SW, Russ WP, Lee H, Gardner KH, Ranganathan R. 2005. Evolutionary information for specifying a protein fold. *Nature* **437:** 512–518. doi:10.1038/nature03991

Stein RR, Marks DS, Sander C. 2015. Inferring pairwise interactions from biological data using maximum-entropy probability models. *PLoS Comput Biol* **11:** e1004182. doi:10.1371/journal.pcbi.1004182

Süel GM, Lockless SW, Wall MA, Ranganathan R. 2003. Evolutionarily conserved networks of residues mediate allosteric communication in proteins. *Nat Struct Biol* **10:** 59–69. doi:10.1038/nsb881

Sun HP, Huang Y, Wang XF, Zhang Y, Shen HB. 2015. Improving accuracy of protein contact prediction using balanced network deconvolution: protein residue contact map prediction. *Proteins* **83:** 485–496. doi:10.1002/prot.24744

The UniProt Consortium. 2023. Uniprot: the universal protein knowledgebase in 2023. *Nucleic Acids Res* **51:** D523–D531. doi:10.1093/nar/gkac1052

Thompson S, Zhang Y, Ingle C, Reynolds KA, Kortemme T. 2020. Altered expression of a quality control protease in E. coli reshapes the in vivo mutational landscape of a model enzyme. *eLife* **9:** e53476. doi:10.7554/eLife.53476

Tian P, Louis JM, Baber JL, Aniana A, Best RB. 2018. Coevolutionary fitness landscapes for sequence design. *Angew Chem Int Ed Engl* **57:** 5674–5678. doi:10.1002/anie.201713220

Trinquier J, Uguzzoni G, Pagnani A, Zamponi F, Weigt M. 2021. Efficient generative modeling of protein sequences using simple autoregressive models. *Nat Commun* **12:** 5800. doi:10.1038/s41467-021-25756-4

Uguzzoni G, John Lovis S, Oteri F, Schug A, Szurmant H, Weigt M. 2017. Large-scale identification of coevolution signals across homo-oligomeric protein interfaces by direct coupling analysis. *Proc Natl Acad Sci* **114:** E2662–E2671. doi:10.1073/pnas.1615068114

Wang H, Feng S, Liu S, Ovchinnikov S. 2022. Disentanglement of entropy and coevolution using spectral regularization. bioRxiv. doi:10.1101/2022.03.04.483009v1

Weinreb C, Riesselman AJ, Ingraham JB, Gross T, Sander C, Marks DS. 2016. 3D RNA and functional interactions from evolutionary couplings. *Cell* **165:** 963–975. doi:10.1016/j.cell.2016.03.030

Weng C, Faure AJ, Lehner B. 2022. The energetic and allosteric landscape for KRAS inhibition. bioRxiv doi:10.1101/2022.12.06.519122

Xie WJ, Asadi M, Warshel A. 2022. Enhancing computational enzyme design by a maximum entropy strategy. *Proc Natl Acad Sci* **119:** e2122355119. doi:10.1073/pnas.2122355119

Yang KK, Wu Z, Arnold FH. 2019. Machine-learning-guided directed evolution for protein engineering. *Nat Methods* **16:** 687–694. doi:10.1038/s41592-019-0496-6

Zhang H, Gao Y, Deng M, Wang C, Zhu J, Li SC, Zheng WM, Bu D. 2016. Improving residue–residue contact prediction via low-rank and sparse decomposition of residue correlation matrix. *Biochem Biophys Res Commun* **472:** 217–222. doi:10.1016/j.bbrc.2016.01.188

Zhang H, Bei Z, Xi W, Hao M, Ju Z, Saravanan KM, Zhang H, Guo N, Wei Y. 2021. Evaluation of residue-residue contact prediction methods: from retrospective to prospective. *PLoS Comput Biol* **17:** e1009027. doi:10.1371/journal.pcbi.1009027

Petabase-Scale Homology Search for Structure Prediction

Sewon Lee,[1,10] Gyuri Kim,[1,10] Eli Levy Karin,[2] Milot Mirdita,[1] Sukhwan Park,[3] Rayan Chikhi,[4] Artem Babaian,[5,6] Andriy Kryshtafovych,[7] and Martin Steinegger[1,3,8,9]

[1]School of Biological Sciences, Seoul National University, Gwanak-gu, Seoul 08826, South Korea

[2]ELKMO, Copenhagen 2720, Denmark

[3]Interdisciplinary Program in Bioinformatics, Seoul National University, Seoul 08826, South Korea

[4]Institut Pasteur, Université Paris Cité, G5 Sequence Bioinformatics, 75015 Paris, France

[5]Department of Molecular Genetics, University of Toronto, Toronto, Ontario M5S 1A8, Canada

[6]Donnelly Centre for Cellular and Biomolecular Research, University of Toronto, Toronto, Ontario M5S 3E1, Canada

[7]Genome Center, University of California, Davis, California 95616, USA

[8]Artificial Intelligence Institute, Seoul National University, Seoul 08826, South Korea

[9]Institute of Molecular Biology and Genetics, Seoul National University, Seoul 08826, South Korea

Correspondence: martin.steinegger@snu.ac.kr

The recent CASP15 competition highlighted the critical role of multiple sequence alignments (MSAs) in protein structure prediction, as demonstrated by the success of the top AlphaFold2-based prediction methods. To push the boundaries of MSA utilization, we conducted a petabase-scale search of the Sequence Read Archive (SRA), resulting in gigabytes of aligned homologs for CASP15 targets. These were merged with default MSAs produced by ColabFold-search and provided to ColabFold-predict. By using SRA data, we achieved highly accurate predictions (GDT_TS > 70) for 66% of the non-easy targets, whereas using ColabFold-search default MSAs scored highly in only 52%. Next, we tested the effect of deep homology search and ColabFold's advanced features, such as more recycles, on prediction accuracy. While SRA homologs were most significant for improving ColabFold's CASP15 ranking from 11th to 3rd place, other strategies contributed too. We analyze these in the context of existing strategies to improve prediction.

Determining the 3D structure of proteins is of great importance to many research fields, encompassing cancer drug discovery (Borkakoti and Thornton 2023; Ren et al. 2023), pesticide development, and crop improvement (Koesoema 2022). Additionally, it plays a crucial role in the design of sensors and enzymes (Pereira et al. 2021), as well as numerous other applications, as reviewed by Pearce and Zhang (2021).

---

[10]These authors contributed equally to this work.

Traditionally, protein structures have been solved using laborious techniques, such as X-ray crystallography, resulting in just under 200,000 structures in more than 50 years of communal effort (Berman et al. 2000; Subramaniam and Kleywegt 2022). Resolved structures are routinely deposited in the Protein Data Bank (wwPDB consortium 2019). The demanding experimental process has motivated the development of computational tools as a less burdensome alternative for structure prediction. Since 1994, the Critical Assessment of protein Structure Prediction (CASP) has aimed to identify state-of-the-art computational methods by competition (Moult et al. 1995). The organizers of CASP provide the participants with protein sequences, whose structures were experimentally solved but not yet deposited in the PDB. The solved structures are unknown to the organizers, assessors as well as to the participants. Also, the group identities are kept anonymous from the assessors, therefore the competition is considered double-blinded, ensuring fairness.

Originally, computational prediction methods could be divided into two main groups: template-based modeling (TBM) and free modeling (FM). However, in the past decade, the lines between the groups have been blurred (Bertoline et al. 2023). TBM is a broad category in which known structures are used as templates to predict the structure of query proteins, based on the sequence similarity between them. In its simplest form, a similarity between a single query and a match from the PDB serves as the base for projecting the match's structure onto the query. The inaugural software MODELLER (Šali and Blundell 1993) and other tools that followed have made use of this principle (see Pearce and Zhang 2021 for a review).

Given the limited size of the PDB and its bias toward model organisms (Orlando et al. 2016), detecting remote sequence homology is crucial. To that end, increasingly sensitive search methods have been developed. The first step forward was taken by algorithms like BLAST (Altschul et al. 1990), which directly compare the query sequence to the reference database. PSI-BLAST (Altschul et al. 1997) improved upon this by computing a multiple sequence alignment (MSA) of the query and its best BLAST hits and calculating a position-specific scoring matrix from that. This generalization of the query is used for a sensitive search of the reference. This approach was further refined by using probabilistic hidden Markov models (HMMs) (Krogh et al. 1994) in tools like HMMer (Eddy 2011). Another significant advancement came with HHsearch (Söding 2005), which expressed both query and reference as HMMs, markedly improving search sensitivity. This underpinned the success of HHpred (Hildebrand et al. 2009) in the CASP9 challenge (Moult et al. 2011). A further development, HHblits (Remmert et al. 2012; Steinegger et al. 2019a), accelerated the HMM-HMM comparison allowing to query databases with millions of HMMs like the Uniclust30 (Mirdita et al. 2017), a clustered version of the Uniprot (UniProt Consortium 2023), to generate diverse query MSAs.

Due to its unprecedented sensitivity, HHpred has transformed CASP in two ways. First, many methods competing in CASP have incorporated HHblits/HHsearch or other tools to identify distant structural homologs (Bertoline et al. 2023). Second, CASP has started using it for classifying target domains in subsequent competitions (Kinch et al. 2011). Specifically, domains in targets for which HHpred could identify a homolog in the PDB, are considered by CASP as "TBM target domains," while the others—as "FM target domains."

Recent advances in deep learning have been harnessed by various methods for protein structure prediction (Torrisi et al. 2020). Undoubtedly, the most revolutionary of these is AlphaFold2 (Jumper et al. 2021), which won the CASP14 challenge by a significant margin (Kryshtafovych et al. 2021), reaching experimental accuracy for over two-thirds of the targets. Despite its success, AlphaFold2's prediction accuracy is not without its limitations. Most notably, it relies on its input MSA diversity (Mirdita et al. 2022), experiencing a significant drop in prediction accuracy when the median number of diverse sequences in the MSA is 30 or less (Jumper et al. 2021). This finding is in agreement with previous studies on the importance of distant homologs to structure prediction (Ashkenazy et al. 2009;

Kuhlman and Bradley 2019). However, the ability to construct a deep MSA depends not only on the sensitivity of search algorithms, such as HHblits, but also on the potential pool of sequences (i.e., the reference database).

Metagenomics allows for sequencing uncultivable organisms directly from the environment, significantly expanding the repertoire of protein sequences deposited in scientific databases. In recent years, metagenomic sequences have shown great potential in increasing the fraction of proteins, whose structure can be modeled accurately (Ovchinnikov et al. 2017; Söding 2017; Wang et al. 2019; Yang et al. 2021). Of note, the largest metagenomic database used in these studies is the IMG/M, which contains 27 terabase pairs (Chen et al. 2023b).

It is therefore not surprising that the top scoring servers in the most recent CASP15 challenge were based on AlphaFold2 and included metagenomic sequences in their constructed MSAs (Table 1). An example for such a server is ColabFold (Mirdita et al. 2022). ColabFold takes as input a query protein sequence(s), whose structure is to be predicted. Its first step, denoted here as CF-search, implements a procedure for collecting homologs of the query using MMseqs2 (Steinegger and Söding 2017). CF-search starts by querying the input against the

UniRef30 database (Mirdita et al. 2017) and computing profiles from the hits. Next, CF-search queries these profiles against one of two metagenomic databases, which were constructed as part of the ColabFold release: BFD/MGnify and ColabFoldDB (the default reference database). As detailed in Table 2, BFD/MGnify contains 513 million nonredundant proteins from the union of the BFD (Jumper et al. 2021) and MGnify (Richardson et al. 2023) databases. ColabFoldDB expanded the BFD/MGnify with various environmental proteins, resulting in ∼740 million proteins. Following the search, an MSA is computed from the detected homologs and, finally, in a step denoted here as CF-predict, the MSA is provided as input to the AlphaFold2 models.

In this study we examined different strategies to improve protein structure prediction along three axes. The first two focused on adding homologs to MSAs used for protein structure prediction and the third on utilizing advanced features of CF-predict. The first and main axis is the breadth of the search, where we studied the impact of a much more systematic inclusion of metagenomic sequences on prediction accuracy. Over 37 petabase pairs are publicly available through the Sequence Read Archive (SRA), the world's largest metagenomic database (Katz

Table 1. Use of homology algorithms and databases among leading CASP15 servers[a]

Name	Rank[b]	Homology search algorithm	Ref. sequence databases (DBs)[c]
Yang-Server	1	HHblits, MMseqs2, Jackhmmer	Uniclust30, UniRef30, BFD, ColabFoldDB, manual
UM-TBM	2	DeepMSA, LOMETS	BFD, IMG/M, Metaclust, MetaSource, Uniclust30, UniRef90, Tara
Manifold-E	3	HHblits, hmmsearch, Jackhmmer	BFD, Uniprot, UniRef30, UniRef90
DFolding	4	CRFalign, Jackhmmer, HHblits, HHpred, Kalign	BFD, Uniclust30, UniRef90, MGnify
MULTICOM	5–7, 9	DeepMSA (modified), Foldseek, HHblits, Jackhmmer, MMseqs2	BFD, ColabFoldDB, IMG/M, MGnify, Uniclust30, UniRef90, Uniprot
RaptorX	8	HHblits, Jackhmmer	BFD, SMAG + MetaEuk + TOPAZ + MGV + GPD + IMG/M (in-house HHblits DB), MGnify, Uniclust30, UniRef90
MultiFOLD	10	MMseqs2, UniRef30	ColabFoldDB
ColabFold	11	MMseqs2	ColabFoldDB, UniRef30

[a]The information about the servers was extracted from the CASP15 abstract book.
[b]The rank refers to the "server only" performance on protein targets, excluding the TBM-easy category.
[c]A detailed overview of the reference sequence DBs is provided in Table 2.

Table 2. Size and composition of reference databases used by leading CASP15 servers

Database name	References[a]	Source; processing	Type	ca. No. sequences	Most y environmental proteins
UniProt/Swiss-Prot	UniProt Consortium 2023	Experiments; manual annotation + redundancy reduction	Proteins	$<10^6$	No
RefSeq	UniProt Consortium 2023	NCBI; annotation + redundancy reduction	Proteins	$>100 \times 10^6$	No
UniProt/TrEMBL	UniProt Consortium 2023	EMBL-Bank/GenBank/DDBJ; annotation	Proteins	$>100 \times 10^6$	No
UniParc	UniProt Consortium 2023	UniProt + RefSeq + other sources	Proteins	$>500 \times 10^6$	No
UniRef100	UniProt Consortium 2023	UniProt + selected UniParc; redundancy reduction	Proteins	$>100 \times 10^6$	No
UniRef90	UniProt Consortium 2023	UniProt; clustering at 90%	Proteins	$>100 \times 10^6$	No
UniRef30	Mirdita et al. 2017	UniProt; clustering at 30%	Proteins	$>10 \times 10^6$	No
Uniclust30	Mirdita et al. 2017	UniProt; clustering at 30%	Proteins	$>10 \times 10^6$	No
GPD	Camarillo-Guerrero et al. 2021	Human gut bacteriophages; prediction + clustering	Predicted proteins	$<10^6$	Yes
MGV	Nayfach et al. 2021	Human gut viruses, mostly DNA viruses; prediction + clustering	Predicted proteins	$<10^6$	Yes
TOPAZ	Alexander et al. 2023	TARA oceans expedition: enriched for Eukaryotes; prediction	Predicted Proteins	$>5 \times 10^6$	Yes
SMAG	Delmont et al. 2022	TARA oceans expedition: sunlit ocean; prediction + annotation	Predicted proteins	$\sim 10 \times 10^6$	Yes
MetaEukDB	Levy Karin et al. 2020	TARA oceans expedition: eukaryotic metagenomics; prediction	Predicted proteins	$>5 \times 10^6$	Yes
SRC	Steinegger et al. 2019b	Metagenomic soil samples; protein assembly	Predicted proteins	$>500 \times 10^6$	Yes
MERC	Steinegger et al. 2019b	TARA oceans expedition: metatranscriptomic data; protein assembly	Predicted proteins	$>100 \times 10^6$	Yes
Metaclust	Steinegger and Söding 2018	JGI's metagenomic and metatranscriptomic data; prediction + clustering	Predicted proteins	$>500 \times 10^6$	Yes
MGnify	Richardson et al. 2023	ENA + microbiome studies; EMBL-EBI pipeline	Predicted proteins	$>1000 \times 10^6$	Yes
BFD	Jumper et al. 2021	Various sources: TrEMBL + Swissprot + SRC + MERC + Metaclust; clustering	predicted proteins	$>500 \times 10^6$	yes
BFD/MGnify	Mirdita et al. 2022	Various sources: BFD + MGnify; clustering	predicted proteins	$>500 \times 10^6$	yes
ColabFoldDB	Mirdita et al. 2022	Various sources: BFD/MGnify + MetaEukDB + TOPAZ + MGV + GPD + SMAG; clustering	predicted proteins	$>500 \times 10^6$ (ca. 0.5 T base pairs)	yes
IMG/M	Chen et al. 2023b	Various sources: uncultivatable genomes; isolates; metagenomes; metatranscriptomes; JGI's pipeline	Predicted proteins	$>50,000 \times 10^6$ (27 T base pairs)	Yes

[a]References are to the latest version of each DB. Of note, the servers detailed in Table 1 may have used older versions of these resources. In addition, some resources rely on others for their construction, meaning an older version was used.

Cite this article as *Cold Spring Harb Perspect Biol* doi: 10.1101/cshperspect.a041465

et al. 2022). Recently, Serratus, a cloud-based pipeline for high-throughput homology search of the whole SRA, was introduced by Edgar et al. (2022). It allows searching through one million SRA read sets for about $5000 ($0.005 per set). Here, we constructed MSAs based on Serratus-mined homologs of the CASP15 targets and merged them with the default MSAs produced by CF-search. In the second axis of this study, we further enhanced the merged MSAs by searching for distant homologs of their sequences using HHblits against the BFD. The third axis concerns tuning the advanced parameters of Colab-Fold to control the use of templates and multimer models and the number of recycles. We then provided MSAs produced by each of these strategies as input to CF-predict and compared the resulting prediction accuracy to that measured with default CF-search MSAs as well as to the leading CASP15 servers. For fair comparison, we ensured all databases used in this study excluded any sequences deposited after the start of CASP15 competition (May 2022).

Our results show that adding SRA-mined homologs improves prediction accuracy for 61% of the examined targets. Tuning advanced features of CF-predict, especially adding more recycles, also contributed to better prediction. By combining the different strategies, Colab-Fold's CASP15 ranking among servers on non-easy template targets increased from 11th to 3rd place, indicating the vast potential of large-scale sequence exploration for better structure prediction.

HOMOLOG SEARCH AND MSA CONSTRUCTION

The entry point to this study was a list of 126 targets provided by CASP15. We excluded from this list all targets, which were RNA, heteromers, canceled by CASP, or indicated as auxiliary structure for ligand prediction, leaving 77 targets. Each of these targets had one or more domains, which are divided into categories by CASP as follows: FM, FM/TBM, TBM-hard, and TBM-easy. We used CF-search to query the targets against ColabFoldDB (Fig. 1①), resulting in 77 MSAs, denoted here as *cfdb* MSAs.

Thousand Times Broader Search

Our next goal was to expand the search beyond ColabFoldDB and explore the SRA. With its 37 petabase pairs of publicly available data, the SRA is orders of magnitude bigger than any previously used metagenomic resource, including Colab-FoldDB (Table 2). We queried the 77 CASP15 targets using Serratus. By using search terms, such as virome, metagenome, and eukaryotes, we set the Serratus search space to cover more than half of the publicly available SRA, comprising 22 petabase pairs, organized in over five million SRA runs. In each run, reads that aligned to the CASP15 protein sequence queries were assembled using rnaviralSpades (Meleshko et al. 2021), mounting to over a hundred gigabytes of assembled data. Each CASP15 target was then queried against a reference protein database created from its Serratus-produced assembled proteins using MMseqs2 (Steinegger and Söding 2017). The identified homologs were then aligned by MMseqs2 to create an MSA (Fig. 1②), denoted as *sra* MSA. As a first indication of the tremendous capacity of the SRA, we found that more than half of the targets, which were composed solely of easy template targets (TBM-easy) domains, could be matched with at least 1,000,000 homologs, some even exceeding 100,000,000 (Fig. 2A). Processing MSAs with millions of sequences poses a heavy computational burden. Thus, we opted to exclude from this study targets that only contained TBM-easy domains, focusing on the remaining 46 targets, which had 62 non-TBM-easy domains: 39 FM, 8 FM/TBM, and 15 TBM-hard. On average, the number of homologs detected per domain doubled from 106,586 in the CFDB to 274,231 when including the SRA results (Fig. 2A).

Diving Deeper

Serratus' ability to scan the SRA in feasible time comes at the expense of its sensitivity to detect remote homologs. Specifically, it is limited in its ability to detect sequences with less than 50% identity to the query (Edgar et al. 2022). This prompted us to search deeply for remote homologs. To that end, we merged for each target its

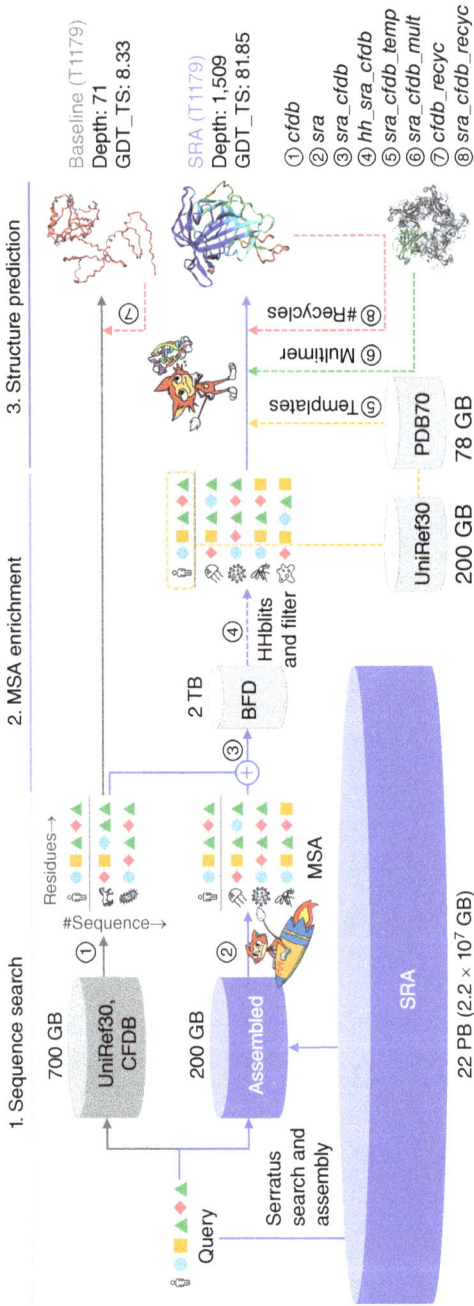

Figure 1. Multiple sequence alignment (MSA) enrichment using the Sequence Read Archive (SRA) and other strategies to improve protein structure prediction. Workflow of the different strategies examined in this study, ①~⑧. All strategies construct an MSA (but differ in the homology databases [DBs] they utilize) and provide it to CF-predict (but differ in the way they tune its parameters). The size of each homology DB is denoted close to it. The baseline MSA (*cfdb* MSA, ①) is constructed by CF-search. The SRA-detected homologs are aligned to create ② using MMseqs2. The *sra_cfdb* MSA (③) is constructed by combining ① and ②. The *hh_sra_cfdb* MSA (④) is constructed by querying ③ against UniRef30 and BFD using HHblits. Strategies ⑤, ⑥, ⑦, and ⑧ refer to the following CF-predict options applied on top of strategy ① or ③: use of templates, multimer (homo-oligomer) modeling, and 12 recycles (instead of the default 3). Before being provided to CF-predict, each MSA is filtered based on the sequence identity between its members and the query.

cfdb and *sra* MSAs (Fig. 1③), resulting in an *sra_cfdb* MSA. We then ran HHblits with each *sra_cfdb* MSA as input against UniRef30 and BFD (Fig. 1④), setting parameters to include all sequences in the output without any filtering and ensure maximum sensitivity (for a full parameter description, see Supplemental information in Jumper et al. 2021). The resulting MSAs, denoted as *hh_sra_cfdb* MSAs, contained the input sequences as well as the homologs detected by HHblits.

To filter each input MSA based on the sequence identity (seqid) between its members and the query, we utilized a newly introduced filter module called `filtera3m` in MMseqs2, which implements ColabFold's MSA filtering strategy. We removed unlikely homologs (seqid < 0.2), noninformative homologs (seqid > 0.95) and redundant sequences, keeping the most diverse and informative set of sequences in the MSA.

The filtered *cfdb* MSAs had on average 2395 sequences, *sra_cfdb* MSAs-5731 and *hh_sra_cfdb* MSAs-8133. We used HHmake (Steinegger et al. 2019a) to compute the number of effective sequences (N_{eff}), where higher values indicate less similarity between the sequences and more diverse MSAs. Here, a more moderate increase was observed with the average N_{eff} score, rising from 4.87 for *cfdb* MSA to 5.43 for *sra_cfdb* MSA and to 7.26 for *hh_sra_cfdb* MSA (Fig. 2B).

THE EFFECT OF HOMOLOGS ON STRUCTURE PREDICTION

To measure the effect of the various homolog collection strategies on structure prediction, we provided CF-predict with different input MSAs. These included the *sra_cfdb* and *hh_sra_cfdb* MSAs as well as their controls, which did not include Serratus-detected homologs from the SRA: *cfdb* and *hh_cfdb* MSAs. The *hh_cfdb* MSAs were produced in a similar manner to *hh_sra_cfdb* MSAs, using *cfdb* MSAs, rather than *sra_cfdb* MSAs, as the input to HHblits.

For each strategy, five protein models were produced and the best one was selected according to its computed predicted local distance difference test (pLDDT) score (Jumper et al. 2021). For the selected models, we measured the do-

main level accuracy, using the GDT_TS score (Zemla 2003). Compared to using *cfdb* MSAs, *sra_cfdb* MSAs significantly improved GDT_TS scores (Table 3), increasing scores for 38 out of 62 domains (Fig. 2C). On the other hand, using *hh_sra_cfdb* MSAs did not lead to a significant improvement over *sra_cfdb* MSAs (Table 3) and is therefore omitted from the figure.

We further examined the relative performance of each strategy compared to other CASP15 servers using Z-scores, as follows. For each strategy, we deducted from its GDT_TS scores the mean servers' GDT_TS score and divided it by the servers' standard deviation. The sum of nonnegative Z-scores and the average GDT_TS of all evaluated target domains were used as representative scores for each strategy (Table 3). As expected, the addition of *sra* to *cfdb* MSAs substantially improved the performance, increasing the sum of Z-scores from 17.09 to 30.30 (Table 3). On average, no significant improvement was observed when adding HHblits-detected homologs (Table 3). However, specific domains gained a substantial improvement by including these homologs, indicating a variable effect for each target. A notable example with the highest improvement is target T1178-D1, where running HHblits increased GDT_TS from 28.61 for *sra_cfdb* to 84.17 for *hh_sra_cfdb* MSAs.

TUNING PARAMETERS

In addition to homolog collection strategies, we examined the impact of three advanced features of CF-predict: using templates, multimer models, and increasing the number of recycles. The strategies corresponding to these features were built on the MSAs constructed in previous steps and are denoted *sra_cfdb_temp*, *sra_cfdb_recyc*, and *sra_cfdb_mult* and their control *sra_cfdb*. Other strategies, taken by the leading CASP15 servers, are detailed in Table 4.

Leveraging Templates

Two out of the five AlphaFold2 models require structural features as input. Setting the "templates" parameter (Fig. 1⑤) changes the default behavior of CF-predict from using mock templates to querying the PDB70 (Steinegger et al.

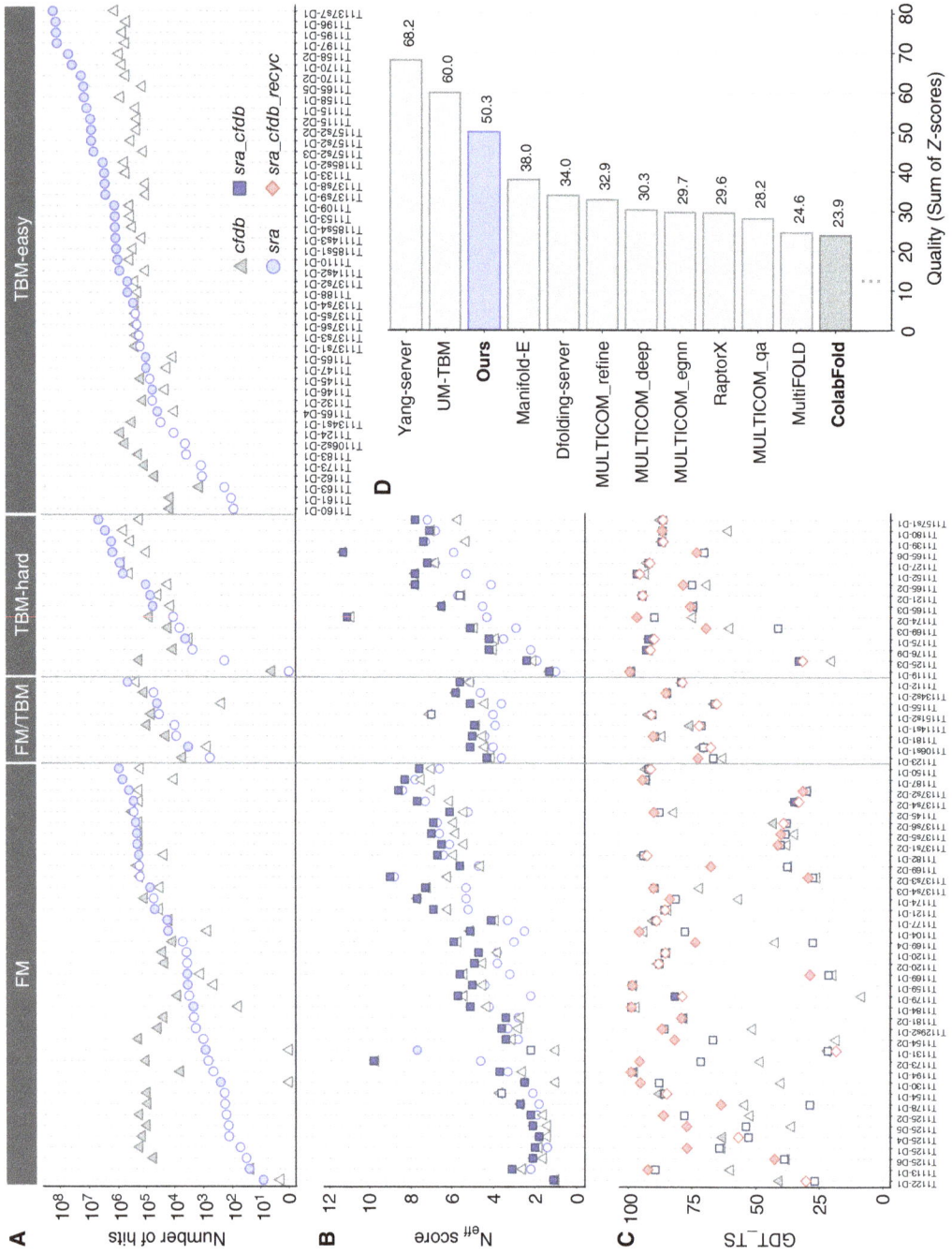

Figure 2. (*See following page for legend.*)

Table 3. Effects of different strategies on protein structure prediction performance

CFDB	SRA	HHblits	Templates	Multimer	12 recycles	GDT_TS[b]	Sum Z (>0.0)[c]
✓						65.80 ± 25.58	17.09
✓		✓				67.36 ± 25.15	15.82
✓					✓	68.12 ± 25.25[d]	17.69
✓	✓					70.97 ± 24.40[e]	30.30
✓	✓	✓				71.44 ± 23.30[f]	26.32
✓	✓		✓			70.94 ± 23.99	31.60
✓	✓			✓		70.98 ± 24.12	30.97
✓	✓				✓	75.54 ± 22.17[g]	40.59
Model1[a]						77.85 ± 20.59	50.31

[a]Model1 is the set of predicted-best strategies among all examined strategies, chosen based on pLDDT for each target domain.
[b]GDT_TS scores are represented as mean ± standard deviation.
[c]Z-scores are calculated from each target domain's GDT_TS, based on the server groups' mean and standard deviation. Sum Z (>0.0) refers to the sum of positive Z-scores. All strategies were compared to their controls using the Wilcoxon signed-rank test. No significant improvement ($P > 0.05$) was found except for those in footnotes d–g.
[d]*sra_cfdb_recyc* over *cfdb_recyc* MSAs ($P = 0.0107$); *cfdb_recyc* over *cfdb* MSAs ($P = 0.0272$).
[e]*sra_cfdb* over *cfdb* MSAs ($P = 0.0116$).
[f]*hh_sra_cfdb* over *hh_cfdb* MSAs ($P = 0.0126$).
[g]*sra_cfdb_recyc* over *sra_cfdb* MSAs ($P = 0.0003$).

2019a) using the UniRef30-based constructed profile as input and returning hits, which are later aligned by HHsearch.

Adding templates did not improve the accuracy compared to using default parameters (Table 3). However, it should be noted that our examined targets are FM, FM/TBM, and TBM-hard, which are classified as difficult targets to find templates for. Applying templates to TBM-easy targets might have different effects on prediction performance.

Multimer Modeling

We also tested the impact of using ColabFold's "multimer" option (Fig. 1⑥). This changes the default behavior of CF-predict from treating the input query sequence as a monomer to considering it to be a part of a complex. This has the potential of stabilizing the structure, thereby improving prediction accuracy. We applied multimer modeling only for the homo-oligomer targets, based on the stoichiometry provided by CASP and used the multimer model weights from AlphaFold-multimer version 2 (Evans et al. 2022).

Using multimer modeling had a diverse range of effects. Overall, it did not make a significant improvement over *sra_cfdb* MSAs (Table 3). However, it drastically increased or decreased prediction performance for certain targets. Two notable examples stand out: T1178-D1 achieved the highest improvement, with its GDT_TS score soaring from 28.61 to 61.02 after incorporating multimer modeling. Conversely, the GDT_TS score of target

Figure 2. Effect of ColabFold parameters on structure prediction accuracy. (*A*) Comparison of homology search of 109 domains of 77 CASP15 targets. Each mark denotes the number of hits found for each target domain using either CF-search against CFDB (triangle) or MMseqs2 against SRA-mined and assembled proteins (circle) before the MSA filtering step. (*B*) N_{eff} scores of the different MSAs computed for each domain. The multiple sequence alignment (MSA) with the most homologs and the highest N_{eff} is indicated with a filled mark in panels *A* and *B*, respectively. (*C*) Structure prediction of 62 target domains in the categories: FM, FM/TBM, and TBM-hard was evaluated based on GDT_TS scores of three prediction strategies: *cfdb* MSA, *sra_cfdb* MSA, and *sra_cfdb_recyc*. The best-scoring strategy for each target domain is indicated with filled marks. (*D*) Prediction performance comparison between server groups in CASP15. The *x*-axis refers to the Sum Z (>0.0) in Table 3. The score of this study is from the Model1 in Table 3. Here, ColabFold refers to the performance of the server group submitted in CASP15.

Table 4. Use of prediction algorithms and strategies among leading CASP15 servers[a]

Name	Monomer prediction algorithm	Monomer strategies	Strategy category	Multimer prediction algorithm	Multimer strategies	Comment
Yang-Server	trRosettaX2, AF2	Use AF2 when trRosettaX2 is not satisfactory	Modeler selection	AF-Multimer	Replace template search with HHsearch	
		Use AF2's Evoformer in trRosettaX2	Modeler selection		Disable MSA pairing	
		Generate MSAs using HHblits and MMseqs2	MSA			
UM-TBM	AF2, I-TASSER	Generate MSAs using DeepMSA	MSA	NA	NA	Multistep pipeline
		Detect multidomain templates using LOMETS	Templates			
		Construct model using I-TASSER REMC simulation	Model refinement			
		Refine model using MD simulation	Model refinement			
Manifold-E	AF2 (reimplemented)	Generate MSAs using HHblits and JackHMMER	MSA	AF-Multimer (reimplemented)	Not specified separately from monomer	Reimplemented in PyTorch
		Train models with different configs: No. sequences, No. templates, etc.	AF2 configuration			
		Modify predict params: MSA sampling, templates, No. recycles	AF2 configuration			
		Relax with OpenMM and AMBER99 force field	Model relaxation			
DFolding	AF2 (modified)	Modify AF2's torsion and FAPE loss functions	AF2 modification	AF2Complex (modified)	Use nonpaired MSA for features generated by DeepFold	Multistep pipeline DeepFold
		Introduce loss functions for sidechain and secondary structure	AF2 modification		Generate sub-complexes by domains for large targets and combine using Modeller	
		Train models using uni-fold and protein chains from PDB40	AF2 configuration			
		Search MSA features using HHpred, kalign, and HHblits	MSA			

Continued

Table 4. *Continued*

Name	Monomer prediction algorithm	Monomer strategies	Strategy category	Multimer prediction algorithm	Multimer strategies	Comment
MULTICOM	AF2	Replace template features by using CRFalign	Templates	AF-Multimer	Rank multimer models using MMalign	Different strategies tested in the different MULTICOM servers
		Refine model using MD simulation	Model refinement			
		Generate MSAs using HHblits, JackHMMER, MMseqs2, and DeepMSA	MSA			
		Augment AF2's templates with one found by searching an in-house template DB	Templates			
		Augment AF2's MSAs with IMG/M homologs if depth <200	MSA			
		Rank AF2's models using APOLLO, DeepRank, and EnQA	Model ranking			
		Refine models using a method based on FoldSeek	Model refinement			
RaptorX	AF2 (modified)	Generate MSAs like AF2 and augment with homologs from in-house metagenomic DB if shallow	MSA	Not specified	Not specified	
		Use three template DBs: PDB70, PDB100, and DistillPDB (predicted decoy structures by AF2)	Templates			
		Use TMalign and DeepAlign to find and align templates based on highest pLDDT (iterative)	Templates			
		Modify AF2: use a linear layer to integrate scalar, point, and pair attention values in IPA model	AF2 modification			

Continued

Table 4. *Continued*

Name	Monomer prediction algorithm	Monomer strategies	Strategy category	Multimer prediction algorithm	Multimer strategies	Comment
		Modify AF2 feature module and train four combinations: MSA, MSA + template, MSA + MSAtransformer + template, MSA + MSAtransformer + template + AF2 model	AF2 configuration			
		Train single sequence model with included ESM-1b protein language model embeddings	Custom model			
		Top five highest pLDDT and replace one model with TM-based model clustering center model	Model ranking			
MultiFOLD	AF2 (modified ColabFold)	Use templates if seq length <1000	Templates	AF-Multimer		LocalColabFold extends ColabFold and AF-Multimer
		Use ModFOLDdockR to score and rank models	Model ranking			
		Use AMBER if seq length <1000	Model relaxation			
		Use 12 recycles if seq length <1000	AF2 configuration			
ColabFold	AF2	Use MMseqs2 to search and align	MSA	AF-Multimer-v2		
		Use 12 recycles, templates, ensembles	AF2 configuration			
		Relax with OpenMM/Amber	Model relaxation			

[a]The information about the servers was extracted from the CASP15 abstract book.

T1174-D1 experienced the largest decrease, dropping from 81.48 to 59.84.

Adding More Recycles

Through the "recycle" parameter (Fig. 1⑦,⑧), CF-predict allows setting the number of iterations in which a prediction will be re-fed to the AlphaFold2 models. By default, this value is set to 3, but additional recycles have the potential to improve prediction accuracy (Mirdita et al. 2022). Thus, when exploring this option, we set it to 12. The recently released version 3 of Alpha-Fold-multimer (Evans et al. 2022) uses up to 20 recycle iterations, with early stopping if a model has already converged.

Increasing recycles significantly improved the prediction accuracy compared to the control sra_cfdb with default parameters (Table 3). As depicted in Figure 2C, sra_cfdb_recyc MSAs scored higher than cfdb MSAs and sra_cfdb MSAs in 34 domains and achieved high accuracy (GDT_TS > 70) in 72% of the 62 domains. To further examine this effect, we measured the performance of another strategy, cfdb_recyc MSAs, which does not include SRA-mined homologs, serving as another control to sra_cfdb_recyc MSAs. We found significant improvements in prediction accuracy both when comparing cfdb_recyc MSAs to cfdb MSAs and when comparing sra_cfdb_recyc MSAs to cfdb_recyc MSAs (Table 3), suggesting that both SRA homologs and the additional recycles contribute to the improved performance and that the SRA contributes most.

STRATEGY SELECTION AND COMPARISON WITH CASP15 SERVERS

Finally, among the eight examined strategies, we selected the one with the highest pLDDT for each target domain, denoted here as Model1. We then compared the performance of Model1 with the leading server groups in CASP15, including the original ColabFold server, which is similar to this study's cfdb MSAs (Fig. 2D). Model1 resulted in an average GDT_TS of 77.85 and sum of Z-scores of 50.31, increasing 12.05 and 33.22 units from the baseline cfdb MSAs, respectively. Notably, 49 out of 62 domains (79%) of Model1

achieved high accuracy scores (GDT_TS > 70), compared to 52% of the cfdb MSAs. When comparing with other server groups based on the sum of Z-scores, Model1 would have ranked third, outperforming the ColabFold original server, which ranked 11th in CASP15 among server-only groups on non-easy targets.

To examine the validity of using pLDDT as selection criterion, we compared for each domain the GDT_TS score of Model1 and the highest GDT_TS score (Model_best), across all strategies. In case of perfect agreement between pLDDT and GDT_TS, these values should be equal. However, we observed a disagreement for 18 out of 62 target domains, resulting in an increase in Model_best with the average GDT_TS reaching 78.26 and the sum of Z-scores reaching 52.84. This disparity between pLDDT and GDT_TS highlights the challenge in selecting the best model. For instance, choosing the strategy with the highest pLDDT for target T1125-D3 yielded a GDT_TS score of 31.52 (sra_cfdb_recyc MSA), while the actual best GDT_TS score was 61.52 (cfdb_hh MSA).

To address this discrepancy, there were attempts to use alternative model selection (ranking) methods in CASP15, instead of relying solely on pLDDT (Table 4). For instance, the MULTICOM servers (Liu et al. 2023) utilized APOLLO (Wang et al. 2011), DeepRank (Renaud et al. 2021), and EnQA (Chen et al. 2023a) for ranking models, and MultiFOLD (McGuffin et al. 2023) employed ModFOLD-dockR (Edmunds et al. 2023) for both scoring and ranking purposes. Further developments of ranking methods are needed to improve the accuracy and reliability of model selection for structure prediction.

CONCLUDING REMARKS

In this study, we have shown the importance of a comprehensive inclusion of metagenomic sequences from the SRA for improving protein structure prediction. These results accentuate protein structure prediction as a data-driven discipline, gaining from countless communal contributions to databases, such as the SRA and the PDB.

Our results highlight the large variation in the number of homologs found for different targets. For example, over 100 million environmental homologs were found for T1137s7, T1195, T1196, and T1197—the same order of magnitude as the entire UniProt database. However, there are some targets for which few matches were detected, possibly due to the limited sensitivity of the mining procedure.

Serratus, the tool used for mining the SRA has impressive capabilities, but also significant constraints. It is limited to detecting homologs, which have about 50% sequence identity to the query, missing the full potential of homologs from the twilight zone (Rost 1999). Fast and more sensitive search methods are thus required to further improve our ability to exploit the SRA. Additionally, using Serratus in a similar manner to this study is likely to cost thousands of dollars (Edgar et al. 2022) and this cost could become limiting with the expected continued exponential growth of the SRA. Current scalable search methods (Bingmann et al. 2019; Bradley et al. 2019; Camacho et al. 2023; Shiryev and Agarwala 2023) face a challenging trade-off between how much they cost and how informative they are. Due to this trade-off, they either only indicate the presence or absence of a specific sequence fragment in an SRA experiment at a low cost or return full alignments at a high cost. Therefore, it is critical to develop new homology search methods that scale to the size of the SRA through efficient compression and indexing. We see this as having the highest potential to further improve structure prediction in the AF2 era and call upon the community to tackle this challenge of making the SRA searchable, alignable, accessible, and affordable for everyone.

We further investigated the impact of advanced CF-predict features on structure prediction performance. While adding more recycles led to improvement, using multimer models and templates did not contribute significantly. Nonetheless, each feature may have varying effects on different targets, as demonstrated by some notable examples. Thus, it is highly recommended to experiment with different combinations of these features to optimize performance.

In conclusion, while limitations persist, advancements in metagenomic data mining tools, coupled with a blend of automated and human-guided predictions, promise exciting prospects for the future. Further, the results of this research underline the necessity for diversification in methodologies used in protein structure prediction. Finally, the disparity in MSA coverage among different targets stresses the importance of individual target evaluation and tailored approaches. As the field continues to evolve, we anticipate these findings to contribute to the ongoing quest for accurate protein structure prediction.

DATA AVAILABILITY

The data that support the findings of this study are available at doi.org/10.5281/zenodo.8126538.

ACKNOWLEDGMENTS

M.S. acknowledges the support by the National Research Foundation of Korea, grants [2020M3-A9G7-103933, 2021-R1C1-C102065, 2021-M3A9-I4021220], Samsung DS research fund and the Creative-Pioneering Researchers Program through Seoul National University. M.M. acknowledges support by the National Research Foundation of Korea (grant RS-2023-00250 470). A.B. is supported by a Project Grant from the Canadian Institutes for Health Research (CIHR PJT - 190150). Computing resources were provided by the University of British Columbia Community Health and Wellbeing Cloud Innovation Centre, powered by AWS. R.C. was supported by ANR grants [ANR-19-CE45-0008, ANR-22-CE45-0007, PIA/ANR16-CONV-0005, ANR-19-P3IA-0001], and H2020 Marie Skłodowska-Curie grants [Nos. 956229 and 872539]. A.K. acknowledges the support of the US National Institute of General Medical Sciences (NIGMS/NIH), grant R01GM100482. We would like to acknowledge everyone who contributes data to public reference databases, making our work possible, and extend our thanks to Kresten Lindorff-Larsen for pointing this out (Lindorff-Larsen 2023).

Cite this article as *Cold Spring Harb Perspect Biol* doi: 10.1101/cshperspect.a041465

REFERENCES

Alexander H, Hu SK, Krinos AI, Pachiadaki M, Tully BJ, Neely CJ, Reiter T. 2023. Eukaryotic genomes from a global metagenomic data set illuminate trophic modes and biogeography of ocean plankton. *mBio* e01676. doi:10.1128/mbio.01676-23

Altschul SF, Gish W, Miller W, Myers EW, Lipman DJ. 1990. Basic local alignment search tool. *J Mol Biol* **215**: 403–410. doi:10.1016/S0022-2836(05)80360-2

Altschul SF, Madden TL, Schäffer AA, Zhang J, Zhang Z, Miller W, Lipman DJ. 1997. Gapped BLAST and PSI-BLAST: a new generation of protein database search programs. *Nucleic Acids Res* **25**: 3389–3402. doi:10.1093/nar/25.17.3389

Ashkenazy H, Unger R, Kliger Y. 2009. Optimal data collection for correlated mutation analysis. *Proteins* **74**: 545–555. doi:10.1002/prot.22168

Berman HM, Westbrook J, Feng Z, Gilliland G, Bhat TN, Weissig H, Shindyalov IN, Bourne PE. 2000. The protein data bank. *Nucleic Acids Res* **28**: 235–242. doi:10.1093/nar/28.1.235

Bertoline LMF, Lima AN, Krieger JE, Teixeira SK. 2023. Before and after AlphaFold2: an overview of protein structure prediction. *Front Bioinform* **3**: 1120370. doi:10.3389/fbinf.2023.1120370

Bingmann T, Bradley P, Gauger F, Iqbal Z. 2019. COBS: A compact bit-sliced signature index. In *String processing and information retrieval: 26th international symposium, SPIRE 2019*, Segovia, Spain, October 7–9, 2019, pp. 285–303. Springer, New York.

Borkakoti N, Thornton JM. 2023. Alphafold2 protein structure prediction: implications for drug discovery. *Curr Opin Struct Biol* **78**: 102526. doi:10.1016/j.sbi.2022.102526

Bradley P, Den Bakker HC, Rocha EP, McVean G, Iqbal Z. 2019. Ultrafast search of all deposited bacterial and viral genomic data. *Nat Biotechnol* **37**: 152–159. doi:10.1038/s41587-018-0010-1

Camacho C, Boratyn GM, Joukov V, Vera Alvarez R, Madden TL. 2023. ElasticBLAST: accelerating sequence search via cloud computing. *BMC Bioinform* **24**: 1–16. doi:10.1186/s12859-023-05245-9

Camarillo-Guerrero LF, Almeida A, Rangel-Pineros G, Finn RD, Lawley TD. 2021. Massive expansion of human gut bacteriophage diversity. *Cell* **184**: 1098–1109.e9. doi:10.1016/j.cell.2021.01.029

Chen C, Chen X, Morehead A, Wu T, Cheng J. 2023a. 3D-equivariant graph neural networks for protein model quality assessment. *Bioinformatics* **39**: btad030. doi:10.1093/bioinformatics/btad030

Chen I-MA, Chu K, Palaniappan K, Ratner A, Huang J, Huntemann M, Hajek P, Ritter SJ, Webb C, Wu D, et al. 2023b. The IMG/M data management and analysis system v.7: content updates and new features. *Nucleic Acids Res* **51**: D723–D732. doi:10.1093/nar/gkac976

Delmont TO, Gaia M, Hinsinger DD, Frémont P, Vanni C, Fernandez-Guerra A, Eren AM, Kourlaiev A, d'Agata L, Clayssen Q, et al. 2022. Functional repertoire convergence of distantly related eukaryotic plankton lineages abundant in the sunlit ocean. *Cell Genom* **2**: 100123. doi:10.1016/j.xgen.2022.100123

Eddy SR. 2011. Accelerated profile HMM searches. *PLoS Comput Biol* **7**: e1002195. doi:10.1371/journal.pcbi.1002195

Edgar RC, Taylor J, Lin V, Altman T, Barbera P, Meleshko D, Lohr D, Novakovsky G, Buchfink B, Al-Shayeb B, et al. 2022. Petabase-scale sequence alignment catalyses viral discovery. *Nature* **602**: 142–147. doi:10.1038/s41586-021-04332-2

Edmunds NS, Alharbi SMA, Genc AG, Adiyaman R, McGuffin LJ. 2023. Estimation of model accuracy in CASP15 using the ModFOLDdock server. *Proteins* doi:10.1002/prot.26532

Evans R, O'Neill M, Pritzel A, Antropova N, Senior A, Green T, Žídek A, Bates R, Blackwell S, Yim J, et al. 2022. Protein complex prediction with AlphaFold-multimer. bioRxiv doi:10.1101/2021.10.04.463034

Hildebrand A, Remmert M, Biegert A, Söding J. 2009. Fast and accurate automatic structure prediction with HHpred. *Proteins* **77**(Suppl. 9): 128–132. doi:10.1002/prot.22499

Jumper J, Evans R, Pritzel A, Green T, Figurnov M, Ronneberger O, Tunyasuvunakool K, Bates R, Žídek A, Potapenko A, et al. 2021. Highly accurate protein structure prediction with AlphaFold. *Nature* **596**: 583–589. doi:10.1038/s41586-021-03819-2

Katz K, Shutov O, Lapoint R, Kimelman M, Brister JR, O'Sullivan C. 2022. The sequence read archive: a decade more of explosive growth. *Nucleic Acids Res* **50**: D387–D390. doi:10.1093/nar/gkab1053

Kinch LN, Shi S, Cheng H, Cong Q, Pei J, Mariani V, Schwede T, Grishin NV. 2011. CASP9 target classification. *Proteins* **79**: 21–36. doi:10.1002/prot.23190

Koesoema AA. 2022. Protein structure determination as a powerful tool for the sustainable development of agriculture field (and its potential relevance in Indonesia). *IOP Conf Ser Earth Environ Sci* **978**: 012021. doi:10.1088/1755-1315/978/1/012021

Krogh A, Brown M, Mian IS, Sjölander K, Haussler D. 1994. Hidden Markov models in computational biology. applications to protein modeling. *J Mol Biol* **235**: 1501–1531. doi:10.1006/jmbi.1994.1104

Kryshtafovych A, Schwede T, Topf M, Fidelis K, Moult J. 2021. Critical assessment of methods of protein structure prediction (CASP)—round XIV. *Proteins* **89**: 1607–1617. doi:10.1002/prot.26237

Kuhlman B, Bradley P. 2019. Advances in protein structure prediction and design. *Nat Rev Mol Cell Biol* **20**: 681–697. doi:10.1038/s41580-019-0163-x

Levy Karin E, Mirdita M, Söding J. 2020. Metaeuk—sensitive, high-throughput gene discovery, and annotation for large-scale eukaryotic metagenomics. *Microbiome* **8**: 48. doi:10.1186/s40168-020-00808-x

Lindorff-Larsen K. 2023. Nice example of how AlphaFold—like other structure determination methods—is driven by experimental data, and the more/better data you collect, the better the structure. *Twitter.* https://twitter.com/LindorffLarsen/status/1679010235560673280

Liu J, Guo Z, Wu T, Roy RS, Chen C, Cheng J. 2023. Improving AlphaFold2-based protein tertiary structure prediction with MULTICOM in CASP15. *Commun Chem* **6**: 188. doi:10.1038/s42004-023-00991-6

McGuffin LJ, Edmunds NS, Genc AG, Alharbi SMA, Salehe BR, Adiyaman R. 2023. Prediction of protein structures, functions and interactions using the IntFOLD7, Multi-FOLD and ModFOLDdock servers. *Nucleic Acids Res* 51: W274–W280. doi:10.1093/nar/gkad297

Meleshko D, Hajirasouliha I, Korobeynikov A. 2021. coronaSPAdes: from biosynthetic gene clusters to RNA viral assemblies. *Bioinformatics* 38: 1–8. doi:10.1093/bioinformatics/btab597

Mirdita M, von den Driesch L, Galiez C, Martin MJ, Söding J, Steinegger M. 2017. Uniclust databases of clustered and deeply annotated protein sequences and alignments. *Nucleic Acids Res* 45: D170–D176. doi:10.1093/nar/gkw1081

Mirdita M, Schütze K, Moriwaki Y, Heo L, Ovchinnikov S, Steinegger M. 2022. Colabfold: making protein folding accessible to all. *Nat Methods* 19: 679–682. doi:10.1038/s41592-022-01488-1

Moult J, Pedersen JT, Judson R, Fidelis K. 1995. A large-scale experiment to assess protein structure prediction methods. *Proteins* 23: ii–v. doi:10.1002/prot.340230303

Moult J, Fidelis K, Kryshtafovych A, Tramontano A. 2011. Critical assessment of methods of protein structure prediction (CASP)—round IX. *Proteins* 79(Suppl. 10): 1–5. doi:10.1002/prot.23200

Nayfach S, Páez-Espino D, Call L, Low SJ, Sberro H, Ivanova NN, Proal AD, Fischbach MA, Bhatt AS, Hugenholtz P, et al. 2021. Metagenomic compendium of 189,680 DNA viruses from the human gut microbiome. *Nat Microbiol* 6: 960–970. doi:10.1038/s41564-021-00928-6

Orlando G, Raimondi D, Vranken WF. 2016. Observation selection bias in contact prediction and its implications for structural bioinformatics. *Sci Rep* 6: 36679. doi:10.1038/srep36679

Ovchinnikov S, Park H, Varghese N, Huang PS, Pavlopoulos GA, Kim DE, Kamisetty H, Kyrpides NC, Baker D. 2017. Protein structure determination using metagenome sequence data. *Science* 355: 294–298. doi:10.1126/science.aah4043

Pearce R, Zhang Y. 2021. Toward the solution of the protein structure prediction problem. *J Biol Chem* 297: 100870. doi:10.1016/j.jbc.2021.100870

Pereira JM, Vieira M, Santos SM. 2021. Step-by-step design of proteins for small molecule interaction: a review on recent milestones. *Protein Sci* 30: 1502–1520. doi:10.1002/pro.4098

Remmert M, Biegert A, Hauser A, Söding J. 2012. HHblits: lightning-fast iterative protein sequence searching by HMM-HMM alignment. *Nat Methods* 9: 173–175. doi:10.1038/nmeth.1818

Ren F, Ding X, Zheng M, Korzinkin M, Cai X, Zhu W, Mantsyzov A, Aliper A, Aladinskiy V, Cao Z, et al. 2023. Alphafold accelerates artificial intelligence powered drug discovery: efficient discovery of a novel CDK20 small molecule inhibitor. *Chem Sci* 14: 1443–1452. doi:10.1039/D2SC05709C

Renaud N, Geng C, Georgievska S, Ambrosetti F, Ridder L, Marzella DF, Réau MF, Bonvin AMJJ, Xue LC. 2021. Deeprank: a deep learning framework for data mining 3D protein-protein interfaces. *Nat Commun* 12: 7068. doi:10.1038/s41467-021-27396-0

Richardson L, Allen B, Baldi G, Beracochea M, Bileschi ML, Burdett T, Burgin J, Caballero-Pérez J, Cochrane G, Colwell LJ, et al. 2023. MGnify: the microbiome sequence data analysis resource in 2023. *Nucleic Acids Res* 51: D753–D759. doi:10.1093/nar/gkac1080

Rost B. 1999. Twilight zone of protein sequence alignments. *Protein Eng* 12: 85–94. doi:10.1093/protein/12.2.85

Šali A, Blundell TL. 1993. Comparative protein modelling by satisfaction of spatial restraints. *J Mol Biol* 234: 779–815. doi:10.1006/jmbi.1993.1626

Shiryev SA, Agarwala R. 2023. Indexing and searching petabyte-scale nucleotide resources. bioRxiv doi:10.1101/2023.07.09.547343

Söding J. 2005. Protein homology detection by HMM-HMM comparison. *Bioinformatics* 21: 951–960. doi:10.1093/bioinformatics/bti125

Söding J. 2017. Big-data approaches to protein structure prediction. *Science* 355: 248–249. doi:10.1126/science.aal4512

Steinegger M, Söding J. 2017. MMseqs2 enables sensitive protein sequence searching for the analysis of massive data sets. *Nat Biotechnol* 35: 1026–1028. doi:10.1038/nbt.3988

Steinegger M, Söding J. 2018. Clustering huge protein sequence sets in linear time. *Nat Commun* 9: 2542. doi:10.1038/s41467-018-04964-5

Steinegger M, Meier M, Mirdita M, Vöhringer H, Haunsberger SJ, Söding J. 2019a. HH-suite3 for fast remote homology detection and deep protein annotation. *BMC Bioinform* 20: 473. doi:10.1186/s12859-019-3019-7

Steinegger M, Mirdita M, Söding J. 2019b. Protein-level assembly increases protein sequence recovery from metagenomic samples manyfold. *Nat Methods* 16: 603–606. doi:10.1038/s41592-019-0437-4

Subramaniam S, Kleywegt GJ. 2022. A paradigm shift in structural biology. *Nat Methods* 19: 20–23. doi:10.1038/s41592-021-01361-7

Torrisi M, Pollastri G, Le Q. 2020. Deep learning methods in protein structure prediction. *Comput Struct Biotechnol J* 18: 1301–1310. doi:10.1016/j.csbj.2019.12.011

UniProt Consortium. 2023. Uniprot: the universal protein knowledgebase in 2023. *Nucleic Acids Res* 51: D523–D531. doi:10.1093/nar/gkac1052

Wang Z, Eickholt J, Cheng J. 2011. APOLLO: a quality assessment service for single and multiple protein models. *Bioinformatics* 27: 1715–1716. doi:10.1093/bioinformatics/btr268

Wang Y, Shi Q, Yang P, Zhang C, Mortuza SM, Xue Z, Ning K, Zhang Y. 2019. Fueling ab initio folding with marine metagenomics enables structure and function predictions of new protein families. *Genome Biol* 20: 229. doi:10.1186/s13059-019-1823-z

wwPDB consortium. 2019. Protein data bank: the single global archive for 3D macromolecular structure data. *Nucleic Acids Res* 47: D520–D528. doi:10.1093/nar/gky949

Yang P, Zheng W, Ning K, Zhang Y. 2021. Decoding the link of microbiome niches with homologous sequences enables accurately targeted protein structure prediction. *Proc Natl Acad Sci* 118: e2110828118. doi:10.1073/pnas.2110828118

Zemla A. 2003. LGA: a method for finding 3D similarities in protein structures. *Nucleic Acids Res* 31: 3370–3374. doi:10.1093/nar/gkg571

 Cite this article as *Cold Spring Harb Perspect Biol* doi: 10.1101/cshperspect.a041465

Variant Effect Prediction in the Age of Machine Learning

Yana Bromberg,[1,2] R. Prabakaran,[1,*] Anowarul Kabir,[3,*] and Amarda Shehu[3]

[1]Department of Biology; [2]Department of Computer Science, Emory University, Atlanta 30322, Georgia, USA
[3]Department of Computer Science, George Mason University, Fairfax 22030, Virginia, USA

Correspondence: yana.bromberg@emory.edu

Over the years, many computational methods have been created for the analysis of the impact of single amino acid substitutions resulting from single-nucleotide variants in genome coding regions. Historically, all methods have been supervised and thus limited by the inadequate sizes of experimentally curated data sets and by the lack of a standardized definition of variant effect. The emergence of unsupervised, deep learning (DL)-based methods raised an important question: Can machines learn the language of life from the unannotated protein sequence data well enough to identify significant errors in the protein "sentences"? Our analysis suggests that some unsupervised methods perform as well or better than existing supervised methods. Unsupervised methods are also faster and can, thus, be useful in large-scale variant evaluations. For all other methods, however, their performance varies by both evaluation metrics and by the type of variant effect being predicted. We also note that the evaluation of method performance is still lacking on less-studied, nonhuman proteins where unsupervised methods hold the most promise.

Prediction of genomic variant effect is arguably the holy grail of precision medicine, evolutionary analysis, and molecular function annotation. Over the last 20 years, machine learning techniques have captured the leading role in this field. This is primarily because of the enormous number of variables that go into establishing the effect of a specific variant. For example, evaluating the change in a protein function due to a single amino acid substitution, that is, the result of a missense single-nucleotide variant (SNV), may require evaluating all concomitant structural changes, proximity to active sites, and evolutionary history/conservation measures of the protein as a whole and at the specific position in question. Furthermore, the number of variants that have yet to be analyzed is large and growing. That is, every human genome differs from the reference by ~0.1% in SNVs, leading to roughly 3.5 million variants per individual (Pang et al. 2010; Pelak et al. 2010) and at least 84.7 million unique variants in the human population (1000 Genomes Project Consortium et al. 2015). Note that if we commit to the task of evaluating the variant effect on the abilities of other species, particularly bacteria, the number of variants needing annotation would grow exponentially.

---

*These authors contributed equally to this work.

Cite this article as *Cold Spring Harb Perspect Biol* doi: 10.1101/cshperspect.a041467

A [NOT SO] PERFECT CASE
FOR MACHINE LEARNING

Using machine learning methods for human variant analysis, however, has been historically complicated by the difficulty of defining measurable variant effect classes. For example, "Does this variant cause disease?" may be a yes/no question only for some cancer drivers and monogenic disorders; note that the variant effect may need to be further modified by the allelic dominance. Nevertheless, many methods have claimed the ability to predict variant pathogenicity, without explicitly defining the term's meaning. This reductionist paradigm that has long dominated the field of genetics (Gibson 2012; Katsanis 2016) answers a very different question than whether the complex interplay of all individual genome variants with the environment is likely to give rise to a particular disease.

Existing experimentally derived, quantified, and validated variant annotations that could be used for method training are few in the scientific databases and have, so far, been hard to extract from the literature. For example, the manual curation efforts to collect monogenic disease/phenotype-associated variants in OMIM (Hamosh et al. 2000) have only annotated ~23,000 (0.027% of 84.7M) SNVs (Savojardo et al. 2021). Additional disease-associated and putatively neutral missense SNVs were collected in HumVar (Capriotti et al. 2006), from the work of SwissProt curators (Boeckmann et al. 2003; Bairoch et al. 2004), and in the more recent HuVarBase (Ganesan et al. 2019). The latter parses a large number of resources to extract 774,863 disease-annotated variants (718,590 are missense)—a "staggering" 0.92% of all known human SNVs.

Non-disease-relevant variant functional effect is even harder to come by. The Protein Mutant Database (PMD) (Nishikawa et al. 1994; Kawabata et al. 1999), a project started in 1989, required reading over 42,000 relevant scientific articles to extract the subjective, qualitatively binned measurements of the variant effect of over 200K variants in different species on protein structure, stability, and/or function; note that human variants make up less than a fifth of this data set. While recent small studies have been more proactive in distributing their data (see, e.g., the VariBench collection [Sarkar et al. 2020]), study reporting, data accessibility, and cross-study standardization, remain limited. Given the resulting disproportionately wide and short training data tables (i.e., relatively few variants and many features that can describe a variant), the field has gotten very creative in data collection for prediction method development/training.

NOT ALL VARIANT IMPACT IS THE SAME

Here lies the root of the confusion—understanding of exactly which variant effect is being predicted is limited across the nearly 200 currently available, supervised methods (Hu et al. 2019; Zhu et al. 2020). For example, SNAP (Bromberg and Rost 2007; Hecht et al. 2015) is trained using (all species) single amino acid substitutions collected from the PMD, where functional effect refers to the result of experimental evaluations of the activity of wild-type versus mutated proteins. As expected, SNAP aims to predict the variant functional effect. One version of PolyPhen-2 (Adzhubei et al. 2010) is trained on the HumVar data set, that is, pathogenic variants versus those with no such annotation. Curiously, PolyPhen is also meant to predict the possible effect of the variant on the function and structure of the affected protein. CADD (Kircher et al. 2014) contrasts observed coding and noncoding variants with simulated mutations, predicting variant functional effect and both disease causation and association. Furthermore, meta-methods (e.g., REVEL [Ioannidis et al. 2016]) rely on input from a variety of disparate techniques to summarize some fluid definition of variant effect. The recent advances in deep learning (DL) and unsupervised methods (e.g., DeepSequence [Riesselman et al. 2018], EVE [Frazer et al. 2021], and ESM1V [Meier et al. 2021]) have given rise to even less-explainable effect predictions. While these models are better suited to evaluate variants on a large scale, the specific type of effect that their variant scores capture is unclear. Numerous efforts have been undertaken to evaluate the performance of the various methods (e.g., The Critical Assessment

Cite this article as *Cold Spring Harb Perspect Biol* doi: 10.1101/cshperspect.a041467

of Genome Interpretation Consortium 2022), but these are also limited by the lack of correspondence between the questions being asked and the type of answers that these tools have been developed to provide.

Excluding the prediction of measurable structural or stability changes, there are three broad kinds of effects that are relevant: (evolutionary) fitness effect, pathogenicity (disease causation), and (molecular) function change. We argue that existing computational methods have largely failed to recognize the difference between these three types of effect.

IMPACT OF VARIANT EFFECT TYPE ON METHOD DEVELOPMENT

While closely related, the three types of variant effects are not identical and, in fact, are very different in non-edge cases. Evolutionary fitness, for example, is often evaluated in terms of population frequency or conservation of the mutated site across species; that is, variants are expected to be more or less common in human population or across species because of the impact they have on their carrier ability to survive and reproduce. Evolutionary history, however, does not guarantee current success. Thus, for individual human genomes, variant population frequency or site conservation alone is unlikely to lead to a precise conclusion about variant functional effect or involvement in causing a disease. Similarly, a slight change in the function of a given protein may be insufficient to make a fitness difference on the human population level or to bring on a clearly definable disease phenotype.

The relationship of evolutionary fitness with disease is complicated by the polygenic nature of most disorders, making it impossible to infer truly causative variants. However, even in the case of monogenic disorders, the selective disadvantage of causative variants is not always guaranteed, and population-specific frequency is paramount. For example, the sickle cell hemoglobin allele is exceedingly rare (∼0% minor allele frequency [MAF]) in Europe and the Americas, but more common (MAF > 0.5%) throughout most of the African continent, and pervasive (MAF > 9%) across the large area stretching from southern

Ghana to northern Zambia (Piel et al. 2010). Thus, elucidating the specific relationship between disease and fitness requires a much deeper understanding of the nature of the disease and of the features of the affected populations.

Technological problems, for example, insufficient sequencing across populations, limited experimental resolution in establishing functional effect, and even statistical inference parameters, complicate matters further. For example, genome-wide association studies (GWAS), require that a given variant be present in some significant fraction of the population (e.g., 5%). This frequency, however, would be improbable, if not impossible, for a true disease-causing, selectively disadvantageous variant.

Furthermore, outside of human populations, the words fitness, pathogenicity, and population frequency are measured using different scales and, thus, methods pretrained with human data cannot be expected to produce similar results across organisms.

The lack of effect type differentiation in literature could be attributed to the variants historically analyzed in wet-lab experiments and used for, first, establishing the theoretical framework for effect identification and, later, for computational method development. Due to scientific interest and time/money constraints, scientists aimed to study variants likely to cause a disease; after all, why not start with a variant that is likely to cause diabetes or trigger cancer development? At the same time, due to experimental limitations, functional variant effect could only be measured reliably for high-effect variants with visible phenotypic displays, such as a drastic reduction in catalytic activity, reduced number/size of cell colonies, or even misshapen red blood cells. On the other hand, many experimentally annotated negative/no-effect findings were incidental and often not explicitly or rigorously evaluated.

This bias toward hunting for high-effect variants has been recently elucidated through the development of deep scanning mutagenesis (DMS) methods that quantitatively, for one type of function per experiment, evaluate many, if not all, possible variants per given protein (Araya and Fowler 2011). DMS experiments demonstrated that the scale of variant effects is continuous and, by sampling from only the high-effect variants, we forgo the understanding of what "effect" means (Miller et al. 2017, 2019; Pejaver et al. 2019). As such, even a perfectly trained supervised classifier would tend to underestimate the number of function effect-carrying variants in a given protein or gene, while method evaluations, of either supervised or unsupervised methods, may overestimate the correlation between the three types of variant effect.

In evaluating variant effect, one readily available representation of evolutionary fitness—conservation of the mutated site—has been exceedingly useful and used by many methods. The reasoning behind the use of this feature is clear —conserved positions are conserved for a reason and, thus, should not change. The use of conservation across methods has, however, ensured that most predictions are largely nonorthogonal, that is, they tend to predict the same conservation-evidenced outcomes. This conclusion is supported by the minimal variation in performance across methods and minimal performance improvement of meta-methods. The recent development of single-sequence-based, unsupervised, protein language models (pLMs) appears to bypass the explicit need for conservation information. These models may capture a different effect signal and, alone or in concert with other techniques, may be helpful in distinguishing variant effect types.

CAN THE TOOLS OF TODAY ANNOTATE VARIANT EFFECT?

In this work, we evaluate the advances in variant effect prediction across the various models and effect types and sizes. We first evaluate the contribution of variant frequency and position conservation to variant effect predictor scores. We suggest that these characteristics of variants are the best currently available, although indirect, representations of variant "fitness." We then use experimentally validated sets of variants to evaluate the method's ability to predict variant functional effects. Here, we assess both a selection of variants explored in the scientific literature (reported in the PMD) and an exhaustive collection of variants from DMS experiments. Finally, we assess the per-

formance of methods in annotating variants as being pathogenic (i.e., likely disease-causing [reported in ClinVar; Landrum et al. 2018]).

We consider three types of methods (SOM methods): classical methods and supervised and unsupervised DL methods. Classical methods use computational (e.g., SIFT [Ng and Henikoff 2003; Hu and Ng 2013]) and machine learning–based (e.g., REVEL) techniques, where input features (e.g., conservation, structure, or other method predictions) are selected on the basis of biological relevance. Their training/development data classifies variants into two groups—effect or no effect—however defined. Supervised DL methods may similarly use selected biological variant or protein features (e.g., MetaRNN uses variant frequency), conservation scores, and output of other prediction methods, or they may rely simply on protein sequence and structure (e.g., Seq-UNet [Dunham et al. 2023]). They are also trained to reflect on variant effect (e.g., pathogenicity in the case of MetaRNN [Li et al. 2022]) and variant "rare-ness" (probability that the variant is rare), or the corresponding position-specific scoring matrix (PSSM) of amino acids (in the case of Seq-UNet). As opposed to classical methods, however, all methods in this category use DL architectures to achieve their task. Finally, unsupervised DL methods use sequence information alone without an attached variant classifier. Their outputs are then evaluated to explore the difference between wild-type and mutant sequences (SOM methods in Supplemental Material).

To evaluate the performance of all methods, we scaled each method's prediction scores into the [0, 1] range using min–max scaling across all variants in all data sets and made all scores unidirectional (1 = effect, 0 = no effect; SOM methods in Supplemental Material). Furthermore, we developed a standardized, population frequency-based means for score threshold selection. For each method, we identified a threshold to separate effect versus no-effect variants as the one below in which 95% of the variants common in the population (allele frequency ≥0.01) were observed (Zeng and Bromberg 2019). We argue that this approach allows for high precision in identifying variants of high effect, even if the recall of smaller effect variants is limited. Finally, to represent the conservation baseline, we used phastCons (Siepel et al. 2005), computed on multiple sequence alignments of 16 primate genomes to the human genome. Earlier work found that using primate genome alignments was more informative of variant effect than all vertebrate alignment (Sun and Yu 2019). Genome-alignment-based methods may not be as sensitive to protein structure/function changes as the more protein-focused methods. However, the conservation of genomic sites in protein-coding regions may be driven by non-protein-sequence-specific needs (e.g., splice sites, mRNA stability, tRNA binding, etc.). We thus argue that these estimates are better suited for baseline assessments of variant fitness and report phastCons performance for comparison with all data sets.

VARIANT EFFECT PREDICTORS AND LANGUAGE MODELS RECOGNIZE POPULATION FREQUENCY

Variants in the human genome are not evenly distributed across coding genes. In our analysis of the ALFA data (Phan et al. 2020), only slightly more than half (58%) of the 16,671 proteins considered in this study corresponded to genes that carried a common variant (allele frequency ≥0.01), while nearly all had a singleton (allele count = 1) or an ultra-rare variant (frequency <0.001). Rare variants were only present in roughly three-quarters of the proteins (0.01> allele frequency ≥0.001; 74%). Although the uneven distribution of variants in some of the proteins can be explained by insufficient data, disagreement between sequencing projects, which could be expected if individual project data were not representative, is rare. Where projects do disagree, it is often one, usually smaller project that assigns higher frequency, while all others do not. For example, the FLYWCH-type protein (NP_001294997) contains no common variants according to ALFA (sample size = 48,480) or any of the population-specific projects but has one SNP (rs61747748; ALFA MAF = 0.0014) labeled as common by the Simons genome diversity project (sample size = 12). Thus, we expect that common variants are indeed limited to a subset of genes. For these, the absence of common variants may identify evolutionary novelty and thus less time to perpetuate variants throughout the population. Alternatively, these genes can be essential and thus resist variance altogether. We note that in-depth evaluation of the reasons for this variant disparity across genes/proteins is beyond the scope of this work.

Here, we set out to evaluate the relationship of the variant frequency with the predicted variant effect. We considered the genes lacking common variants to be somewhat biologically unique and, possibly, misrepresented in training data of computational methods. We thus retained only the variants in those genes that carried both common and other variant types together and sampled 35,082 variants in 9,142 proteins (Supplemental Fig. S1; Supplemental Table S1).

The nature of training/development data used by many supervised methods had, due to experimental limitations, mostly labeled rare human pathogenic variants as deleterious (i.e., pathogenic and/or of large negative functional effect), while common variants were largely deemed neutral (i.e., not pathogenic nor bearing functional effect). For example, very common (i.e., >5% allele frequency) variants are defined as benign by ACMG–AMP experts (i.e., stand-alone evidence of benign effect [BA1]). We thus expected that classical methods for prediction of variant effect, represented here by CADD, REVEL, PolyPhen, and SIFT, as well as supervised DL methods (i.e., MetaRNN, MVP [Qi et al. 2021], Seq-UNet, VESPA [Marquet et al. 2022], and AlphaMissense [Cheng et al. 2023]), would identify less frequent variants as having a larger effect.

We observed the expected method behavior (Fig. 1A), although the difference between scores of common versus rare-type variants was not large for most methods. All classical methods performed roughly the same, but, curiously, supervised DL methods spanned the range of abilities in this analysis; they ranged from identifying almost no disparity in population frequency (Seq-UNet and MVP) to classic-like performance (VESPA) to nearly perfect frequency classification (MetaRNN). Note that while all methods were better at differentiating common from ultra-rare variants than from rare ones, MetaRNN was sig-

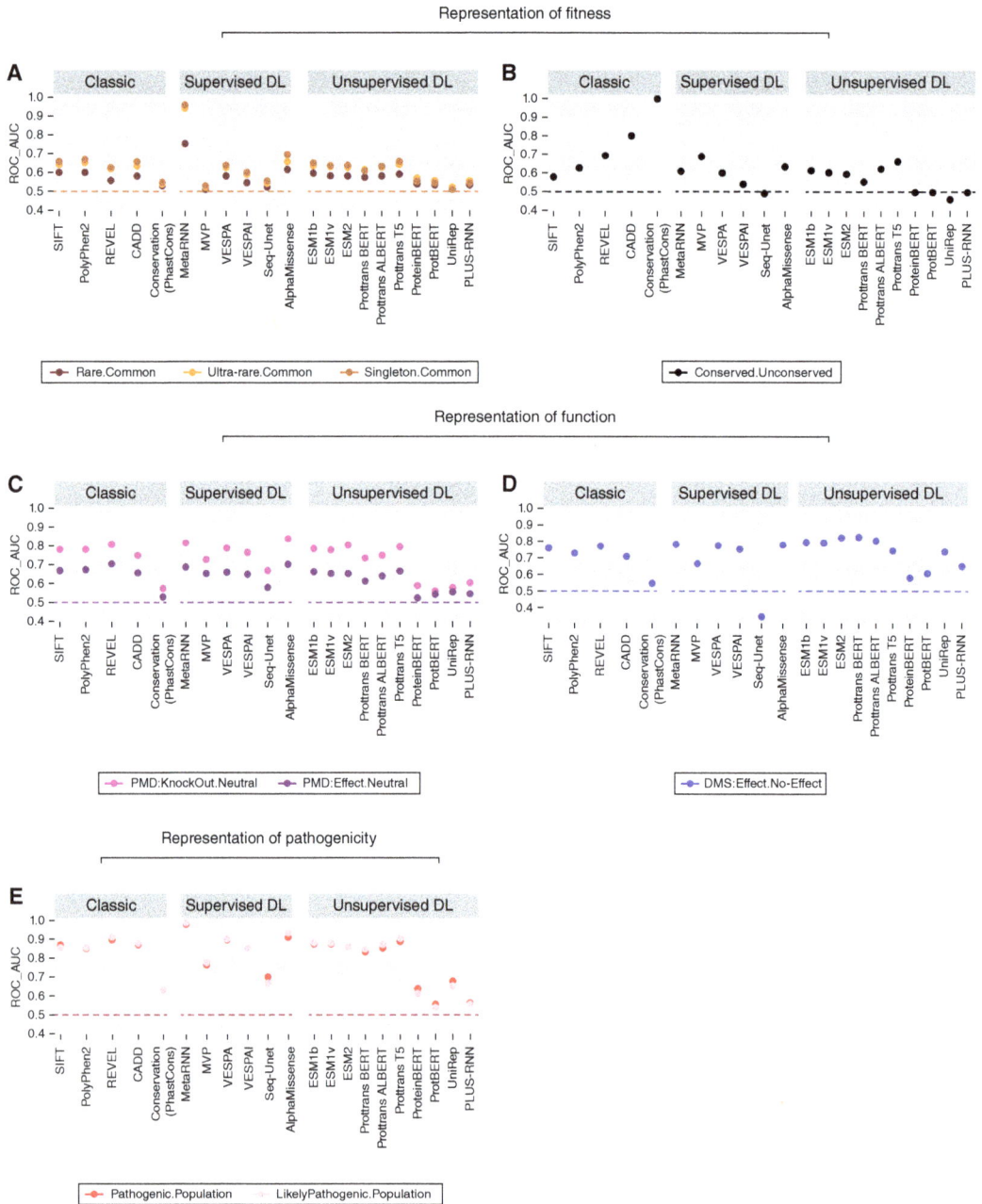

Figure 1. Evaluation of performance of various methods on data sets. Performance is reported as the ROC AUC for all data sets and methods used in this study. Higher ROC indicates better ability to differentiate (*A*) common variants from all other population frequency variants, (*B*) conserved from unconserved variants, (*C*) functionally neutral variants in the Protein Mutant Database (PMD) from variants assigned an effect, (*D*) effect/no-effect deep scanning mutagenesis (DMS) variants, and (*E*) pathogenic variants from putatively benign baseline. Dashed lines indicate the performance of an empirical random classifier.

nificantly more so, suggesting that the pathogenicity signal that it learned is primarily frequency-driven. Note that at our selected scoring threshold most methods included the majority of all variants —common or not—indicating the rarity of variants with high effect (threshold line is higher than the bulk of variant scores) (Fig. 2; Supplemental Fig. S2).

We expected that unsupervised models, trained on most available protein sequences, would capture global sequence signals rather than specific population frequencies. In fact, common human variants may be expected to have some level of evolutionary significance for phenotypically distinct human populations (Bromberg et al. 2013; Mahlich et al. 2017) but are unlikely to disturb the global protein language patterns. Furthermore, while rare and ultra-rare variants carry the bulk of evolutionarily deleterious variation, these drastically damaging changes make up only a small fraction of the massive total number of rare variants. That is, we expected to see little difference between the scores of common and rare

variants. Indeed, for some models (e.g., protein-BERT [Brandes et al. 2022] or UniRep [Alley et al. 2019]), scores were not a major indicator of variant frequency. For others, however (e.g., ESM1b [Rives et al. 2021] and ProtTransT5 [Elnaggar et al. 2022]), we observed that scores were as different across variant frequency classes as for some classical methods (Fig. 1A). Note AlphaMissense was trained using frequency labels; thus, its improved performance in capturing frequency signals was expected if underwhelming.

Our findings thus indicate that, on average, a given variant's population frequency is only moderately indicative of its predicted effect, as captured by (most) methods regardless of predictor class.

VARIANT PREDICTIONS CORRELATE ACROSS MANY METHODS

In the presence of many methods for predicting variant effect, it is natural for scientists to evaluate individual variants using a number of tools to

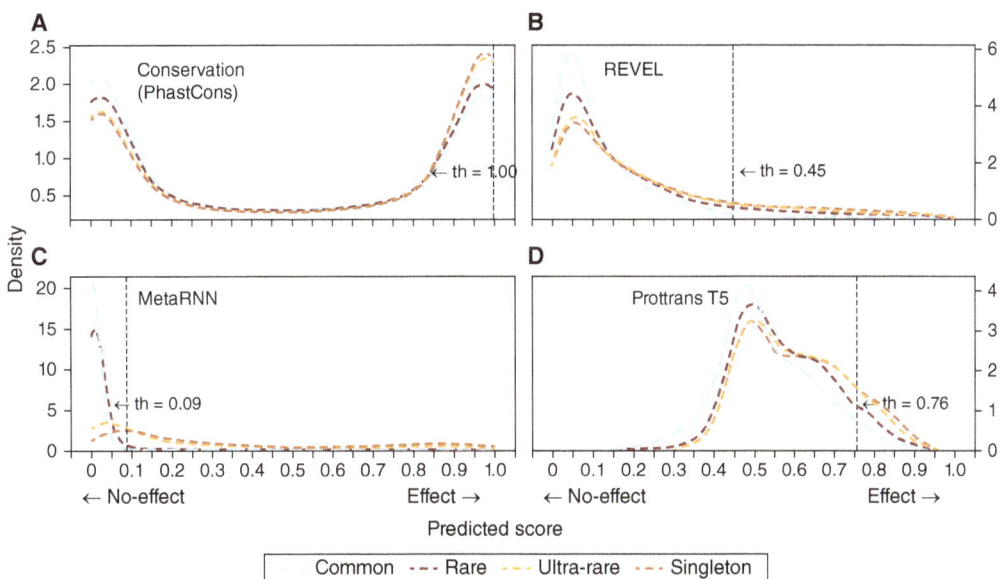

Figure 2. Distribution of scores for different variant population frequencies. Scores attained by variants in different population frequency classes (line colors) from (A) phastCons, our conservation score, (B) REVEL, an example of a classical method in our set, (C) MetaRNN an example of a supervised deep learning (DL), and (D) ProtTransT5, an example of an unsupervised predictor. Thresholds indicate scores below which 95% of common variants were observed. Representative methods attained the best ROC AUCs in Figure 1A; other methods are in Supplemental Figure S2.

establish agreement. The above findings, however, confirm that many methods are nonorthogonal and will often produce similar results—regardless of the actual effect that the variant may have. Indeed, it is informative to evaluate the performance of different methods on the set of variants where their predictions disagree (Bromberg and Rost 2007). We thus asked what is the relationship between method scores? (Fig. 3).

We observed that in binary variant assessment at the established threshold (Fig. 3, red), all methods, except MetaRNN, were in nearly perfect agreement. We also found that conserva-

tion (phastCons) scores correlated with method predictions, suggesting that conservation alone may have been sufficiently informative of these predictions.

The relationship between method scores, however, was somewhat more informative (Fig. 3, blue). As expected from binary comparisons, many predictor scores correlated to some extent, UniRep and Seq-UNet being the exceptions. However, conservation was no longer as well representative of other methods. There was a significant level of correlation between most supervised methods, whether DL-based or not. ESM

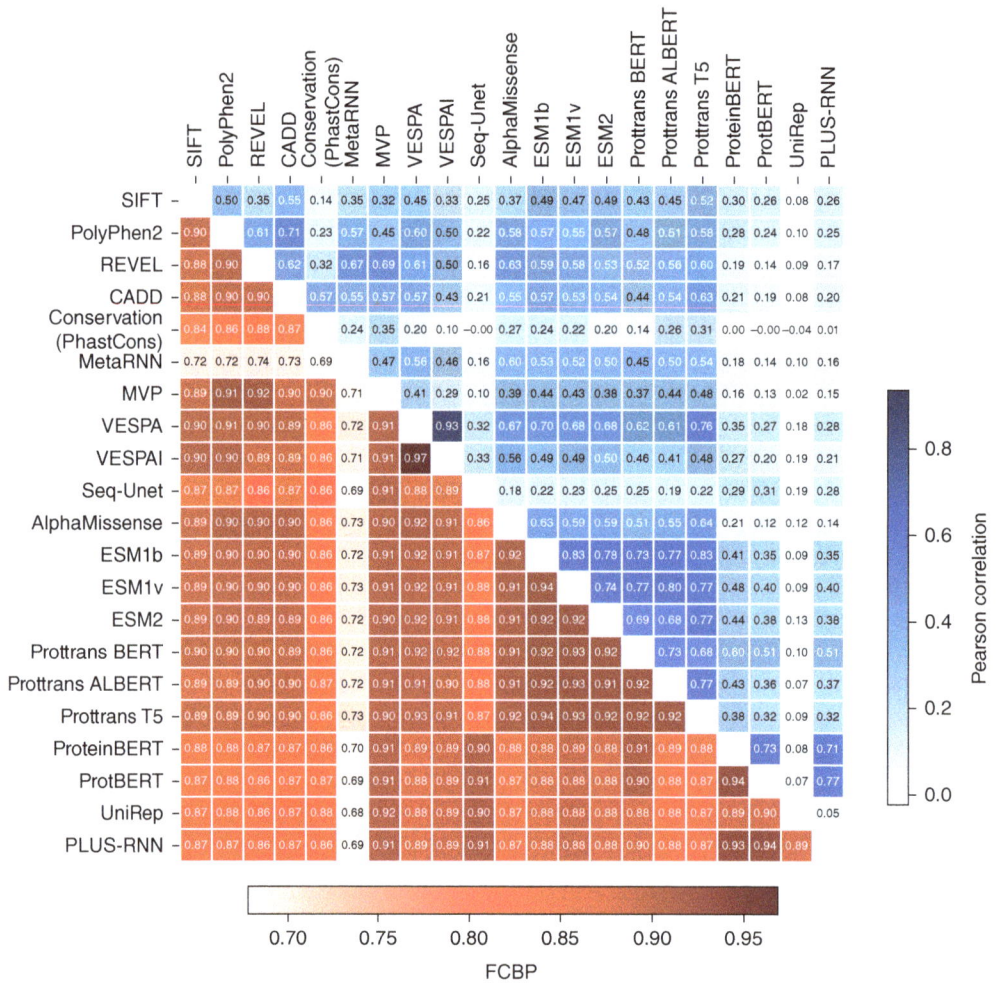

Figure 3. Correlation of variant predictor scores. Correlation of predictor scores for the population frequency data set is reported as the Pearson correlation coefficient (blue) and the fraction of consensus binary predictions (FCBPs) (red). Higher scores (darker colors) indicate a better correlation of predictor outputs.

and ProtTrans scores were correlated among themselves and also with supervised methods. ProteinBert (Brandes et al. 2022), ProtBert (Rao et al. 2019), and PLUS-RNN (Min et al. 2021) agreed in their predictions, but only somewhat correlated with other unsupervised methods, suggesting that these pLMs captured a somewhat different signal.

Importantly, PLUS-RNN sequence representations were contextualized in protein structure, as were Seq-UNet predictions, allowing these methods to capture longer-range residue contacts. We thus suggest that Seq-UNet, PLUS-RNN and, by correlation, ProteinBert and ProtBert, predictions may be orthogonal to ESM and ProtTrans unsupervised DL methods. That is, the agreement of these methods on variant effect may be more informative than either one of the methods alone. We further explore this notion with functionally annotated PMD variants below.

Conservation Is Orthogonal to Frequency and Recognized by All Predictors

Many, if not most, of the classical variant effect prediction methods heavily rely on the conservation of variant position across protein homologs. This is warranted as significantly deleterious substitutions could be expected to be eliminated in evolution. In our work, this observation is also supported by the agreement between binary assessments of variants using phastCons versus other methods (Fig. 3, red). However, conservation scores alone are limited in the prediction of nuances of variant effect, as is evident from lower corresponding score correlations (Fig. 3, blue). There are multiple reasons for this observation. First, sufficiently descriptive estimates of per residue conservation are complicated and often limited to large gene/protein families (Triant and Pearson 2015; Malhis et al. 2019). Second, co-occurring mutations across multiple positions may dampen or increase individual variant effects (Holcomb et al. 2021). Finally, position conservation does not easily translate into quantitative descriptions of the severity of the effect (Miller et al. 2017). Here, the nature of a particular variant substitution may

ameliorate the impact of affecting a conserved position or worsen the impact of tweaking an unconserved one. Thus, an aspartic to glutamic acid substitution in a conserved negatively charged site may be acceptable, while a serine to tryptophan change in a variable but buried position may be severely disruptive.

We used the population frequency variant set to evaluate how well variant effect predictors capture conservation. We labeled as "conserved" all variants with a phastCons score ≥ 0.5, and "unconserved" otherwise. We then asked whether variant conservation and population frequency were related terms. That is, we evaluated whether common variants are differently conserved than rare ones. In a discretized comparison, common variants were indeed less conserved than rare and ultra-rare ones; that is, common, rare, ultra-rare, and singleton phastCons score medians were = 0.43, 0.59, 0.73, 0.76, respectively (Fig. 2A). However, there was no significant correlation between conservation scores and frequency of variants in the population (Pearson correlation = −0.09, Spearman correlation = −0.12). These observations suggest that the signal describing the conservation of variant sites is orthogonal to that describing variant frequency.

We further observed that most effect prediction methods distinguished between variants in differentially conserved positions (Fig. 1B). However, classical method performance greatly varied, highlighting the differential emphasis on conservation (as reported by genome-based alignments) in evaluating variant effect—some surprisingly low (SIFT) and some high (CADD). Supervised DL methods also displayed significantly varied performance. As expected, MetaRNN conservation prediction performance was drastically lower than for prediction of variant frequency (Fig. 1A,B), while MVP's performance improved, likely due to the latter's reliance on conservation scores.

We then asked whether conservation can be predicted by methods that do not use it in training. In earlier work, Marquet et al. (2022) and Dunham et al. (2023) found this to be possible. For our data, pLMs appeared to be as good at differentiating variant conservation as they were for variant frequency, but their performance was not as good as that of some classical methods (Fig. 1B) that use conservation for making predictions. Notably, the supervised DL VESPA method performed worse than its baseline language model ProtTransT5, highlighting the fact that recognizing variant effect is not equivalent to recognizing conservation. Also note that ProtTransT5 was significantly better at this task than other pLMs, suggesting that some language models may produce more biologically interpretable embeddings than others.

If predictions of the methods in our study could be considered equivalent to experimental, in vitro or in vivo, analysis, these results would indicate that conservation plays a significant, if not all-encompassing, role in explaining variant effect (i.e., it contributes 0.1–0.3 of overall ROC AUC). However, given that method performance in predicting true variant effect is limited and exceedingly varied by the test/evaluation set used to establish performance metrics, our results suggest that the value of conservation alone is unlikely large, as confirmed by phastCons performance on function-relevant and pathogenic variants (Fig. 1C–E). That is, the signal of billions of years of evolution has to be seen through the prism of more information to be interpreted and applied to effect prediction.

Functional Effect Is a Combination of Many Factors Recognizable by Unsupervised Methods

As expected from the previously described bias in experimental studies, functionally significant changes affect conserved sites somewhat more frequently than unconserved ones (phastCons performance; Fig. 1C,D). Thus, given their ability to differentiate variants by conservation, most methods could be expected to perform at least as well or better in differentiating mutations of functional effect from those of no effect. Note that both effect and neutral (no-effect) variants were found across the full range of conservation scores, somewhat complicating the problem (Supplemental Fig. S3).

All methods were indeed significantly better at identifying knockout (large effects) than mild and moderate effect variants (Fig. 2C; Supplemental Fig. S3). Notably, unsupervised methods were as good as earlier techniques in differentiating the knockout/effect versus no-effect variants. However, all method score distributions were sufficiently overlapping as to often "misidentify" experimentally labeled neutral variants as having an effect (Supplemental Fig. S3). Note that experimentally establishing variants as neutral is a difficult task, with literature reports often disproven in later publications (Bromberg et al. 2013; Zeng and Bromberg 2019).

We found that, at the binary effect threshold, REVEL was excellent at labeling all effect/knockout variants (PMD set of human variants, F1 measure = 0.78; Table 1). Similar behavior was observed for MetaRNN (0.84). Note that MetaRNN was trained to predict pathogenic variants, this result suggesting that only variants of high functional effect in disease genes would be identified as impactful. Its high recall confirms that many of the MetaRNN predicted pathogenic variants were also of high functional effect. On the other hand, VESPA (F1 = 0.62) was trained using the PMD data and hence could be expected to perform well. However, our population-based threshold has somewhat altered its performance in favor of very high precision (90%), but lower recall (47%) predictions. At the default threshold (Supplemental Table S4), VESPA F1 was higher (0.66), while MetaRNN performance lower (0.81); for MVP, the difference between thresholds was even greater (F1 = 0.65 at our threshold, F1 = 0.84 at default threshold; Table 1; Supplemental Table S4). These differences highlight the issues of applying scoring thresholds to methods without optimization for specific tasks.

This method performance may also seem higher than expected in light of the small number of no-effect variants in our data. In fact, all methods mislabeled neutral variants to a certain extent (e.g., MetaRNN labeled 86% of all experimental neutrals as having an effect). VESPA, however, only tagged 16% of the neutral variants incorrectly (Table 1). This mislabeling by all supervised methods highlights the effect of the previously mentioned bias in selecting variants for evaluation—variants in disease genes are more likely to be experimentally evaluated and may be erroneously tagged as neutral on the basis of nonexhaustive experimentation. We note that

Table 1. Method performance at the frequency-established threshold

	Predictor	Thrsh	PMD					Pathogenicity					DMS				
			Prec	rec	F1	%at	%nat	Prec	rec	F1	%at	%nat	Prec	rec	F1	%at	%nat
Classic	CADD	0.48	0.87	0.40	0.55	0.34	0.18	0.90	0.55	0.68	0.36	0.09	0.17	**0.93**	0.29	0.75	0.72
	PolyPhen2	0.99	**0.89**	0.40	0.55	0.34	0.15	0.90	0.48	0.63	0.32	0.08	**0.26**	0.50	**0.35**	0.26	0.22
	REVEL	0.45	0.87	**0.70**	**0.78**	0.61	0.32	0.87	**0.89**	**0.88**	0.61	0.20	0.20	0.87	0.33	0.59	0.54
	SIFT	1.00	0.90	0.32	0.47	0.26	**0.11**	**0.91**	0.46	0.61	0.30	**0.06**	0.09	0.01	0.02	0.02	**0.02**
Supervised DL	MetaRNN	0.09	0.77	**0.94**	**0.84**	0.92	0.86	**0.93**	**0.99**	**0.96**	0.64	0.12	0.14	**0.98**	0.25	0.94	0.93
	MVP	0.92	0.86	0.52	0.65	0.46	0.26	0.90	0.70	0.79	0.54	0.18	0.21	0.58	0.31	0.38	0.35
	Seq-UNet	0.92	0.87	0.09	0.16	0.07	**0.04**	0.82	0.19	0.30	0.14	**0.06**	0.00	0.00	0.00	0.00	**0.00**
	VESPA	0.55	**0.90**	0.47	0.62	0.39	0.16	0.92	0.64	0.76	0.42	0.08	0.34	0.78	**0.48**	0.40	0.32
	VESPAl	0.58	**0.90**	0.40	0.56	0.34	0.13	**0.93**	0.49	0.65	0.32	**0.06**	**0.39**	0.53	0.45	0.24	0.18
	AlphaMissense	0.57	0.88	0.65	0.75	0.55	0.26	0.91	0.80	0.86	0.52	0.11	0.28	0.92	0.43	0.57	0.50
Unsupervised DL	ESM1b	0.65	0.89	0.47	0.62	0.40	0.18	0.90	0.69	0.78	0.46	0.11	0.05	0.04	0.04	0.32	0.38
	ESM1v	0.64	0.89	0.45	0.60	0.38	0.17	0.90	0.67	0.77	0.44	0.10	0.05	0.04	0.04	0.23	0.27
	ESM2	0.68	0.90	0.44	0.59	0.37	0.15	**0.91**	0.59	0.72	0.39	0.09	0.02	0.02	0.02	0.13	0.16
	ProtBERT	0.53	0.89	0.13	0.23	0.11	0.05	0.79	0.13	0.22	0.10	0.05	0.80	0.26	0.26	0.91	0.93
	ProteinBERT	0.43	**0.91**	0.13	0.22	0.11	**0.04**	0.88	0.25	0.39	0.17	0.05	**0.98**	**0.30**	**0.30**	0.99	0.99
	ProtTransALBERT	0.66	0.88	0.42	0.57	0.36	0.18	0.89	0.63	0.74	0.42	0.11	0.02	0.03	0.03	0.16	0.19
	ProtTransBERT	0.66	0.89	0.44	0.58	0.37	0.16	0.90	0.63	0.74	0.42	0.10	0.17	0.09	0.09	0.45	0.51
	ProtTransT5	0.76	0.88	**0.50**	**0.64**	0.42	0.20	0.90	**0.74**	**0.81**	0.49	0.12	0.00	0.01	0.01	0.02	**0.02**
	UniRep	0.07	0.85	0.13	0.23	0.12	0.07	0.81	0.10	0.17	0.07	**0.03**	0.20	0.27	0.27	0.09	0.07
	PLUS-RNN	0.41	0.90	0.16	0.27	0.13	0.05	0.82	0.19	0.31	0.14	0.06	0.91	0.28	0.28	0.97	0.99

(Prec) Precision, (rec) recall, (F1) F measure, (%at) percent of all variants predicted above threshold (positive), (%nat) percent of all negatives (neutrals/no effect) above threshold (false positives).

Best performance values in each class of tools and for each set are highlighted in bold. Note that all values are rounded to the second digit, leading to ~0 values in some entries.

these could potentially be successfully recovered as having an effect using computational analysis.

ESM and ProtTrans pLMs did as well as supervised methods in differentiating distributions of knockout/effect versus no-effect variants (Fig. 1C) and worse than supervised models at our selected cutoff (Table 1). Nevertheless, ProtTransT5, the best performer of all unsupervised models (F1 = 0.64; Table 1), mislabeled only 20% of the neutrals—a performance on par with best-supervised methods. We also note that at the standard (nonfrequency-optimized threshold; Supplemental Table S4), all ESM and ProtTrans models improved in performance as measured by the F1 measure (e.g., ProtTransT5 $F1_{nonoptimized} = 0.84$, as the cost of drastically reduced precision [prec = 0.78 vs. = 0.88, respectively]).

Given the bias in available experimental evaluation data toward variants of high effect, we further evaluated predictions of variant annotations extracted by DMS techniques. Specifically, we considered annotations of two proteins (PTEN and TPMT) (Notin et al. 2022; Supplemental Material). For this set of variants, supervised DL methods performed best (highest ROC AUC; Fig. 1D) across the scoring spectrum. Specifically, the best performers were VESPA (and VESPAl), both pLM (ProtTransT5)-based models, closely followed by AlphaMissense (Fig. 1D; Table 1), also pLM-based. Their improved performance (over unsupervised methods) suggests significant value in fine-tuning.

Given the above results, we suggest that while ESM and ProtTrans pLMs could possibly be used in identifying variant effect, large-scale analysis of variants benefits from more fine-tuned method application. Nevertheless, pLMs use may be particularly meaningful for nonhuman variants, where gene/protein large family alignments and experimental annotations are not readily available.

Prediction Methods Capture Different Signals

Given the results of our score correlation experiments, we asked whether combining methods may produce more precise classification of variants. We evaluated the precision of a jury-of-two

method on PMD knockout versus no-effect variants, by only considering an agreement between the two methods as an effect prediction. In fact, asking two methods to agree significantly improved precision, albeit at the cost of recall (Fig. 4).

We observed this behavior for almost all combinations of methods. As expected, methods that performed worse on their own, got more of a boost. phastCons, for example, greatly benefitted from the addition of almost any other method's input—even without major cost to recall. Some interesting combinations were present (e.g., REVEL—an ensemble method including SIFT, PolyPhen, and CADD scores, still benefitted from the addition of either of these methods), albeit at a significant >15% cost to recall.

We did not expect much improvement from combining pLMs with correlated scores. However, adding ESM2 (Lin et al. 2023) to ProtTransT5, did result in an 8% gain in precision and a 12% loss in recall. This observation suggests that scores could be further fine-tuned to eke out only the (few) high-reliability variants in each set. Adding uncorrelated unsupervised model scores (PLUS-RNN or ProteinBERT) was somewhat beneficial for ProtTransALBERT precision (adding 13%), but less so for ProtTransT5 or ESM2 precision (adding 8%–10%), while drastically reducing recall for all (by 38%–47%).

These results suggest that relying on the differences in latent spaces described by individual unsupervised models does not necessarily improve variant effect capture. More analysis of data set selection and parameter optimization choices is necessary to define the unique characteristics of pLMs that care about individual "words" (residues) in protein "sentences" (sequences).

Unsupervised Methods Clearly Capture Signals of Pathogenicity

Identifying variant pathogenicity has long been the focus of human genetics and a major driver of research initiatives. Currently, two major resources provide information about designations of variant pathogenicity—ClinVar and ClinGen (Rehm et al. 2015). ClinVar focuses on variant pathogenicity, as recommended by ACMG–

Cite this article as *Cold Spring Harb Perspect Biol* doi: 10.1101/cshperspect.a041467

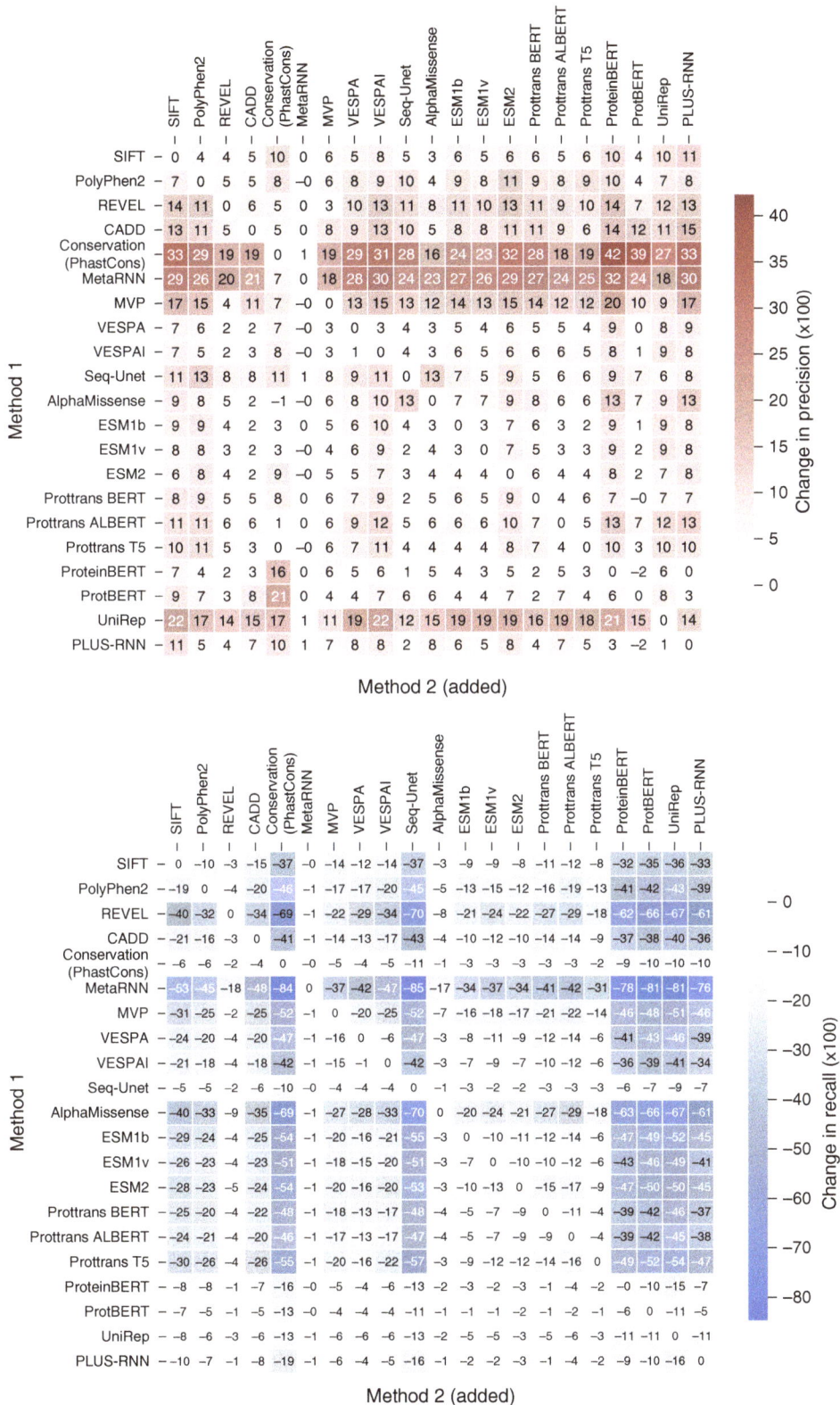

Figure 4. Improved precision at cost of recall by method jury. Most predictors' precision (red) benefits and recall (blue) suffers from the addition of another method in identifying severe (knockout) variants.

AMP (Richards et al. 2015) for variants interpreted for Mendelian conditions. ClinGen, on the other hand, is primarily concerned with establishing gene-disease involvement. Variant pathogenicity is then determined by evaluating genetic variants using the ACMG–AMP guidelines and gene-specific criteria developed by ClinGen expert panels. Note that in a sort of circular logic, disease genes are often defined on the basis of the pathogenic variants they carry.

For Mendelian/monogenic disorders, the process for establishing variant pathogenicity is a relatively well-defined, if laborious, task. For complex disorders, clinical observations and the experimental evidence of variant effect need to be overwhelming. As such, at the time of writing of this article (July 2023), 5312 variants in 2599 genes have been curated by ClinGen (ClinGen Statistics 2023), of which 2546 were designated as pathogenic or likely pathogenic. ClinVar, on the other hand, has collected 15,410 variants labeled by curators and nearly two million with some level of annotation (ClinVar Statistics 2023). Note that even given all precautions, a sufficient number of designated pathogenic variants have been observed in seemingly healthy individuals, suggesting the involvement of supporting and/or alternative molecular pathways (Shah et al. 2018). This biological incongruence is arguably even more pronounced in larger, literature-curated repositories of pathogenicity data (Cassa et al. 2013) (e.g., Human Genome Mutation Database [HGMD] [Stenson et al. 2020]). Nevertheless, in this work, we assumed that, regardless of the possible mislabeling of some variants, the full collection of ACMG–AMP guidelines-verified pathogenic variants is greatly enriched in disease-causing mutations. Note that to evaluate the methods' performance on a set that would not have likely overlapped with the methods' training data, we limited the extraction of pathogenic variants to those identified after 2022 (Supplemental Material).

Identifying putatively non-disease-causing variants for comparison to pathogenic ones was nearly impossible. Given the ACNG requirements for classifying variants as benign, we do not expect that these variants' characteristics (e.g., high population frequency, experimental data showing lack of functional effect, and nonsegregation with known disease) could in any way significantly overlap with those of the putatively pathogenic variants. Moreover, accepting multiple lines of computational evidence as a strong support for the likely benign-ness of variants is logically circuitous—train predictors to recognize benign variants and then label them as likely benign. Thus, in this work, we simply asked whether pathogenic variants can be recognized by the existing methods as different from the rest of the variants observed in the human population.

In our evaluation, methods differentiated likely pathogenic versus baseline variants with similar accuracy as pathogenic versus baseline variants (Fig. 1E; Supplemental Fig. S4). It thus stands to reason that curated likely pathogenic variants are indeed pathogenic according to the current criteria or that the predictor resolution is insufficient to tell the difference between the two. The first inference is more likely, given the inability of the variant and protein characteristic-naive, unsupervised methods to significantly better label pathogenic variants than likely pathogenic ones. This observation further suggests that the current process of pathogenic variant accumulation is either near perfect or heavily biased by the used experimental and clinical techniques.

All classical methods, however designed, explicitly select features of variants and their host protein sequences (e.g., conservation, structure, solvent accessibility, etc.) to attempt capture of variant effect. As the pathogenicity of a variant generally translates into a large effect on protein function, if not vice versa, these tools consistently did better in our hands-on differentiating pathogenic variants from putatively benign ones than functionally impactful variants from neutral ones (higher ROC AUC; Fig. 1C,E). Even simply using conservation was somewhat more informative for the pathogenicity set of variants than for the functional effect set. However, conservation alone was insufficient to precisely differentiate pathogenic variants, suggesting that they are not, contrary to expectations, confined to the strongly conserved sites.

Of all supervised methods, MVP and SeqUNet were the worst performers for the pathogenicity set, but even they attained an ROC AUC = ~0.7 (Fig. 1E). Note that the contribution of

Cite this article as *Cold Spring Harb Perspect Biol* doi: 10.1101/cshperspect.a041467

variant rarity in defining pathogenicity was illustrated by MetaRNN performance, that is, a method that considers population frequency explicitly and was thus able to differentiate baseline (more frequent) variants from pathogenic ones almost perfectly.

Of the unsupervised DL methods, ESM and ProtTrans were able to differentiate clinically significant variants from the general population better than simply using conservation. These models were also as good as or better than many of the supervised methods. At our selected thresholds, all ESM and ProtTrans models did well in recognizing pathogenicity (F1 ≥ 0.72, Table 1), and were better for this set than at predicting functional effect. This observation once again reaffirms that pathogenic variants are of high functional effect and are almost never common. However, as neither of the unsupervised methods captured variant population frequency well (Fig. 1A), the rarity of pathogenic variants is an unlikely cause of these models' pathogenicity classification abilities.

Variant Effects Are Correlated but Not Interchangeable for Effect Prediction

The ability of all methods in our study to differentiate variant effect across effect types is worth exploring further (Fig. 5; Supplemental Fig. S5). For example, all methods note that large-scale functional effect (purple lines) is almost as bad as pathogenicity (red lines). Furthermore, most methods frequently label variants found in the population, regardless of their frequency, as having no effect. In fact, variants that are annotated as having no functional effect (green line) are more frequently predicted to have an effect by all methods than ultra-rare variants (orange dashed line in Fig. 5; Supplemental Fig. S5).

Our graphs also tell us that optimal threshold selection, rarely considered by new methods in the field, is a difficult to capture. However, establishing this threshold is a necessary exercise for allowing the evaluation of individual variants. That is, a higher ROC AUC of the method is

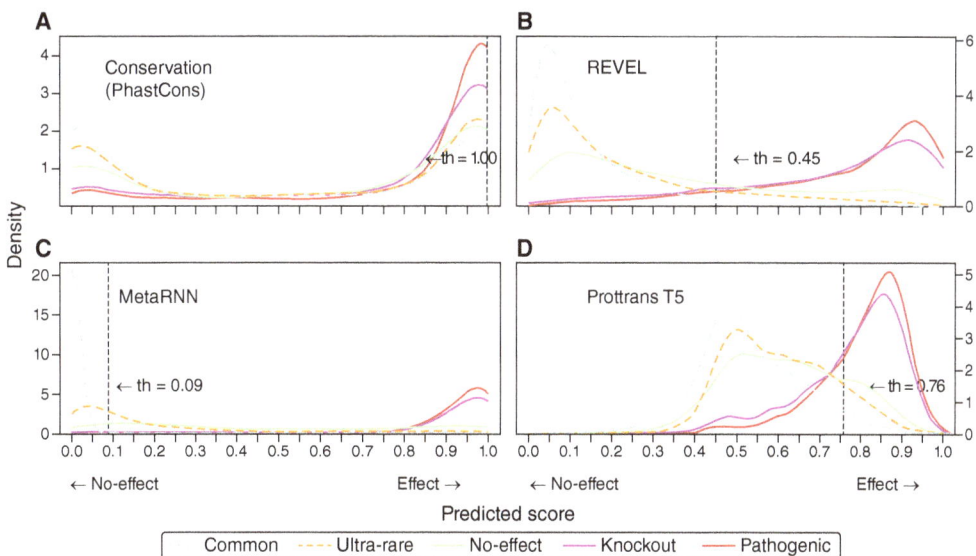

Figure 5. Distribution of scores for different variant classes. Scores attained by variants in different population frequency, functional effect, and pathogenicity classes (line colors) from (A) PhastCons, our conservation score, (B) REVEL, an example of a classical method in our set, (C) MetaRNN, an example of supervised deep learner, and (D) ProtTransT5, an example of an unsupervised predictor. Thresholds indicate scores below which 95% of common variants were observed. Representative methods were selected from the complete set (Supplemental Fig. S2) by best performance (ROC AUC) in differentiating population variant frequencies.

meaningless to a scientist looking for an assessment of the effect of their variant of interest. Furthermore, comparisons of method performance on different data sets and at differently selected cutoffs are bound to bring different performance results, making the selection of the best variant effect predictor nearly impossible (Fig. 1; Table 1; Supplemental Table S4).

Our results indicate that (some) unsupervised methods capture more than variant population frequency or variant site conservation. Instead, they seem to reflect functionally relevant features of variants, learning to extract information directly from the language of life. This ability allows for their correct labeling of pathogenic variants as well. Interpreting their assessments as binary classifications of a particular variant, however, requires a much deeper understanding of what makes an SNV unacceptable to the protein or organism it affects. It is also bound to broaden our horizons, allowing for the evaluation of variants in less well-annotated genes—a significant gain for exploration of, for example, the bacterial world.

CONCLUDING REMARKS

Multiple methods have been developed for annotation of the effect of missense variants. Appearance of unsupervised models has produced yet more of such methods. Our evaluation of method performance suggests that ESM and ProtTrans-based methods are the best performers in this space, exhibiting similar or better performance than specifically trained tools. However, neither of the existing supervised or unsupervised methods is able to evaluate all variant effect correctly.

To improve performance, better definitions of variant effect, as well as larger training sets for model fine-tuning are still necessary. The future "gold standard" predictor should indeed identify variant effect precisely. However, we hope that these models are also able to speed up discovery by labeling impactful variants in less studied spaces before experimental annotations catch up.

ACKNOWLEDGMENTS

We thank Drs. Michael Heinzinger (Technical University of Munich), Celine Marquet (Technical University of Munich), Ethan C. Alley (Alvea), and Mohammed AlQuraishi (Columbia University) for help with running and understanding their methods. A huge thanks goes to Christian Dallago (NVIDIA) for the invitation to participate in the special issue. A particular thanks is due the CSH Press Perspectives editor Barbara Acosta for her patience as this work was being delayed over and over again. Finally, we thank all researchers who deposit their data into public databases and those who maintain these databases. Y.B.'s and R.P.'s work on this manuscript was funded by the Emory Start-up Funding and NASA Astrobiology Institute Cycle 8 (80NSSC18M0093). All gene/protein lists, scores, and method evaluation measures are available from Bromberg et al. (2023).

REFERENCES

1000 Genomes Project Consortium; Auton A, Brooks LD, Durbin RM, Garrison EP, Kang HM, Korbel JO, Marchini JL, McCarthy S, McVean GA, et al. 2015. A global reference for human genetic variation. *Nature* **526:** 68–74. doi:10.1038/nature15393

Adzhubei IA, Schmidt S, Peshkin L, Ramensky VE, Gerasimova A, Bork P, Kondrashov AS, Sunyaev SR. 2010. A method and server for predicting damaging missense mutations. *Nat Methods* **7:** 248–249. doi:10.1038/nmeth0410-248

Alley EC, Khimulya G, Biswas S, AlQuraishi M, Church GM. 2019. Unified rational protein engineering with sequence-based deep representation learning. *Nat Methods* **16:** 1315–1322. doi:10.1038/s41592-019-0598-1

Araya CL, Fowler DM. 2011. Deep mutational scanning: assessing protein function on a massive scale. *Trends Biotechnol* **29:** 435–442. doi:10.1016/j.tibtech.2011.04.003

Bairoch A, Boeckmann B, Ferro S, Gasteiger E. 2004. Swiss-Prot: juggling between evolution and stability. *Brief Bioinform* **5:** 39–55. doi:10.1093/bib/5.1.39

Boeckmann B, Bairoch A, Apweiler R, Blatter MC, Estreicher A, Gasteiger E, Martin MJ, Michoud K, O'Donovan C, Phan I, et al. 2003. The SWISS-PROT protein knowledgebase and its supplement TrEMBL in 2003. *Nucleic Acids Res* **31:** 365–370. doi:10.1093/nar/gkg095

Brandes N, Ofer D, Peleg Y, Rappoport N, Linial M. 2022. ProteinBERT: a universal deep-learning model of protein sequence and function. *Bioinformatics* **38:** 2102–2110. doi:10.1093/bioinformatics/btac020

Bromberg Y, Rost B. 2007. SNAP: predict effect of non-synonymous polymorphisms on function. *Nucleic Acids Res* **35:** 3823–3835. doi:10.1093/nar/gkm238

Bromberg Y, Kahn PC, Rost B. 2013. Neutral and weakly nonneutral sequence variants may define individuality. *Proc Natl Acad Sci* **110:** 14255–14260. doi:10.1073/pnas.1216613110

Bromberg Y, Kabir A, Ramakrishnan P, Shehu A. 2023. Variant prediction in the age of machine learning. figshare doi:10.6084/m9.figshare.c.6746316.v2

Capriotti E, Calabrese R, Casadio R. 2006. Predicting the insurgence of human genetic diseases associated to single point protein mutations with support vector machines and evolutionary information. *Bioinformatics* 22: 2729–2734. doi:10.1093/bioinformatics/btl423

Cassa CA, Tong MY, Jordan DM. 2013. Large numbers of genetic variants considered to be pathogenic are common in asymptomatic individuals. *Hum Mutat* 34: 1216–1220. doi:10.1002/humu.22375

Cheng J, Novati G, Pan J, Bycroft C, Žemgulytė A, Applebaum T, Pritzel A, Wong LH, Zielinski M, Sargeant T, et al. 2023. Accurate proteome-wide missense variant effect prediction with AlphaMissense. *Science* 381: eadg7492. doi:10.1126/science.adg7492

ClinGen Statistics. 2023. https://search.clinicalgenome.org/kb/reports/stats

ClinVar Statistics. 2023. https://www.ncbi.nlm.nih.gov/clinvar/submitters

Dunham AS, Beltrao P, AlQuraishi M. 2023. High-throughput deep learning variant effect prediction with Sequence UNET. *Genome Biol* 24: 110. doi:10.1186/s13059-023-02948-3

Elnaggar A, Heinzinger M, Dallago C, Rehawi G, Wang Y, Jones L, Gibbs T, Feher T, Angerer C, Steinegger M, et al. 2022. Prottrans: toward understanding the language of life through self-supervised learning. *IEEE Trans Pattern Anal Mach Intell* 44: 7112–7127. doi:10.1109/TPAMI.2021.3095381

Frazer J, Notin P, Dias M, Gomez A, Min JK, Brock K, Gal Y, Marks DS. 2021. Disease variant prediction with deep generative models of evolutionary data. *Nature* 599: 91–95. doi:10.1038/s41586-021-04043-8

Ganesan K, Kulandaisamy A, Binny Priya S, Gromiha MM. 2019. Huvarbase: a human variant database with comprehensive information at gene and protein levels. *PLoS ONE* 14: e0210475. doi:10.1371/journal.pone.0210475

Gibson G. 2012. Rare and common variants: twenty arguments. *Nat Rev Genet* 13: 135–145. doi:10.1038/nrg3118

Hamosh A, Scott AF, Amberger J, Valle D, McKusick VA. 2000. Online Mendelian Inheritance in Man (OMIM). *Hum Mutat* 15: 57–61. doi:10.1002/(SICI)1098-1004(200001)15:1<57::AID-HUMU12>3.0.CO;2-G

Hecht M, Bromberg Y, Rost B. 2015. Better prediction of functional effects for sequence variants. *BMC Genomics* 16: S1. doi:10.1186/1471-2164-16-S8-S1

Holcomb D, Hamasaki-Katagiri N, Laurie K, Katneni U, Kames J, Alexaki A, Bar H, Kimchi-Sarfaty C. 2021. New approaches to predict the effect of co-occurring variants on protein characteristics. *Am J Hum Genet* 108: 1502–1511. doi:10.1016/j.ajhg.2021.06.011

Hu J, Ng PC. 2013. SIFT indel: predictions for the functional effects of amino acid insertions/deletions in proteins. *PLoS ONE* 8: e77940. doi:10.1371/journal.pone.0077940

Hu Z, Yu C, Furutsuki M, Andreoletti G, Ly M, Hoskins R, Adhikari AN, Brenner SE. 2019. VIPdb, a genetic variant impact predictor database. *Hum Mutat* 40: 1202–1214. doi:10.1002/humu.23858

Ioannidis NM, Rothstein JH, Pejaver V, Middha S, McDonnell SK, Baheti S, Musolf A, Li Q, Holzinger E, Karyadi D, et al. 2016. REVEL: an ensemble method for predicting the pathogenicity of rare missense variants. *Am J Hum Genet* 99: 877–885. doi:10.1016/j.ajhg.2016.08.016

Katsanis N. 2016. The continuum of causality in human genetic disorders. *Genome Biol* 17: 233. doi:10.1186/s13059-016-1107-9

Kawabata T, Ota M, Nishikawa K. 1999. The protein mutant database. *Nucleic Acids Res* 27: 355–357. doi:10.1093/nar/27.1.355

Kircher M, Witten DM, Jain P, O'Roak BJ, Cooper GM, Shendure J. 2014. A general framework for estimating the relative pathogenicity of human genetic variants. *Nat Genet* 46: 310–315. doi:10.1038/ng.2892

Landrum MJ, Lee JM, Benson M, Brown GR, Chao C, Chitipiralla S, Gu B, Hart J, Hoffman D, Jang W, et al. 2018. Clinvar: improving access to variant interpretations and supporting evidence. *Nucleic Acids Res* 46: D1062–D1067. doi:10.1093/nar/gkx1153

Li C, Zhi D, Wang K, Liu X. 2022. MetaRNN: differentiating rare pathogenic and rare benign missense SNVs and InDels using deep learning. *Genome Med* 14: 115. doi:10.1186/s13073-022-01120-z

Lin Z, Akin H, Rao R, Hie B, Zhu Z, Lu W, Smetanin N, Verkuil R, Kabeli O, Shmueli Y, et al. 2023. Evolutionary-scale prediction of atomic-level protein structure with a language model. *Science* 379: 1123–1130. doi:10.1126/science.ade2574

Mahlich Y, Reeb J, Hecht M, Schelling M, De Beer TAP, Bromberg Y, Rost B. 2017. Common sequence variants affect molecular function more than rare variants? *Sci Rep* 7: 1608. doi:10.1038/s41598-017-01054-2

Malhis N, Jones SJM, Gsponer J. 2019. Improved measures for evolutionary conservation that exploit taxonomy distances. *Nat Commun* 10: 1556. doi:10.1038/s41467-019-09583-2

Marquet C, Heinzinger M, Olenyi T, Dallago C, Erckert K, Bernhofer M, Nechaev D, Rost B. 2022. Embeddings from protein language models predict conservation and variant effects. *Hum Genet* 141: 1629–1647. doi:10.1007/s00439-021-02411-y

Meier J, Rao R, Verkuil R, Liu J, Sercu T, Rives A. 2021. Language models enable zero-shot prediction of the effects of mutations on protein function. *Adv Neural Inf Process Syst* 29287–29303.

Miller M, Bromberg Y, Swint-Kruse L. 2017. Computational predictors fail to identify amino acid substitution effects at rheostat positions. *Sci Rep* 7: 41329. doi:10.1038/srep41329

Miller M, Vitale D, Kahn PC, Rost B, Bromberg Y. 2019. Funtrp: identifying protein positions for variation driven functional tuning. *Nucleic Acids Res* 47: e142. doi:10.1093/nar/gkz818

Min S, Park S, Kim S, Choi HS, Lee B, Yoon S. 2021. Pre-training of deep bidirectional protein sequence representations with structural information. *IEEE Access* 9: 123912–123926. doi:10.1109/ACCESS.2021.3110269

Ng PC, Henikoff S. 2003. SIFT: predicting amino acid changes that affect protein function. *Nucleic Acids Res* 31: 3812–3814. doi:10.1093/nar/gkg509

Nishikawa K, Ishino S, Takenaka H, Norioka N, Hirai T, Yao T, Seto Y. 1994. Constructing a protein mutant database. *Protein Eng* **7**: 733. doi:10.1093/protein/7.5.733

Notin P, Dias M, Frazer J, Hurtado JM, Gomez AN, Marks D, Gal Y. 2022. Tranception: protein fitness prediction with autoregressive transformers and inference-time retrieval. In *International Conference on Machine Learning*, pp. 16990–17017. PMLR.

Pang AW, MacDonald JR, Pinto D, Wei J, Rafiq MA, Conrad DF, Park H, Hurles ME, Lee C, Venter JC, et al. 2010. Towards a comprehensive structural variation map of an individual human genome. *Genome Biol* **11**: R52. doi:10.1186/gb-2010-11-5-r52

Pejaver V, Babbi G, Casadio R, Folkman L, Katsonis P, Kundu K, Lichtarge O, Martelli PL, Miller M, Moult J, et al. 2019. Assessment of methods for predicting the effects of PTEN and TPMT protein variants. *Hum Mutat* **40**: 1495–1506. doi:10.1002/humu.23838

Pelak K, Shianna KV, Ge D, Maia JM, Zhu M, Smith JP, Cirulli ET, Fellay J, Dickson SP, Gumbs CE, et al. 2010. The characterization of twenty sequenced human genomes. *PLoS Genet* **6**: e1001111. doi:10.1371/journal.pgen.1001111

Phan L, Jin Y, Zhang H, Qiang W, Shekhtman E, Shao D, Revoe D, Villamarin R, Ivanchenko E, Kimura M. 2020. ALFA: Allele frequency aggregator. NCBI, U.S. NLM Gatew. http://www.ncbi.nlm.nih.gov/snp/docs/gsr/alfa

Piel FB, Patil AP, Howes RE, Nyangiri OA, Gething PW, Williams TN, Weatherall DJ, Hay SI. 2010. Global distribution of the sickle cell gene and geographical confirmation of the malaria hypothesis. *Nat Commun* **1**: 104. doi:10.1038/ncomms1104

Qi H, Zhang H, Zhao Y, Chen C, Long JJ, Chung WK, Guan Y, Shen Y. 2021. MVP predicts the pathogenicity of missense variants by deep learning. *Nat Commun* **12**: 510. doi:10.1038/s41467-020-20847-0

Rao R, Bhattacharya N, Thomas N, Duan Y, Chen X, Canny J, Abbeel P, Song YS. 2019. Evaluating protein transfer learning with TAPE. *Adv Neural Inf Process Syst* **32**: 9689–9701.

Rehm HL, Berg JS, Brooks LD, Bustamante CD, Evans JP, Landrum MJ, Ledbetter DH, Maglott DR, Martin CL, Nussbaum RL, et al. 2015. Clingen—the clinical genome resource. *N Engl J Med* **372**: 2235–2242. doi:10.1056/NEJMsr1406261

Richards S, Aziz N, Bale S, Bick D, Das S, Gastier-Foster J, Grody WW, Hegde M, Lyon E, Spector E, et al. 2015. Standards and guidelines for the interpretation of sequence variants: a joint consensus recommendation of the American College of Medical Genetics and Genomics and the Association for Molecular Pathology. *Genet Med* **17**: 405–424. doi:10.1038/gim.2015.30

Riesselman AJ, Ingraham JB, Marks DS. 2018. Deep generative models of genetic variation capture the effects of mutations. *Nat Methods* **15**: 816–822. doi:10.1038/s41592-018-0138-4

Rives A, Meier J, Sercu T, Goyal S, Lin Z, Liu J, Guo D, Ott M, Zitnick CL, Ma J, et al. 2021. Biological structure and function emerge from scaling unsupervised learning to 250 million protein sequences. *Proc Natl Acad Sci* **118**: e2016239118: doi:10.1073/pnas.2016239118

Sarkar A, Yang Y, Vihinen M. 2020. Variation benchmark datasets: update, criteria, quality and applications. *Database* **2020**: baz117. doi:10.1093/database/baz117

Savojardo C, Babbi G, Martelli PL, Casadio R. 2021. Mapping OMIM disease–related variations on protein domains reveals an association among variation type, Pfam models, and disease classes. *Front Mol Biosci* **8**: 617016. doi:10.3389/fmolb.2021.617016

Shah N, Hou YC, Yu HC, Sainger R, Caskey CT, Venter JC, Telenti A. 2018. Identification of misclassified ClinVar Variants via disease population prevalence. *Am J Hum Genet* **102**: 609–619. doi:10.1016/j.ajhg.2018.02.019

Siepel A, Bejerano G, Pedersen JS, Hinrichs AS, Hou M, Rosenbloom K, Clawson H, Spieth J, Hillier LW, Richards S, et al. 2005. Evolutionarily conserved elements in vertebrate, insect, worm, and yeast genomes. *Genome Res* **15**: 1034–1050. doi:10.1101/gr.3715005

Stenson PD, Mort M, Ball EV, Chapman M, Evans K, Azevedo L, Hayden M, Heywood S, Millar DS, Phillips AD, et al. 2020. The Human Gene Mutation Database (HGMD): optimizing its use in a clinical diagnostic or research setting. *Hum Genet* **139**: 1197–1207. doi:10.1007/s00439-020-02199-3

Sun H, Yu G. 2019. New insights into the pathogenicity of non-synonymous variants through multi-level analysis. *Sci Rep* **9**: 1667. doi:10.1038/s41598-018-38189-9

The Critical Assessment of Genome Interpretation Consortium. 2022. CAGI, the Critical Assessment of Genome Interpretation, establishes progress and prospects for computational genetic variant interpretation methods. arXiv doi:10.48550/arXiv.2205.05897

Triant DA, Pearson WR. 2015. Most partial domains in proteins are alignment and annotation artifacts. *Genome Biol* **16**: 99. doi:10.1186/s13059-015-0656-7

Zeng Z, Bromberg Y. 2019. Predicting functional effects of synonymous variants: a systematic review and perspectives. *Front Genet* **10**: 914. doi:10.3389/fgene.2019.00914

Zhu C, Miller M, Zeng Z, Wang Y, Mahlich Y, Aptekmann A, Bromberg Y. 2020. Computational approaches for unraveling the effects of variation in the human genome and microbiome. *Annu Rev Biomed Data Sci* **3**: 411–432. doi:10.1146/annurev-biodatasci-030320-041014

Is Novelty Predictable?

Clara Fannjiang and Jennifer Listgarten

Department of Electrical Engineering and Computer Sciences, University of California, Berkeley, California 94720, USA

Correspondence: clarafy@berkeley.edu; jennl@berkeley.edu

Machine learning–based design has gained traction in the sciences, most notably in the design of small molecules, materials, and proteins, with societal applications ranging from drug development and plastic degradation to carbon sequestration. When designing objects to achieve novel property values with machine learning, one faces a fundamental challenge: how to push past the frontier of current knowledge, distilled from the training data into the model, in a manner that rationally controls the risk of failure. If one trusts learned models too much in extrapolation, one is likely to design rubbish. In contrast, if one does not extrapolate, one cannot find novelty. Herein, we ponder how one might strike a useful balance between these two extremes. We focus in particular on designing proteins with novel property values, although much of our discussion is relevant to machine learning–based design more broadly.

CHALLENGES IN FINDING NOVELTY WITH MACHINE LEARNING–BASED DESIGN

How can one find novelty, given only what is known? Focusing on machine learning–based protein design, herein we highlight conceptual challenges and current approaches to solving them, as well as underexplored areas and future directions.

The goal of protein design is to specify the sequence of a protein that satisfies a novel condition. There are three types of novel conditions a protein engineer may seek. One type of novelty is in the sequence—that is, one seeks a protein that differs in sequence from that of known proteins, but not necessarily in its structure or biochemical or biophysical properties. For example, one may seek a novel sequence for an enzyme to avoid a patent, or to serve as a different initial sequence for directed evolution. Much recent work in machine learning–based protein design focuses on sequence novelty, with applications ranging from gene therapy vectors (Bryant et al. 2021; Zhu et al. 2022) and antibodies (Shin et al. 2021) to signal peptides (Wu et al. 2020) and enzymes (Russ et al. 2020; Hawkins-Hooker et al. 2021; Repecka et al. 2021; Madani et al. 2023) A second type of novelty is in the structure—for example, when building a scaffold to support a functional site (Correia et al. 2014; Sesterhenn et al. 2020; Wang et al. 2022).

The third type of novelty is in the biophysical or biochemical properties. That is, one seeks a protein with property values that have yet to be observed, which necessitates that the sequence, and likely the structure, are both also novel. Examples tackled with machine learning include enzymes with enhanced catalytic activity (Fox et al. 2007; Romero et al. 2013; Biswas et al. 2021; Greenhalgh et al. 2021; Fram et al. 2023), brighter fluorescent proteins (Brookes et al. 2019; Biswas et al. 2021; Stanton et al. 2022), optimized channelrhodopsins for optogenetics (Bedbrook et al. 2017, 2019), and cell-type-specific gene therapy vectors (Zhu et al. 2022). Herein, we focus on this third type of novelty. Although machine learning models can facilitate achieving this goal, the pursuit of novelty is also the root cause of unique in silico challenges, the focus of our discussion.

The difficulty of these challenges depends on the extent of novelty sought. Is a protein with a melting point that is one degree higher or that fluoresces slightly brighter at a similar wavelength novel? Novelty is not a binary phenomenon. Rather, it exists on a spectrum, and the further one is on this spectrum the greater the challenges. At the extreme end is what could be considered "radical novelty," which is not just an improvement of an existing phenomenon, but rather a fundamentally different outcome. For example, one might refer to the design of an enzyme that catalyzes a reaction completely differently from what known enzymes can catalyze as radical novelty. If the conditions one is interested in are not radically novel, then machine learning–based design may already be close to providing reliable solutions. Radical novelty is much more elusive.

We frame our discussion around three key tasks for machine learning–based design of novel property values. The first two are (1) learning a trustworthy model that makes predictions for the property of interest given, for example, a protein sequence and/or structure, and (2) choosing what we call a design algorithm: an algorithm that consults the model to propose sequences intended to have the desired property condition. Not all design strategies appear to explicitly comprise these two tasks, such as some conditional generative modeling approaches. Nevertheless, these concepts and corresponding challenges are present under the hood.

Although sometimes overlooked, there is also a third key task, that of (3) uncertainty quantification. Quantifying uncertainty of the model's predictions helps a protein engineer understand what risk portfolio they are adopting when choosing a design strategy.

Our goal herein is not to comprehensively survey existing methods, nor to recast these into a unifying framework. Rather, we identify and discuss fundamental challenges, corresponding strategies, and underexplored areas that arise from the following dilemma: to find novel property values, any design algorithm must consider regions of sequence space away from the training data, but these regions are precisely where any learned model is least trustworthy.

Scope of Protein Design Problems Considered

Sometimes the property of interest is sufficiently mediated by the protein's structure that the goal can be reframed as identifying a sequence that folds into a specific structure or substructure. A common example is when the goal is to bind to a target molecule where the structure of the desired complex is known with high resolution. Such structure-based design has traditionally been performed with biophysics-based modeling (Kuhlman and Bradley 2019), but is now increasingly done with deep generative models (Norn et al. 2021; Dauparas et al. 2022; Hsu et al. 2022b; Watson et al. 2023); we refer the reader to Ovchinnikov and Huang (2021), Pan and Kortemme (2021), and Malbranke et al. (2023) for excellent reviews of these developments. However, there are many settings in which structure-based design is not a viable solution. In particular, how structural changes affect the property of interest is not always known with sufficient precision, with catalytic activity of enzymes being a classic example (Romero and Arnold 2009; Tokuriki and Tawfik 2009). Additionally, the property may be substantially mediated by conformational dynamics or quantum chemistry (Gao and Truhlar 2002; Faheem and Heyden 2014), neither of which can, at present, be readily captured by structure-based design. In these settings, one must rely additionally on directed evolution (Romero and Arnold 2009; Arnold 2018) and/or an approach based on a machine learning model that leverages assay and/or evolutionary data relevant to the property of interest. We focus

Cite this article as *Cold Spring Harb Perspect Biol* doi: 10.1101/cshperspect.a041469

herein on this latter setting, which includes that of machine learning–assisted directed evolution (Wu et al. 2019; Wittmann et al. 2021).

Machine learning models in this setting may be trained in a supervised fashion on sequences labeled with experimental measurements. Alternatively, some models may capture the property more implicitly, such as density models fit to families of protein sequences (Cheng et al. 2016; Figliuzzi et al. 2016; Hopf et al. 2017; Riesselman et al. 2018; Laine et al. 2019; Frazer et al. 2021; Shin et al. 2021; Trinquier et al. 2021) or to natural protein sequences more broadly (Alley et al. 2019; Madani et al. 2020; Meier et al. 2021; Ferruz et al. 2022; Notin et al. 2022a,b). The likelihood of these models or approximations thereof can provide useful information about protein properties. The conditional likelihoods of structure-conditioned generative models can also be used in a similar fashion (Dauparas et al. 2022; Hsu et al. 2022b; Ingraham et al. 2022).

Overview of Challenges

Almost surely, one does not have access to the true causal model for the property of interest, and without the true causal model, model predictions are almost surely wrong. They are particularly likely to be wrong given the shifts between the distributions of training and designed sequences that are necessary when seeking novel property values. Despite the inability to access true causal models, predictive models can still be useful. How can one ensure that such models are useful for design? Motivated by this question, this article is organized into three sections, summarized next.

First, we ask: How can one learn a trustworthy model for pursuing novel properties? We consider how different types of data—such as evolutionary and assay-labeled—contain different types of noise and biases. We then discuss how the search for novel property values induces distribution shifts between the training and designed sequences that jeopardize how much one can trust the model. Finally, we discuss how to potentially mitigate such problems by infusing relevant biological knowledge into the inductive bias of the model, analogous to how the convolution operation in computer vision encodes fundamental knowledge about the translational invariance of objects in images.

Second, we ask: How can one quantify uncertainty for design? Because design necessitates making predictions for sequences far from the training data, the predictions can deviate wildly from the truth. It is therefore desirable to quantify the model's uncertainty, and thereby understand the risk one is incurring, before synthesizing and measuring designed sequences in the laboratory. To do so, one can choose from a variety of technical notions of uncertainty that we discuss, each with its own strengths and limitations.

Finally, we ask: What are the design algorithm considerations? Given a fixed budget of sequences, how should the design algorithm place its bets so as to maximize the chances of finding a protein with the desired property values? One reasonable approach is for the design algorithm to take into account the model's uncertainty. Or, alternatively, perhaps one should quantify uncertainty jointly over the model and the design algorithm.

HOW CAN ONE LEARN A TRUSTWORTHY MODEL FOR PURSUING NOVEL CONDITIONS?

We start by describing how one's choices about the training data impact the model's predictions. We then discuss the distribution shifts between training and test data that emerge from design, and, finally, potential strategies to mitigate problems arising from these shifts.

Trade-Off between Quality and Quantity of Training Data

To frame our discussion of how training data choices impact predictions, we appeal to the recently introduced bias-variance decomposition of prediction error described by Posani et al. (2023). Although bias-variance decompositions have conventionally been used to help understand why different model classes yield different predictive performance (Geman et al. 1992; Hastie et al. 2001), Posani et al. (2023) repurpose the

idea to understand how choices in curating homologous protein sequences affect how well the
likelihoods of a Potts model fit to those sequences can provide a ranking of protein property values. Inspired by their work, we appeal
to a similar framework to discuss the impact of
training data more broadly on prediction error.

More concretely, consider the prediction error for a given test sequence, when averaged over
predictions made from models trained on different random draws of data from a fixed distribution (Geman et al. 1992; Hastie et al. 2001).[1] This
average prediction error can be decomposed into
two parts. The first is a bias component, or the
average difference between the prediction and
true property value of the test sequence; intuitively, this component reflects how relevant the
training data are for the prediction task at hand.
The second, a variance component, captures
how sensitive the prediction is to perturbations
to the training data due to sampling; intuitively,
this component reflects the amount of information content in the data set—more content is
harder to perturb. For example, given equal relevance of the data, larger training data sets result
in lower-variance predictions because the information contained in larger data sets is more robust to perturbations than that contained in
smaller ones. These bias and variance terms exhibit a trade-off (Grenander 1952; Geman et al.
1992): if one chooses the training data in such a
way as to improve one, then typically the other
degrades. For example, increasing the amount of
training data by incorporating data that is less
relevant, such as using homologous protein sequences that are more evolutionarily distant, results in increased bias, but decreased variance.
Consequently, one can generally achieve different points on a trade-off curve by modulating the
quantity versus quality of the training data. Next,
we invoke this quantity–quality trade-off to gain
insights into choosing both assay-labeled and
evolutionary data.

---

[1]Note that for the purposes of our discussion, we do not
consider randomness in the test data, as done in some
bias-variance analyses in the machine learning literature
(Bishop 2007).

Assay-Labeled Data

When choosing a laboratory assay to label protein sequences, there is typically a trade-off between two extremes. On one end, there are high-
quality, low-throughput experiments that yield
measurements of a few dozen to a few thousand
sequences at most, but which directly measure
the biochemical or biophysical property of interest (Acker and Auld 2014; Markin et al. 2021).
Training a model on the small amount of resulting data will result in low predictive bias but high
predictive variance. On the other end, there are
lower-quality, high-throughput assays that can
yield measurements of up to hundreds of thousands or even millions of sequences, but which
only provide a biased proxy of the property of
interest. A common form of the latter are sequencing-based assays known collectively as
deep mutational scanning, in which a pool of
sequences is subject to selection experiments
that indirectly encourage sequences with the desired condition to become more abundant relative to other sequences (Fowler and Fields 2014;
Wrenbeck et al. 2017). For example, to measure
how well different enzyme sequences catalyze the
biosynthesis of a compound necessary for a cell's
survival, those enzymes can be expressed in a
population of cells, whose subsequent survival
rates reflect the catalytic efficiencies of the variants. From such data, one can derive a quantitative label for each sequence reflecting how much
more abundant it became after the selection—for
example, by computing ratios of sequence counts
from before and after the selection (Fowler et al.
2011; Rubin et al. 2017), although recent work
has shown how to improve such quantification
using density ratio estimation (Busia and Listgarten 2023). Even though such labels indirectly
inform us about the biophysical property of interest, they will generally be biased because factors unrelated to the property can drive changes
in sequence abundance (Song et al. 2021). Using
them as training data consequently introduces
predictive bias, albeit with low predictive variance due to the typically large data sets.

In either the high-quality–low-throughput
or low-quality–high-throughput settings, it can
be fruitful to augment assay-labeled data with

Cite this article as *Cold Spring Harb Perspect Biol* doi: 10.1101/cshperspect.a041469

evolutionary data—that is, increasing quantity at the expense of quality (Alley et al. 2019; Biswas et al. 2021; Hsu et al. 2022a). The blend of data types that achieves the optimal trade-off between quantity and quality will be different for each protein engineering campaign. Ultimately, empirical assessments will dictate the choice, although these may be guided at times by theoretical results such as those in Posani et al. (2023), discussed next.

Homologous Sequence Data

One can also understand the effects of curating homologous sequences with respect to the bias-variance trade-off. The goal is to identify a set of sequences known to appreciably exhibit the property of interest, and then fit a density model to those sequences. The likelihood of a sequence under such a model, if correctly specified, should then correlate with its real-valued property of interest. However, there are often only a handful of proteins, if any, that are laboratory-verified to exhibit the property of interest above some threshold. The likelihoods of a density model fit to such a small number of protein sequences would have low bias as property predictions, but extremely high variance. To reduce this variance at the expense of increased bias, one can increase the quantity of training data by using heuristics to identify proteins that are likely—but not known with certainty—to exhibit the property. One popular heuristic is to include homologous protein sequences, or homologs: those that are evolutionarily related to a natural protein that appreciably exhibits the property of interest. If these proteins underwent the same selective pressure that gave rise to that property, then they too should exhibit it. However, one never knows the extent to which they experienced that selective pressure, if at all. Moreover, finding homologs is itself a nontrivial task, as evolution cannot be observed over the relevant timescales. One must instead leverage heuristic search algorithms based on sequence similarity (Altschul et al. 1997; Johnson et al. 2010). Nevertheless, a rich line of work has shown that the likelihoods of various density models (Cheng et al. 2016; Figliuzzi et al. 2016; Hopf et al. 2017; Riesselman et al. 2018; Frazer et al. 2021; Shin et al. 2021; Trinquier et al. 2021) and predictions based on phylogenetic trees (Laine et al. 2019) fit to such putative homologs can be correlated with the property values of protein variants, despite the predictive bias introduced by data of uncertain relevance (Qin and Colwell 2018; Weinstein et al. 2022a; Posani et al. 2023).

Algorithms for heuristically identifying homologs look through databases of natural protein sequences for similar sequences and have various hyperparameters that can also be interpreted as navigating a quantity–quality trade-off. In particular, hyperparameter settings that return a larger set of sequences may include proteins that are not actually homologs, whereas settings that are too conservative may miss some (Pearson 2013). Posani et al. propose methods for selecting optimal subsets of homologs based on their sequence distances to a wild-type. Selection based on auxiliary information, such as what types of species the homologs are from, can also be useful heuristics for reducing predictive bias (Jagota et al. 2023). Thinking creatively about how to address the sources of predictive bias and variance we have discussed may be fruitful for improving "zero-shot" protein property prediction in settings where assay-labeled data are not yet available.

Pan-Protein Data

The protein modeling community has borrowed ideas from the field of natural language processing to train unsupervised sequence models on pan-protein data (i.e., all known natural proteins, spanning all known protein families and beyond). The likelihood of these models, or approximations thereof, can also serve as zero-shot predictions of protein property values (Meier et al. 2021; Hesslow et al. 2022; Nijkamp et al. 2022; Notin et al. 2022a,b). In principle, learning a density over all these proteins could result in an implicit mixture of family-specific "modes," in which each mode captures the distribution of a single protein family, similar to family-specific models such as Potts models. However, depending on the inductive bias of the pan-protein model architecture, such a model is likely to share information between families. The extent of this information sharing can be viewed as implicit navigation of the bias-variance trade-off. Moreover, a single trained pan-protein model likely navigates this trade-off differently for different protein families: the fewer the known homologs for a family, for example, the more information the model may have borrowed from other families. What is the nature of this shared information, and what determines which families contribute to it? Such questions should be investigated to make sense of how, why, and when such models may be providing benefits, which in turn could be used to improve them or other approaches.

As of this writing, the best approach for zero-shot property prediction blends pan-protein and family-specific models (Notin et al. 2022b), which suggests that pan-protein models are not yet achieving an optimal trade-off by themselves. Indeed, the best family-specific method (Laine

et al. 2019) performs similarly well, particularly when homologs are abundant for the family of interest (Fig. 1; Notin et al. 2022a,b). Far more conspicuous than the margin between these two methods' average performance is how dramatically their performances—as well as the magnitude of the difference between their performances—vary over different protein families (Fig. 1), which raises open and underexplored questions. How much do the sources of predictive bias and variance we have described account for this variation? Given a protein family of interest, is it possible to anticipate beforehand how accurate the predictions from family-specific versus pan-protein methods will be?

We anticipate that in the future, it will be possible to forgo curation of homologs entirely and develop strategies for training pan-protein models to automatically find more effective points on the bias-variance trade-off. However, as family-specific models can currently perform

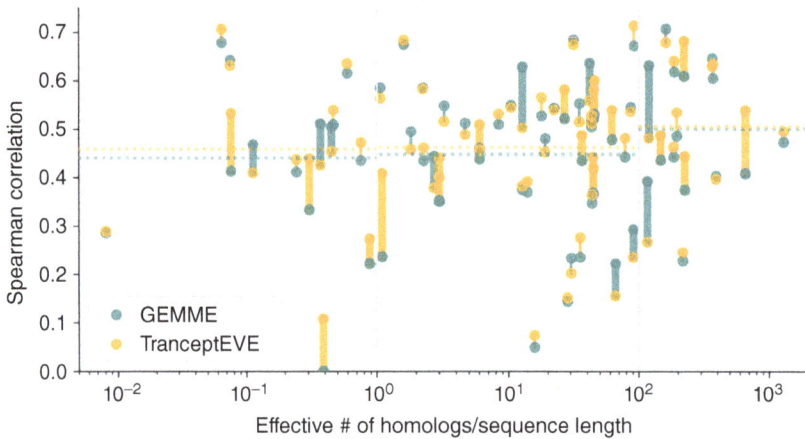

Figure 1. Protein property predictive performance of the top two methods on the ProteinGym substitution benchmark. All data in this plot are from Notin et al. (2022a,b) and the affiliated ProteinGym substitution benchmark (www.proteingym.org/substitutions). As of this writing, the two top-performing methods are one that combines pan-protein and family-specific models (TranceptEVE [Notin et al. 2022b], orange dots) and one family-specific method (GEMME [Laine et al. 2019], teal dots). The methods' performances on each ProteinGym data set (each teal and orange dot) are measured by Spearman correlations between predictions and labels (higher is better). To improve visualization, the magnitude of the difference between the methods is shown by a solid vertical line. Its color denotes the winning method for that data set, and the line is thick if the difference is greater than 0.05 and thin otherwise. For different data sets comprising mutants of the same wild-type protein, the average correlation over those data sets is shown as a single dot. The same thresholds as in Notin et al. (2022a,b) were used to categorize the amount of homologous data as low, medium, or high (demarcated by two solid gray vertical lines), and the average correlation within each of these categories is computed for each of GEMME and TranceptEVE (dashed horizontal teal and orange lines).

 Cite this article as *Cold Spring Harb Perspect Biol* doi: 10.1101/cshperspect.a041469

similarly and are much easier to learn—or require minimal learning (Laine et al. 2019)—their use may continue well into the future.

Accounting for Design-Induced Distribution Shift

So far, our discussion has been about the training data, without considering the specific distribution of test proteins that emerge from design. We now examine this topic more closely.

Because we seek a protein with novel biophysical or biochemical property values, it follows that the designed proteins come from a different distribution than the training proteins. This phenomenon is referred to as a distribution shift, which in many machine learning settings is passively observed rather than purposely induced as in the design setting. One type of distribution shift that emerges in machine learning–based design for novel property values is feedback covariate shift (Fannjiang et al. 2022). As a generalization of the more common covariate shift (Shimodaira 2000), feedback covariate shift additionally encompasses settings where test inputs are chosen based on the training data, such that they are dependent on it, rather than simply drawn independently from a different distribution. Because of this distribution shift, the model's predictions must be trustworthy over regions of protein space "away from" the training proteins—in particular, regions characterized by the distribution of designed proteins, as well as those the model examines en route to finding that distribution.

If one knew these distributions in advance, then one could deploy strategies for learning a model to be accurate over the test rather than training distribution (Shimodaira 2000; Sugiyama et al. 2008; Bickel et al. 2009; Gretton et al. 2009). The chicken-and-egg dilemma, however, is that one does not know the distribution of designed proteins until after the model has been learned; this is the "feedback" described by feedback covariate shift. Nevertheless, it is prudent to try to anticipate and account for plausibly relevant distribution shifts. For example, many design algorithms move through protein space in an iterative fashion to search for promising proteins, where each move induces, either implicitly or explicitly (Brookes et al. 2020), an intermediary distribution of proteins currently under consideration. After each move, one could relearn the model in lockstep with the design algorithm such that it is more accurate over the intermediary distribution (Fannjiang and Listgarten 2020).

Learning strategies that account for distribution shift do have a cost: intuitively, all involve reweighting the training data to better mimic the test distribution. The more the training and test distributions differ, the greater the variance of these weights over the training data (Cortes et al. 2010), which means that the model learns from a smaller effective amount of data. Strategies for mapping the training and test proteins into feature spaces in which the distribution shift is more mild (Rhodes et al. 2020; Choi et al. 2022) or tempering the weights so as to only partially account for the shift (Grover et al. 2019) may provide solutions.

Beyond changing how the model is learned, uncertainty quantification strategies and design algorithms can also take design-induced distribution shifts into account, as we will discuss. In general, problems arising from such shifts disappear to the extent that the model captures the true causal mechanism. One way to move toward this causal model is to incorporate relevant knowledge into the inductive bias of the model, as discussed next.

Incorporating Informative Inductive Biases

In practice, the true causal model is never available. As a practical mitigation strategy, one can imbue the inductive bias of the model with broadly applicable knowledge regarding how amino acid sequences give rise to biophysical or biochemical properties. For example, there is evidence that this relationship is dominated by single-site effects, and that higher-order or epistatic interaction effects are sparse and typically decay with increasing order (Sailer and Harms 2017; Otwinowski et al. 2018; Poelwijk et al. 2019; Yang et al. 2019a; Ballal et al. 2020; Brookes et al. 2022; Ding 2022). Similarly, models that generate or make use of protein structures should respect physical constraints and rotational symmetries (Ingraham et al. 2019; Jing et al. 2021; Ingraham et al. 2022).

Particularly when using high-capacity deep learning models, incorporating such domain-specific, yet task-independent knowledge into the model can reduce its degrees of freedom without ruling out useful parts of parameter space. Doing so can consequently improve the efficiency with which the model distills relevant information from the training data. Intuitively, encoding these types of knowledge is tantamount to having additional high-quality data, enabling a reduction in both predictive bias and variance. Critically, because such knowledge is not based on the training data, a model that incorporates it can better retain its accuracy further away from the training data.

Coherently integrating appropriate knowledge into the model is a case-by-case challenge. Sometimes, as in the case of rotational symmetries, such knowledge is absolute and should be enforced as a hard constraint (Cohen and Welling 2016; Geiger and Smidt 2022), although computational issues can remain challenging. In contrast, the sparsity and decay of epistatic interactions with increasing order are phenomena that hold to different extents for different protein properties, so should only be softly enforced. In one approach, Aghazadeh et al. (2021) leverage ideas and algorithms from compressed sensing to encourage sparse epistatic interactions in neural networks, although they use a binary alphabet and do not discriminate between low- and high-order effects. The main technical challenge in incorporating knowledge about epistasis is how to tractably compute the combinatorially large number of higher-order interactions between protein sites (Erginbas et al. 2023) or devise clever ways to avoid doing so.

So far, we have discussed incorporating knowledge that would be useful for almost any protein property prediction task. However, one can also leverage task-specific biophysical knowledge. Although the accuracy of learned models degrades further away from the training data, biophysics-based models crafted in a relatively data-free manner should have roughly uniform accuracy over protein space, even if learned

models outperform them near their training data. To capitalize on this intuition, Nisonoff et al. (2022) developed an easily instantiable and computationally efficient formalism for blending information from biophysics-based models and Bayesian neural network models of the same properties. A key component of their approach is the uncertainty of the predictions, the topic of our next discussion.

HOW CAN ONE QUANTIFY UNCERTAINTY FOR DESIGN?

Thus far, we have discussed how the quality of the model's predictions, especially away from the training data, is affected by choices regarding the training data and inductive biases of the model. However, it can prove useful to not only improve predictions, but also quantify uncertainty over them, which can help us assess risk in designing new proteins. The precise notion of uncertainty quantification that is most useful for protein design is an open and underexplored question. One must decide what entity needs uncertainty quantification, what notion of uncertainty is suitable, and finally how to go about quantifying it.

First, what is the entity whose uncertainty needs to be quantified? An obvious candidate is the property prediction for any individual test protein, for which we will discuss both Bayesian and frequentist approaches in the remainder of this section. In designing a library of proteins, however, one may alternatively want to quantify uncertainty for the average, median, or other quantiles of the property values contained in the library, to reason about its collective performance. This idea was first proposed and tackled by Wheelock et al. (2022) under a Bayesian lens and will be fruitful to continue investigating.

Bayesian Uncertainty Quantification

In the Bayesian framework, one specifies a prior distribution over the predictive model parameters that encodes the user's current beliefs, as well as a likelihood model that gives, for each possible setting of the model parameters, the probability of the data. The key Bayesian operation is to up-

Cite this article as *Cold Spring Harb Perspect Biol* doi: 10.1101/cshperspect.a041469

date the prior distribution to be more consistent with the evidence contained in the data, by re-weighting each parameter setting according to the likelihood of the data under that setting. This update yields a posterior distribution that encodes updated beliefs about the predictive model parameters, which can in turn be used to derive a posterior distribution over the prediction for any test point, called the posterior predictive distribution. The variance and other measures of dispersion of the posterior predictive distribution are natural and commonly used notions of uncertainty for predictions.

An elegant aspect of the Bayesian framework is that different components can be interpreted as accounting for different sources of uncertainty (Kiureghian and Ditlevsen 2009). In particular, the likelihood model, if correctly specified, captures what is called aleatoric uncertainty, or uncertainty due to inherent stochasticity in the underlying causal mechanism being modeled. Aleatoric uncertainty never vanishes, even with infinite data. In contrast, the process of updating the prior based on the data captures reduction in epistemic uncertainty, or uncertainty due to lack of data, which vanishes with infinite data—at which point the prior has no influence over the posterior. One satisfying consequence is that posterior predictive distributions for test inputs further from the training inputs naturally have higher variance.

For exact, tractable computation of the posterior distribution, the Bayesian framework requires particular pairings of prior distributions and likelihood models, known as conjugate pairs, which may not be the ones that most accurately describe one's beliefs about the system. Nevertheless, these can work well in practice: Bayesian notions of uncertainty from Gaussian process regression models have been put to good use in protein engineering (Romero et al. 2013; Bedbrook et al. 2017, 2019; Greenhalgh et al. 2021; Rapp et al. 2023). Beyond the use of conjugate pairs, accessing the posterior requires sampling methods such as Markov chain Monte Carlo methods (Neal 1996), or variational inference (Gal 2016; Zhang et al. 2019), which learns an approximation to the posterior that affords easy sampling. See Gal (2016) for a thorough study of tools for the latter in deep learning.

In contrast to frequentist approaches, Bayesian notions of uncertainty do not generally satisfy any guarantees about their relationship with the true value of the entity. For example, one might seek a guarantee that the true property value is less than the 90th percentile of the posterior predictive distribution 90% of the time—a notion referred to as calibrated uncertainty (Platt 1999). As the Bayesian framework centers around subjective beliefs (Gelman 2009; Fortuin 2022)—encoding and updating them—it does not yield such guarantees; in fact, such guarantees are antithetical to the Bayesian paradigm (Rubin 1984; Little 2006). In contrast, the frequentist framework, discussed next, explicitly seeks such guarantees.

Frequentist Uncertainty Quantification

The goal of frequentist uncertainty quantification is to produce uncertainty estimates that satisfy some probabilistic notion of correctness. As an example, we will focus on conformal prediction methods, which yield confidence sets for test points that are guaranteed to contain the true labels with high probability, for any predictive model (Vovk et al. 2005; Angelopoulos and Bates 2023). Such coverage guarantees reliance on assumptions about the relationship between the training and test data, such as being exchangeable (e.g., independently and identically distributed) or related by way of particular distribution shifts.

One basic intuition with which to understand conformal prediction is that, if one has "calibration data"—held-out validation data from the same distribution as the test data—then the prediction error on a test point comes from the same distribution as the prediction errors on the calibration data. Consequently, one can obtain guarantees on the test error by assessing the calibration errors. However, we may not have access to such calibration data. Instead, for any test input, we can consider each possible value that its label could take on. For each such candidate label, we ask whether the test input paired with that candidate label together "look like" they come from the same distribution as the training data, according to a frequentist hypothesis test. If they do, then this value is included in the confidence set for that test point. The confidence set that comprises all such values gives coverage: it contains the true test label with a probability determined by the hypothesis test's significance level.

The argument thus far holds when the training and test data are from the same distribution, or, more generally, exchangeable. Fannjiang et al. (2022) extend this framework to obtain coverage under feedback covariate shift, by devising the appropriate hypothesis test, building upon a recent body of work generalizing conformal prediction to various distribution shifts (Tibshirani et al. 2019; Cauchois et al. 2020; Gibbs and Candes 2021; Podkopaev and Ramdas 2021; Barber et al. 2023). Notably, the precise feedback covariate shift that emerges in design is dictated jointly by the design algorithm, predictive model, and training data. Consequently, the confidence set for any given test protein depends on the design algorithm, not just the model and the training data. In contrast, a Bayesian predictive posterior for any given test protein depends on the model and training data but not on the design algorithm (Fig. 2). Regardless, similar to the Bayesian setting, the more the training and design distributions differ, the larger the confidence sets given by conformal prediction, capturing the intuition that the predictions should be more uncertain.

Coverage guarantees are non-asymptotic, meaning they hold for any amount of training data. However, coverage is not a guarantee on the confidence set for any particular test protein —indeed, such conditional coverage guarantees are impossible without making strong assumptions about the sequence–property relationship (Vovk 2012; Foygel Barber et al. 2021). Rather, coverage describes what happens in expectation over random draws of both the training and test data from their respective distributions. Consequently, it may not be appropriate to use these methods to make choices about individual proteins. Rather, these methods are potentially most useful for selecting a design algorithm or

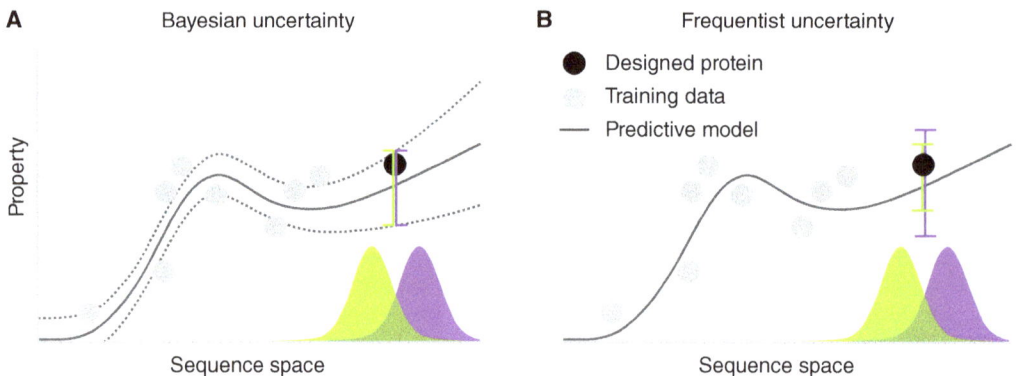

Figure 2. Bayesian versus frequentist uncertainty quantification for design. A predictive model is fit to training data (gray dots). A designed protein (black dot) is drawn from a design distribution; as two examples, here a purple one and a green one are shown. Confidence intervals for the designed protein are shown with corresponding colors. (*A*) Bayesian notions of uncertainty over a designed protein tend to increase further from the training data, as shown by the distance between two standard deviations (dashed gray lines) above and below the mean (solid gray line) of the posterior predictive distribution. Given a designed protein, this estimated uncertainty does not change with the design distribution: the confidence interval, determined by the dashed gray lines, is the same whether it was designed by sampling from the green distribution or the purple distribution. (*B*) In contrast, frequentist notions of uncertainty over a designed protein depend on the distribution it was drawn from. In particular, the confidence interval produced by a conformal prediction method that accommodates distribution shift will generally be smaller if the protein was sampled from a distribution closer to (green), rather than farther from (purple) the training distribution.

   Cite this article as *Cold Spring Harb Perspect Biol* doi: 10.1101/cshperspect.a041469

its hyperparameters, as discussed in the next section.

Density Ratio Estimation: An Important Tool for Handling Distribution Shift

An additional limitation of conformal prediction methods in the design setting is their reliance on the density ratio between the training and the designed input distributions—that is, the ratio of the densities of these two distributions—to characterize the distribution shift (Tibshirani et al. 2019; Fannjiang et al. 2022). In fact, these ratios are also key to a panoply of learning strategies that use them to reweight training data to account for distribution shift (Shimodaira 2000; Sugiyama et al. 2008; Bickel et al. 2009; Gretton et al. 2009).

If both the training and design input distributions have known closed-form densities, then it is straightforward to compute this ratio. For example, training sequences generated with library construction protocols such as error-prone polymerase chain reaction, degenerate codons, or recombination (Neylon 2004) can be framed as sampling from an explicit sequence distribution. Some design algorithms also prescribe sampling from a distribution with a closed-form density, such as a Potts model (Russ et al. 2020; Fram et al. 2023) or a library where parameters have been set such that sequences with desirable predictions are more likely to be sampled (Weinstein et al. 2022b; Zhu et al. 2022; Yang et al. 2023). However, in general this is not the case. For example, if training data comprise homologs, previously designed proteins, proteins gathered from different literature sources, or any combination thereof, then their distribution does not have a closed-form density. Similarly, most design algorithms only implicitly induce a design distribution.

As a naive solution, one could estimate the training and design distributions separately and take the ratio of their densities. However, the modeling choices that are best for estimating individual densities may not be the best for estimating ratios of different densities; in particular, this approach does not account for the nature of how the two distributions differ. Generally, estima-

tion of density ratios is statistically fragile in that it becomes increasingly high-variance the more the distributions differ, particularly given the high dimensionality of protein sequence space. The rich literature on density ratio estimation presents a wealth of alternative strategies mitigating this problem (Sugiyama et al. 2012), which are ripe for further investigation in the context of machine learning–based design (Stanton et al. 2023). These strategies learn parametric forms of the density ratio by, for example, learning a classifier for distinguishing between samples from the two distributions (Qin 1998; Bickel et al. 2009; Gutmann and Hyvärinen 2010), or minimizing various objectives quantifying how well the two distributions agree, after one of them is transformed by the density ratio estimate (Huang et al. 2006; Sugiyama et al. 2008, 2012; Gretton et al. 2009; Kanamori and Hido 2009; Nguyen et al. 2010). These approaches explicitly or implicitly learn which features are most useful for modeling the differences between the two distributions, and which are irrelevant and can be ignored. Consequently, they can be more statistically efficient than the naive approach based on estimating the two densities separately.

Finally, a unique advantage of the design setting is that one can generate as many in silico designed sequences as desired for accurate density ratio estimation, provided the design algorithm is not too computationally costly.

Uncertainty Quantification for Models of Evolutionary Data

As a final note, an open and underexplored question is how to best quantify uncertainty when using the likelihoods of density models fit to evolutionary data as property predictions. Because likelihood lives in entirely different units from the property of interest and is therefore only correlated with it at best, it is unclear what entity one should quantify uncertainty for to facilitate design. Riesselman et al. (2018) invoked Bayesian notions of uncertainty over the model parameters to improve the correlation between likelihoods and property values. It might also be fruitful to quantify notions of uncertainty for this correlation, but it is not clear how a protein engineer could make use of such uncertainty to improve design.

WHAT ARE THE DESIGN ALGORITHM CONSIDERATIONS?

So far, we have discussed considerations for learning a model whose predictions are as trustworthy as possible, and how to quantify its predictive uncertainty, particularly in the face of the distribution shifts induced by machine learning–based design. We now discuss considerations for choosing the design algorithm: the algorithm that leverages the model and any associated uncertainty to propose protein sequences to measure in the laboratory. These may be proposed either as the final set of designed proteins aspiring to achieve the desired novel condition, or to update the predictive model in an iterative manner.

As a concrete example, one simple design algorithm is as follows: pick some initial protein sequence; out of all possible single mutants of this sequence, choose the one with the most desirable predicted property value; and repeat this step with the new mutant until a computational budget is exhausted or the predictions suggest that the desired condition has been achieved. Alternatively, a design algorithm could entail sampling from a generative model—conditional or otherwise—that puts more weight on sequences with promising predicted property values (Brookes et al. 2019; Weinstein et al. 2022b; Zhu et al. 2022), a density model fit to homologs, such as a Potts model or variational autoencoder (Russ et al. 2020; Hawkins-Hooker et al. 2021; Fram et al. 2023), or a pan-protein model (Ferruz et al. 2022; Madani et al. 2023). These examples show that the design algorithm can be either intertwined with or decoupled from the model. Either way, the design algorithm dictates how the design distribution will shift relative to the distribution of the training data. As such, how should one select the design algorithm to have the greatest chance of success?

Because design algorithms employed to seek novel property values must consider sequence space away from the training data, an inherent trade-off arises between "exploration"—proposing sequences whose predictions appear to meet the desired condition but have high uncertainty, versus "exploitation"—proposing sequences with more confident but less promising predictions. Fundamentally, novelty-seeking is about navigating this dilemma as effectively as possible.

Bayesian Optimization

Bayesian optimization is one richly studied framework for tackling this dilemma in the context of iterative rounds of design, so-called because the predictive model is updated in a Bayesian manner after every round of data collection (Snoek et al. 2012; Shahriari et al. 2016). Instantiated in protein engineering terms, Bayesian optimization requires that one specify an acquisition function, a function over protein space whose maximum dictates which protein should be measured next in the laboratory. Acquisition functions typically incorporate both a protein's predicted property and the associated uncertainty. For example, the commonly used upper confidence bound algorithm can be understood as Bayesian optimization with an acquisition function comprising a weighted sum of the posterior predictive mean and standard deviation, where the weight is specified by the user and controls the exploration–exploitation trade-off. After using an optimization algorithm to maximize the acquisition function and identify which protein to measure next in the laboratory, the new data can be used to update the predictive model, and the process is iterated. In batch Bayesian optimization, one can design protein libraries using an acquisition function that is over a set of protein sequences (Azimi et al. 2010; Shah and Ghahramani 2015; Gonzalez et al. 2016; Wu and Frazier 2016; Daxberger and Low 2017; Yang et al. 2019b).

Bayesian optimization approaches, especially the upper confidence bound algorithm instantiated with Gaussian process regression models (GP-UCB), have been successfully used in a variety of protein engineering campaigns (Romero et al. 2013; Bedbrook et al. 2017, 2019; Greenhalgh et al. 2021; Rapp et al. 2023). Justifications for the use of GP-UCB and its batched variants sometimes invoke their theoretical properties, such as efficient rates of convergence in finding

Cite this article as *Cold Spring Harb Perspect Biol* doi: 10.1101/cshperspect.a041469

the global optimum for many functions (Srinivas et al. 2010; Desautels et al. 2014). However, it is unclear how practically informative such rates are, particularly because protein engineers typically want to design proteins in as few rounds as possible, aspirationally in just a single round. Consequently, further analysis and development of algorithms tailored specifically for the low- or single-round setting could be fruitful (Chan et al. 2021).

The form of many common acquisition functions assumes that the predictive model yields a conditional density of the label given the input, such as of the binding affinity to a target molecule given the protein sequence. However, when there is no assay-labeled data to begin with, one often uses the likelihood of a density model fit to homologous or pan-protein data to instead rank proteins. In such cases, it is not straightforward to instantiate common acquisition functions. Moreover, maximizing the acquisition function—which is not in general concave or otherwise friendly to optimization— is itself a nontrivial task (Wilson et al. 2018). Finally, if one wants frequentist-style guarantees on the output of the design algorithm, then Bayesian optimization may not be the most amenable paradigm (although see Stanton et al. (2023), who integrate the two by modulating the Bayesian predictive posterior with conformal confidence sets). For any of these reasons, alternative design algorithms can be employed, discussed next.

Beyond Bayesian Optimization

A number of design algorithms that do not subscribe to the Bayesian optimization paradigm have been developed and used successfully. Many of these can be framed as finding protein sequences that optimize some sort of acquisition function; however, because they do not involve any update of the predictive model, Bayesian or otherwise, they are neither performing Bayesian optimization nor intended for multiround design. The goal of the "acquisition function" in these approaches is to navigate the same exploration–exploitation trade-off already discussed, albeit in the setting where collected data will be

final and cannot be used to inform another round of proposed proteins.

Such design algorithms include those that sample designed proteins using an estimation of distribution algorithm (Brookes et al. 2019, 2020; Fannjiang and Listgarten 2020), a blackbox optimization strategy that can be used to find the parameters of a sequence distribution with, say, maximal expected property value. Others use gradient-based methods with differentiable predictive models to design sequences with desirable predicted values (Killoran et al. 2017; Bogard et al. 2019), or start with an initial set of proteins and iteratively introduce and accept mutations based on their predicted property values (Fox et al. 2007; Sinai et al. 2020; Bryant et al. 2021). When mutations are chosen in a suitable manner, the latter approach is equivalent to Markov chain Monte Carlo sampling from some distribution (Biswas et al. 2021). Although these methods cannot be analyzed through the theoretical lens of Bayesian optimization, one can wrap frequentist formalisms for uncertainty quantification around them to get guarantees on the designed proteins' property values. For example, as mentioned in the previous section, Fannjiang et al. (2022) generalize conformal prediction to the setting of machine learning-based design. By framing any design algorithm as a mapping from training data to a design distribution—even if only implicitly—this method can provide frequentist uncertainty guarantees for the output of any combination of design algorithm, predictive model, and (independently and identically distributed) training data. This flexibility means that one can incorporate whatever heuristics, intuitions, or constraints one desires, such as various mechanisms for encouraging designed sequences to remain in regions where the predictive model is trusted (Brookes et al. 2019; Linder et al. 2020; Biswas et al. 2021; Fram et al. 2023), and still obtain the same type of guarantees.

Although not the original motivation for conformal prediction, these methods could be used as a tool for design algorithm selection, including how to set hyperparameters of a design algorithm in an informed manner. For example, many design algorithms have a hyperparameter that can be thought of as navigating the exploration–exploitation trade-off. Common examples include a "temperature" hyperparameter that controls the entropy of the design distribution (Russ et al. 2020; Biswas et al. 2021; Zhu et al. 2022), and a prediction threshold hyperparameter in genetic algorithms (Sinai et al. 2020). In general, it is unclear how to set these and other hyperparameters. However, the tool of conformal prediction can help one gauge how different values of the hyperparameter trade-off between desirable but conflicting goals—for example, the trade-off between how high the average predicted value is, and frequentist uncertainty about the predictions. Plotting such trade-offs can guide protein engineers in selecting a hyperparameter value that they believe achieves an acceptable point on the trade-off, or multiple such values to achieve a risk portfolio (Fig. 3).

As one moves along the trade-off in Figure 3, not only do the predicted property values and their uncertainties change, but also the diversity of the designed sequences: the smaller the value of the hyperparameter, the greater the entropy of the design distribution. We next discuss the topic of sequence diversity in protein design.

Should Sequence Diversity Be an Explicit Consideration?

In any setting where a batch of sequences, rather than just a single sequence, is designed at once, a common concern might be how to propose a sufficiently diverse set of sequences so as to maximize the chance of achieving the desired condition. It turns out that this explicit goal of proposing a diverse set of sequences is often beside the point. If the real goal is to achieve novel property values,[2] then, provided there is a suitable notion of uncertainty for the property predictions, choosing sequences in a way that accounts for that uncertainty is the principled strategy for exploring sequence space. For exam-

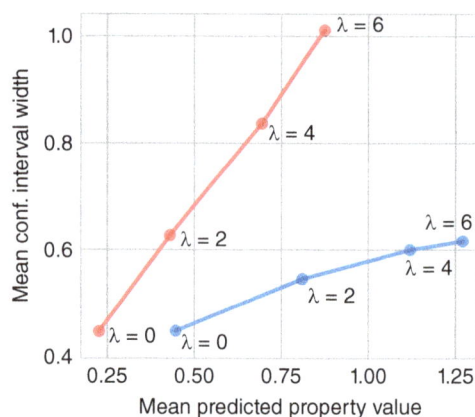

Figure 3. Design algorithm hyperparameter selection with conformal prediction. The "inverse temperature" hyperparameter, λ, controls the entropy of the design distribution (higher values lead to lower entropy). Higher values also correspond to higher average predicted property values (x-axis) and greater predictive uncertainty on the designed proteins, as measured by the width of confidence intervals produced by conformal prediction (y-axis, higher is more uncertain). Two different protein design goals are shown, for achieving brighter red and blue fluorescence (red and blue lines). One can access higher predicted brightness for blue fluorescence without incurring as much predictive uncertainty, compared to red fluorescence (i.e., the blue line is less steep than the red line). One explanation for this difference is that replicate measurements from the red fluorescence assay in Poelwijk et al. (2019) are noisier and hence inherently harder to predict. (Figure reprinted from Fannjiang et al. 2022, courtesy of the authors © 2022; published by PNAS.)

ple, in the batch Bayesian optimization setting, diversity is effectively baked into the acquisition function. Once the acquisition function has been decided, it implicitly determines the appropriate notion of diversity in the resulting proposed batch of sequences. Analogously, if one uses conformal prediction to select hyperparameter settings to navigate the exploration–exploitation trade-off (Fig. 3), then the appropriate notion of sequence diversity is dictated by the selected hyperparameter value. From these two examples, we see that in a design framework that sensibly accounts for the user's notion of uncertainty about the property predictions, the correspondingly appropriate notion of sequence diversity emerges naturally.

---

[2]In contrast, as noted in the introduction, if the goal is to find proteins with novel sequences—but not necessarily novel properties—then sequence diversity must be explicitly considered.

As just described, ideally the design algorithm accounts for predictive uncertainty such that it implicitly yields an effective notion of sequence diversity. However, in practice, there may not be a clear notion of predictive uncertainty that can be leveraged—for example, when using the likelihoods of a density model fit to evolutionary data to rank protein sequences. It is in such cases, or when one is unsure of what uncertainty quantification approach to use, but finds notions of sequence diversity easier to specify, that it may be effective to introduce property-agnostic sequence diversity metrics into the design algorithm. Examples include the entropy of the designed sequence distribution or Hamming or BLOSUM-based distances between protein sequences (Angermueller et al. 2020; Linder et al. 2020). Furthermore, one might consider generalizing such distances to account for higher-order interactions between sequence positions, or otherwise incorporate knowledge about the sequence–property relationship into the metric. As a side note, one should not use entropy to assess the diversity of a finite set of sequences, as the entropy of any set of unique sequences is the same regardless of how different they are from each other.

PREDICTION AND UNCERTAINTY ABOUT THE FUTURE OF MACHINE LEARNING– BASED PROTEIN DESIGN

With the recent rise of large models for "designing" written compositions and images with astonishing quality and apparent novelty, protein engineers have been motivated to try to follow a similar path by training models on large and diverse sets of protein sequences and/or structures. Will such efforts mean that the challenges discussed herein will soon become irrelevant? Important distinctions exist between the setting of protein design and that of language and image generation, although the practical implications of such distinctions remain to be seen. Still, let us consider them.

In large language models and their corresponding dialogue systems (OpenAI 2023), the training data, queries, and outputs all live in the same space: they are all instantiated in terms of language tokens, or, roughly, words. Correspondingly, the information that the model needs can be extracted entirely from textual data. Now consider a similar type of model, trained only on protein sequences and no other information. Could such a model know which proteins fluoresce at a particular wavelength with a particular brightness? No, that information lies in a space unfamiliar to the model—that of real-valued wavelength and brightness, not sequence. The fact that the number of known natural protein sequences is steadily rising because of plummeting sequencing costs does not reveal this information. For a model to know about the biochemical and biophysical properties of any given sequence, information from this space, such as measurements from laboratory experiments—a different modality altogether—must also be provided to the model during training. Databases of natural protein sequences do contain (mostly qualitative) annotations about biophysical and biochemical properties, but the quality of such annotations varies greatly. Although some are from published experiments, many are based on heuristics such as propagating annotations from similar sequences, which can be error-prone in a way not attenuated by scale (Brenner 1999; Schnoes et al. 2009; Radivojac et al. 2013). Consequently, fully automated protein design is currently bottlenecked by a lack of informative data from the relevant space. These data could come from laboratory experiments, or sufficiently accurate predictive models or simulations thereof.

In recent work, Madani et al. (2023) incorporated annotations from a variety of sources into training a pan-protein generative model, and then conditioned on protein family annotations based on sequence homology (Finn et al. 2016) to generate functional proteins from specific families. However, they also fine-tuned the model on abundant homologous sequences from those families. Notably, when such homologous sequences are available, functional proteins have been successfully designed without access to pan-protein data (Russ et al. 2020; Hawkins-Hooker et al. 2021; Repecka et al. 2021; Fram et al. 2023). Although Madani et al. (2023) find that a particular family-specific method (Figliuzzi et al. 2018; Russ et al. 2020) did not yield functional proteins for their families of interest, we discussed earlier how the relative performance of pan-protein and family-specific methods varies dramatically across protein families (Fig. 1). Until we better understand the factors that drive this variation, general advantages of the use of pan-protein data and models remain unclear.

Even if there should eventually be enough annotated sequence data to mimic the successes of large, multimodal models on images and text, there remains a question of how well such models could achieve novel property values, or "radically novel" properties altogether. Claims of dialogue agents exhibiting "emergent behaviors" suggest that these models may be capable of finding novelty, although such claims have also been questioned (Schaeffer et al. 2023) because of the fact that emergent behavior has not been well-defined at a technical level. Consequently, it is difficult to reason about how relevant these supposed emergent behaviors are to the world of protein design. Let us see what emerges—we cannot predict the future!

AUTHOR CONTRIBUTIONS

Whatever you agree with, C.F. wrote. Whatever you disagree with, J.L. wrote.

ACKNOWLEDGMENTS

We are grateful to Akosua Busia, Albert Fannjiang, Hanlun Jiang, Hunter Nisonoff, and Micah Olivas for providing compelling feedback and literature pointers.

REFERENCES

Acker MG, Auld DS. 2014. Considerations for the design and reporting of enzyme assays in high-throughput screening applications. *Perspect Sci* 1: 56–73. doi:10.1016/j.pisc.2013.12.001

Aghazadeh A, Nisonoff H, Ocal O, Brookes DH, Huang Y, Koyluoglu OO, Listgarten J, Ramchandran K. 2021. Epistatic net allows the sparse spectral regularization of deep neural networks for inferring fitness functions. *Nat Commun* 12: 5225. doi:10.1038/s41467-021-25371-3

Alley EC, Khimulya G, Biswas S, AlQuraishi M, Church GM. 2019. Unified rational protein engineering with sequence-based deep representation learning. *Nat Methods* 16: 1315–1322. doi:10.1038/s41592-019-0598-1

Altschul SF, Madden TL, Schäffer AA, Zhang J, Zhang Z, Miller W, Lipman DJ. 1997. Gapped BLAST and PSI-BLAST: a new generation of protein database search programs. *Nucleic Acids Res* 25: 3389–3402. doi:10.1093/nar/25.17.3389

Angelopoulos AN, Bates S. 2023. Conformal prediction: a gentle introduction. *Found Trends Mach Learn* 16: 494–591. doi:10.1561/2200000101

Angermueller C, Belanger D, Gane A, Mariet Z, Dohan D, Murphy K, Colwell L, Sculley D. 2020. Population-based black-box optimization for biological sequence design. *PMLR* 119: 324–334.

Arnold FH. 2018. Directed evolution: bringing new chemistry to life. *Angew Chem Int Ed Engl* 57: 4143–4148. doi:10.1002/anie.201708408

Azimi J, Fern A, Fern X. 2010. Batch Bayesian optimization via simulation matching. *Adv Neural Inf Process Syst* https://proceedings.neurips.cc/paper/2010/hash/e702e51da2c0f5be4dd354bb3e295d37-Abstract.html

Ballal A, Laurendon C, Salmon M, Vardakou M, Cheema J, Defernez M, O'Maille PE, Morozov AV. 2020. Sparse epistatic patterns in the evolution of terpene synthases. *Mol Biol Evol* 37: 1907–1924. doi:10.1093/molbev/msaa052

Barber RF, Candes EJ, Ramdas A, Tibshirani RJ. 2023. Conformal prediction beyond exchangeability. *Ann Statist* 51: 816–845.

Bedbrook CN, Yang KK, Rice AJ, Gradinaru V, Arnold FH. 2017. Machine learning to design integral membrane channelrhodopsins for efficient eukaryotic expression and plasma membrane localization. *PLoS Comput Biol* 13: e1005786. doi:10.1371/journal.pcbi.1005786

Bedbrook CN, Yang KK, Robinson JE, Mackey ED, Gradinaru V, Arnold FH. 2019. Machine learning-guided channelrhodopsin engineering enables minimally invasive optogenetics. *Nat Methods* 16: 1176–1184. doi:10.1038/s41592-019-0583-8

Bickel S, Bruckner M, Scheffer T. 2009. Discriminative learning under covariate shift. *J Mach Learn Res* 10: 2137–2155.

Bishop CM. 2007. *Pattern recognition and machine learning.* Springer, New York.

Biswas S, Khimulya G, Alley EC, Esvelt KM, Church GM. 2021. Low-N protein engineering with data-efficient deep learning. *Nat Methods* 18: 389–396. doi:10.1038/s41592-021-01100-y

Bogard N, Linder J, Rosenberg AB, Seelig G. 2019. A deep neural network for predicting and engineering alternative polyadenylation. *Cell* 178: 91–106.e23. doi:10.1016/j.cell.2019.04.046

Brenner SE. 1999. Errors in genome annotation. *Trends Genet* 15: 132–133. doi:10.1016/S0168-9525(99)01706-0

Brookes D, Park H, Listgarten J. 2019. Conditioning by adaptive sampling for robust design. *PMLR* 97: 773–782.

Brookes D, Busia A, Fannjiang C, Murphy K, Listgarten J. 2020. A view of estimation of distribution algorithms through the lens of expectation-maximization. In *Proceedings of the 2020 Genetic and Evolutionary Computation Conference Companion, GECCO '20*, pp. 189–190. Association for Computing Machinery, New York.

Cite this article as *Cold Spring Harb Perspect Biol* doi: 10.1101/cshperspect.a041469

Brookes DH, Aghazadeh A, Listgarten J. 2022. On the sparsity of fitness functions and implications for learning. *Proc Natl Acad Sci* **119:** e2109649118. doi:10.1073/pnas.2109649118

Bryant DH, Bashir A, Sinai S, Jain NK, Ogden PJ, Riley PF, Church GM, Colwell LJ, Kelsic ED. 2021. Deep diversification of an AAV capsid protein by machine learning. *Nat Biotechnol* **39:** 691–696. doi:10.1038/s41587-020-00793-4

Busia A, Listgarten J. 2023. MBE: model-based enrichment estimation and prediction for differential sequencing data. *Genome Biol* **24:** 218.

Cauchois M, Gupta S, Ali A, Duchi JC. 2020. Robust validation: confident predictions even when distributions shift. arXiv [statML]. http://arxiv.org/abs/2008.04267

Chan J, Pacchiano A, Tripuraneni N, Song YS, Bartlett P, Jordan MI. 2021. Parallelizing contextual linear bandits. arXiv [statML]. http://arxiv.org/abs/2105.10590

Cheng RR, Nordesjö O, Hayes RL, Levine H, Flores SC, Onuchic JN, Morcos F. 2016. Connecting the sequence-space of bacterial signaling proteins to phenotypes using coevolutionary landscapes. *Mol Biol Evol* **33:** 3054–3064. doi:10.1093/molbev/msw188

Choi K, Meng C, Song Y, Ermon S. 2022. Density ratio estimation via infinitesimal classification. *PMLR* **151:** 2552–2573.

Cohen T, Welling M. 2016. Group equivariant convolutional networks. *PMLR* **48:** 2990–2999.

Correia BE, Bates JT, Loomis RJ, Baneyx G, Carrico C, Jardine JG, Rupert P, Correnti C, Kalyuzhniy O, Vittal V, et al. 2014. Proof of principle for epitope-focused vaccine design. *Nature* **507:** 201–206. doi:10.1038/nature12966

Cortes C, Mansour Y, Mohri M. 2010. Learning bounds for importance weighting. In *Advances in neural information processing systems 23* (ed. Lafferty JD, et al.), pp. 442–450. Curran Associates, Red Hook, NY.

Dauparas J, Anishchenko I, Bennett N, Bai H, Ragotte RJ, Milles LF, Wicky BIM, Courbet A, de Haas RJ, Bethel N, et al. 2022. Robust deep learning-based protein sequence design using ProteinMPNN. *Science* **378:** 49–56. doi:10.1126/science.add2187

Daxberger EA, Low BKH. 2017. Distributed batch Gaussian process optimization. *PMLR* **70:** 951–960.

Desautels T, Krause A, Burdick JW. 2014. Parallelizing exploration-exploitation tradeoffs in Gaussian process bandit optimization. *J Mach Learn Res* **15:** 4053–4103.

Ding D. 2022. Independent mutation effects enable design of combinatorial protein binding mutants. bioRxiv doi:10.1101/2022.10.31.514613

Erginbas YE, Kang JS, Aghazadeh A, Ramchandran K. 2023. Efficiently computing sparse fourier transforms of q-ary functions. arXiv [eessSP]. http://arxiv.org/abs/2301.06200

Faheem M, Heyden A. 2014. Hybrid quantum mechanics/molecular mechanics solvation scheme for computing free energies of reactions at metal-water interfaces. *J Chem Theory Comput* **10:** 3354–3368. doi:10.1021/ct500211w

Fannjiang C, Listgarten J. 2020. Autofocused oracles for model-based design. In *Advances in neural information processing systems* (ed. Larochelle H, et al.), Vol. 33, pp. 12945–12956. Curran Associates, Red Hook, NY.

Fannjiang C, Bates S, Angelopoulos AN, Listgarten J, Jordan MI. 2022. Conformal prediction under feedback covariate shift for biomolecular design. *Proc Natl Acad Sci* **119:** e2204569119. doi:10.1073/pnas.2204569119

Ferruz N, Schmidt S, Höcker B. 2022. ProtGPT2 is a deep unsupervised language model for protein design. *Nat Commun* **13:** 4348. doi:10.1038/s41467-022-32007-7

Figliuzzi M, Jacquier H, Schug A, Tenaillon O, Weigt M. 2016. Coevolutionary landscape inference and the context-dependence of mutations in beta-lactamase TEM-1. *Mol Biol Evol* **33:** 268–280. doi:10.1093/molbev/msv211

Figliuzzi M, Barrat-Charlaix P, Weigt M. 2018. How pairwise coevolutionary models capture the collective residue variability in proteins? *Mol Biol Evol* **35:** 1018–1027. doi:10.1093/molbev/msy007

Finn RD, Coggill P, Eberhardt RY, Eddy SR, Mistry J, Mitchell AL, Potter SC, Punta M, Qureshi M, Sangrador-Vegas A, et al. 2016. The Pfam protein families database: towards a more sustainable future. *Nucleic Acids Res* **44:** D279–D285. doi:10.1093/nar/gkv1344

Fortuin V. 2022. Priors in Bayesian deep learning: a review. *Int Stat Rev* **90:** 563–591. doi:10.1111/insr.12502

Fowler DM, Fields S. 2014. Deep mutational scanning: a new style of protein science. *Nat Methods* **11:** 801–807. doi:10.1038/nmeth.3027

Fowler DM, Araya CL, Gerard W, Fields S. 2011. Enrich: software for analysis of protein function by enrichment and depletion of variants. *Bioinformatics* **27:** 3430–3431. doi:10.1093/bioinformatics/btr577

Fox RJ, Davis SC, Mundorff EC, Newman LM, Gavrilovic V, Ma SK, Chung LM, Ching C, Tam S, Muley S, et al. 2007. Improving catalytic function by ProSAR-driven enzyme evolution. *Nat Biotechnol* **25:** 338–344. doi:10.1038/nbt1286

Foygel Barber R, Candès EJ, Ramdas A, Tibshirani RJ. 2021. The limits of distribution-free conditional predictive inference. *Inf Inference* **10:** 455–482. doi:10.1093/imaiai/iaaa017

Fram B, Truebridge I, Su Y, Riesselman AJ, Ingraham JB, Passera A, Napier E, Thadani NN, Lim S, Roberts K, et al. 2023. Simultaneous enhancement of multiple functional properties using evolution-informed protein design. bioRxiv doi:10.1101/2023.05.09.539914

Frazer J, Notin P, Dias M, Gomez A, Min JK, Brock K, Gal Y, Marks DS. 2021. Disease variant prediction with deep generative models of evolutionary data. *Nature* **599:** 91–95. doi:10.1038/s41586-021-04043-8

Gal Y. 2016. "Uncertainty in deep learning." PhD thesis, University of Cambridge, Cambridge.

Gao J, Truhlar DG. 2002. Quantum mechanical methods for enzyme kinetics. *Annu Rev Phys Chem* **53:** 467–505. doi:10.1146/annurev.physchem.53.091301.150114

Geiger M, Smidt T. 2022. E3nn: Euclidean neural networks. arXiv doi:10.48550/arXiv.2207.09453

Gelman A. 2009. Bayes, jeffreys, prior distributions and the philosophy of statistics. *Stat Sci* **24:** 176–178. doi:10.1214/09-STS284D

Geman S, Bienenstock E, Doursat R. 1992. Neural networks and the bias/variance dilemma. *Neural Comput* **4:** 1–58. doi:10.1162/neco.1992.4.1.1

Gibbs I, Candes E. 2021. Adaptive conformal inference under distribution shift. *Adv Neural Inf Process Syst* **34:** 1660–1672.

Gonzalez J, Dai Z, Hennig P, Lawrence N. 2016. Batch Bayesian optimization via local penalization. *PMLR* **51:** 648–657.

Greenhalgh JC, Fahlberg SA, Pfleger BF, Romero PA. 2021. Machine learning-guided acyl-ACP reductase engineering for improved in vivo fatty alcohol production. *Nat Commun* **12:** 5825. doi:10.1038/s41467-021-25831-w

Grenander U. 1952. On empirical spectral analysis of stochastic processes. *Arkiv för Matematik* **1:** 503–531. doi:10.1007/BF02591360

Gretton A, Smola A, Huang J, Schmittfull M, Borgwardt K, Schölkopf B. 2009. Covariate shift by kernel mean matching. In *Dataset shift in machine learning.* MIT Press, Cambridge, MA.

Grover A, Song J, Kapoor A, Tran K, Agarwal A, Horvitz EJ, Ermon S. 2019. Bias correction of learned generative models using likelihood-free importance weighting. In *Advances in neural information processing systems 32* (ed. Wallach H, et al.), pp. 11056–11068. Curran Associates, Red Hook, NY.

Gutmann M, Hyvärinen A. 2010. Noise-contrastive estimation: a new estimation principle for unnormalized statistical models. *PMLR* **9:** 297–304.

Hastie T, Friedman J, Tibshirani R. 2001. *The elements of statistical learning.* Springer, New York.

Hawkins-Hooker A, Depardieu F, Baur S, Couairon G, Chen A, Bikard D. 2021. Generating functional protein variants with variational autoencoders. *PLoS Comput Biol* **17:** e1008736. doi:10.1371/journal.pcbi.1008736

Hesslow D, Zanichelli N, Notin P, Poli I, Marks D. 2022. RITA: a study on scaling up generative protein sequence models. arXiv doi:10.48550/arXiv.2205.05789

Hopf TA, Ingraham JB, Poelwijk FJ, Schärfe CPI, Springer M, Sander C, Marks DS. 2017. Mutation effects predicted from sequence co-variation. *Nat Biotechnol* **35:** 128–135. doi:10.1038/nbt.3769

Hsu C, Nisonoff H, Fannjiang C, Listgarten J. 2022a. Learning protein fitness models from evolutionary and assay-labeled data. *Nat Biotechnol* **40:** 1114–1122. doi:10.1038/s41587-021-01146-5

Hsu C, Verkuil R, Liu J, Lin Z, Hie B, Sercu T, Lerer A, Rives A. 2022b. Learning inverse folding from millions of predicted structures. *PMLR* **162:** 8946–8970.

Huang J, Gretton A, Borgwardt K, Schölkopf B, Smola A. 2006. Correcting sample selection bias by unlabeled data. *Adv Neural Inf Process Syst* **19:** 601–608.

Ingraham J, Garg V, Barzilay R, Jaakkola T. 2019. Generative models for graph-based protein design. In *Advances in neural information processing systems* (ed. Wallach H, et al.), Vol. 32. Curran Associates, Red Hook, NY.

Ingraham J, Baranov M, Costello Z, Frappier V, Ismail A, Tie S, Wang W, Xue V, Obermeyer F, Beam A, et al. 2022. Illuminating protein space with a programmable generative model. bioRxiv doi:10.1101/2022.12.01.518682

Jagota M, Ye C, Albors C, Rastogi R, Koehl A, Ioannidis N, Song YS. 2023. Cross-protein transfer learning substantially improves disease variant prediction. *Genome Biol* **24:** 182.

Jing B, Eismann S, Suriana P, Townshend RJL, Dror R. 2021. Learning from protein structure with geometric vector perceptrons. *9th International Conference on Learning Representations.* https://openreview.net/forum?id=1YLJDvSx6J4

Johnson LS, Eddy SR, Portugaly E. 2010. Hidden Markov model speed heuristic and iterative HMM search procedure. *BMC Bioinformatics* **11:** 431. doi:10.1186/1471-2105-11-431

Kanamori T, Hido S. 2009. A least-squares approach to direct importance estimation. *J Mach Learn Res* **10:** 1391–1445.

Killoran N, Lee LJ, Delong A, Duvenaud D, Frey BJ. 2017. Generating and designing DNA with deep generative models. arXiv doi:10.48550/arXiv.1604.04173

Kiureghian AD, Ditlevsen O. 2009. Aleatory or epistemic? Does it matter? *Struct Saf* **31:** 105–112. doi:10.1016/j.strusafe.2008.06.020

Kuhlman B, Bradley P. 2019. Advances in protein structure prediction and design. *Nat Rev Mol Cell Biol* **20:** 681–697. doi:10.1038/s41580-019-0163-x

Laine E, Karami Y, Carbone A. 2019. GEMME: a simple and fast global epistatic model predicting mutational effects. *Mol Biol Evol* **36:** 2604–2619. doi:10.1093/molbev/msz179

Linder J, Bogard N, Rosenberg AB, Seelig G. 2020. A generative neural network for maximizing fitness and diversity of synthetic DNA and protein sequences. *Cell Syst* **11:** 49–62.e16. doi:10.1016/j.cels.2020.05.007

Little RJ. 2006. Calibrated bayes: a bayes/frequentist roadmap. *Am Stat* **60:** 213–223. doi:10.1198/000313006X117837

Madani A, McCann B, Naik N, Keskar NS, Anand N, Eguchi RR, Huang PS, Socher R. 2020. Progen: language modeling for protein generation. bioRxiv doi:10.48550/arXiv.2004.03497

Madani A, Krause B, Greene ER, Subramanian S, Mohr BP, Holton JM, Olmos JL, Xiong C, Sun ZZ, Socher R, et al. 2023. Large language models generate functional protein sequences across diverse families. *Nat Biotechnol* **41:** 1099–1106. doi:10.1038/s41587-022-01618-2; http://paperpile.com/b/Xd89E6/AtYM

Malbranke C, Bikard D, Cocco S, Monasson R, Tubiana J. 2023. Machine learning for evolutionary-based and physics-inspired protein design: current and future synergies. *Curr Opin Struct Biol* **80:** 102571. doi:10.1016/j.sbi.2023.102571

Markin CJ, Mokhtari DA, Sunden F, Appel MJ, Akiva E, Longwell SA, Sabatti C, Herschlag D, Fordyce PM. 2021. Revealing enzyme functional architecture via high-throughput microfluidic enzyme kinetics. *Science* **373:** eabf8761. doi:10.1126/science.abf8761

Meier J, Rao R, Verkuil R, Liu J, Sercu T, Rives A. 2021. Language models enable zero-shot prediction of the effects of mutations on protein function. *Adv Neural Inf Process Syst* **34:** 29287–29303.

Neal RM. 1996. *Bayesian learning for neural networks.* Springer, New York.

Neylon C. 2004. Chemical and biochemical strategies for the randomization of protein encoding DNA sequences: library construction methods for directed evolution. *Nucleic Acids Res* **32:** 1448–1459. doi:10.1093/nar/gkh315

Nguyen X, Wainwright MJ, Jordan M. 2010. Estimating divergence functionals and the likelihood ratio by convex risk minimization. *IEEE Trans Inf Theory* **56:** 5847–5861.

Nijkamp E, Ruffolo J, Weinstein EN, Naik N, Madani A. 2022. Progen2: exploring the boundaries of protein language models. arXiv doi:10.48550/arXiv.2206.13517

Nisonoff H, Wang Y, Listgarten J. 2022. Augmenting neural networks with priors on function values. arXiv doi:10.48550/arXiv.2202.04798

Norn C, Wicky BIM, Juergens D, Liu S, Kim D, Tischer D, Koepnick B, Anishchenko I, Players F, Baker D, et al. 2021. Protein sequence design by conformational landscape optimization. *Proc Natl Acad Sci* **118:** e2017228118. doi:10.1073/pnas.2017228118

Notin P, Dias M, Frazer J, Hurtado JM, Gomez AN, Marks D, Gal Y. 2022a. Tranception: protein fitness prediction with autoregressive transformers and inference-time retrieval. *PMLR* **162:** 16990–17017.

Notin P, Van Niekerk L, Kollasch AW, Ritter D, Gal Y, Marks DS. 2022b. TranceptEVE: combining family-specific and family-agnostic models of protein sequences for improved fitness prediction. bioRxiv doi:10.1101/2022.12.07.519495

OpenAI. 2023. GPT-4 technical report. arXiv doi:10.48550/arXiv.2303.08774

Otwinowski J, McCandlish DM, Plotkin JB. 2018. Inferring the shape of global epistasis. *Proc Natl Acad Sci* **115:** E7550–E7558. doi:10.1073/pnas.1804015115

Ovchinnikov S, Huang PS. 2021. Structure-based protein design with deep learning. *Curr Opin Chem Biol* **65:** 136–144. doi:10.1016/j.cbpa.2021.08.004

Pan X, Kortemme T. 2021. Recent advances in de novo protein design: principles, methods, and applications. *J Biol Chem* **296:** 100558. doi:10.1016/j.jbc.2021.100558

Pearson WR. 2013. An introduction to sequence similarity ("homology") searching. *Curr Protoc Bioinformatics* **42:** 3.1.1–3.1.8. doi:10.1002/0471250953.bi0301s42

Platt J. 1999. Probabilistic outputs for support vector machines and comparisons to regularized likelihood methods. *Adv Large Margin Class* **10:** 61–74.

Podkopaev A, Ramdas A. 2021. Distribution-free uncertainty quantification for classification under label shift. *PMLR* **161:** 844–853.

Poelwijk FJ, Socolich M, Ranganathan R. 2019. Learning the pattern of epistasis linking genotype and phenotype in a protein. *Nat Commun* **10:** 4213. doi:10.1038/s41467-019-12130-8

Posani L, Rizzato F, Monasson R, Cocco S. 2023. Infer global, predict local: quantity-relevance trade-off in protein fitness predictions from sequence data. *PLoS Comput Biol* **19:** e1011521. doi:10.1371/journal.pcbi.1011521

Qin J. 1998. Inferences for case-control and semiparametric two-sample density ratio models. *Biometrika* **85:** 619–630. doi:10.1093/biomet/85.3.619

Qin C, Colwell LJ. 2018. Power law tails in phylogenetic systems. *Proc Natl Acad Sci* **115:** 690–695. doi:10.1073/pnas.1711913115

Radivojac P, Clark WT, Oron TR, Schnoes AM, Wittkop T, Sokolov A, Graim K, Funk C, Verspoor K, Ben-Hur A, et al. 2013. A large-scale evaluation of computational protein function prediction. *Nat Methods* **10:** 221–227. doi:10.1038/nmeth.2340

Rapp JT, Bremer BJ, Romero PA. 2023. Self-driving laboratories to autonomously navigate the protein fitness landscape. bioRxiv doi:10.1101/2023.05.20.541582

Repecka D, Jauniskis V, Karpus L, Rembeza E, Rokaitis I, Zrimec J, Poviloniene S, Laurynenas A, Viknander S, Abuajwa W, et al. 2021. Expanding functional protein sequence spaces using generative adversarial networks. *Nat Mach Intell* **3:** 324–333. doi:10.1038/s42256-021-00310-5

Rhodes B, Xu K, Gutmann MU. 2020. Telescoping density-ratio estimation. *Adv Neural Inf Process Syst* **33:** 4905–4916.

Riesselman AJ, Ingraham JB, Marks DS. 2018. Deep generative models of genetic variation capture the effects of mutations. *Nat Methods* **15:** 816–822. doi:10.1038/s41592-018-0138-4

Romero PA, Arnold FH. 2009. Exploring protein fitness landscapes by directed evolution. *Nat Rev Mol Cell Biol* **10:** 866–876. doi:10.1038/nrm2805

Romero PA, Krause A, Arnold FH. 2013. Navigating the protein fitness landscape with Gaussian processes. *Proc Natl Acad Sci* **110:** E193–E201. doi:10.1073/pnas.1215251110

Rubin DB. 1984. Bayesianly justifiable and relevant frequency calculations for the applied statistician. *Ann Statist* **12:** 1151–1172. doi:10.1214/aos/1176346785

Rubin AF, Gelman H, Lucas N, Bajjalieh SM, Papenfuss AT, Speed TP, Fowler DM. 2017. A statistical framework for analyzing deep mutational scanning data. *Genome Biol* **18:** 150. doi:10.1186/s13059-017-1272-5

Russ WP, Figliuzzi M, Stocker C, Barrat-Charlaix P, Socolich M, Kast P, Hilvert D, Monasson R, Cocco S, Weigt M, et al. 2020. An evolution-based model for designing chorismate mutase enzymes. *Science* **369:** 440–445. doi:10.1126/science.aba3304

Sailer ZR, Harms MJ. 2017. Detecting high-order epistasis in nonlinear genotype-phenotype maps. *Genetics* **205:** 1079–1088. doi:10.1534/genetics.116.195214

Schaeffer R, Miranda B, Koyejo S. 2023. Are emergent abilities of large language models a mirage? arXiv doi:10.48550/arXiv.2304.15004

Schnoes AM, Brown SD, Dodevski I, Babbitt PC. 2009. Annotation error in public databases: misannotation of molecular function in enzyme superfamilies. *PLoS Comput Biol* **5:** e1000605. doi:10.1371/journal.pcbi.1000605

Sesterhenn F, Yang C, Bonet J, Cramer JT, Wen X, Wang Y, Chiang CI, Abriata LA, Kucharska I, Castoro G, et al. 2020. De novo protein design enables the precise induction of RSV-neutralizing antibodies. *Science* **368:** eaay5051. doi:10.1126/science.aay5051

Shah A, Ghahramani Z. 2015. Parallel predictive entropy search for batch global optimization of expensive objective functions. *Adv Neural Inf Process Syst* **28:** 3330–3338.

Shahriari B, Swersky K, Wang Z, Adams RP, de Freitas N. 2016. Taking the human out of the loop: a review of Bayesian optimization. *Proc IEEE* **104:** 148–175. doi:10.1109/JPROC.2015.2494218

Shimodaira H. 2000. Improving predictive inference under covariate shift by weighting the log-likelihood function. *J*

Stat Plan Inference **90**: 227–244. doi:10.1016/S0378-3758(00)00115-4

Shin JE, Riesselman AJ, Kollasch AW, McMahon C, Simon E, Sander C, Manglik A, Kruse AC, Marks DS. 2021. Protein design and variant prediction using autoregressive generative models. *Nat Commun* **12**: 2403. doi:10.1038/s41467-021-22732-w

Sinai S, Wang R, Whatley A, Slocum S, Locane E, Kelsic ED. 2020. Adalead: a simple and robust adaptive greedy search algorithm for sequence design. arXiv doi:10.48550/arXiv.2010.02141

Snoek J, Larochelle H, Adams RP. 2012. Practical Bayesian optimization of machine learning algorithms. In *Advances in neural information processing systems* (ed. Pereira F, et al.), Vol. 25. Curran Associates, Red Hook, NY.

Song H, Bremer BJ, Hinds EC, Raskutti G, Romero PA. 2021. Inferring protein sequence-function relationships with large-scale positive-unlabeled learning. *Cell Syst* **12**: 92–101.e8. doi:10.1016/j.cels.2020.10.007

Srinivas N, Krause A, Kakade SM, Seeger MW. 2010. Gaussian process optimization in the bandit setting: no regret and experimental design. In *Proceedings of the 27th International Conference on Machine Learning (ICML-10)*, June 21–24, 2010, Haifa, Israel (ed. Fürnkranz J, Joachims T), pp. 1015–1022. Omnipress, Madison, WI.

Stanton S, Maddox W, Gruver N, Maffettone P, Delaney E, Greenside P, Wilson AG. 2022. Accelerating Bayesian optimization for biological sequence design with denoising autoencoders. *PMLR* **162**: 20459–20478.

Stanton S, Maddox W, Wilson AG. 2023. Bayesian optimization with conformal prediction sets. *PMLR* **206**: 959–986.

Sugiyama M, Suzuki T, Nakajima S, Kashima H, von Bünau P, Kawanabe M. 2008. Direct importance estimation for covariate shift adaptation. *Ann Inst Stat Math* **60**: 699–746. doi:10.1007/s10463-008-0197-x

Sugiyama M, Suzuki T, Kanamori T. 2012. *Density ratio estimation in machine learning*. Cambridge University Press, Cambridge.

Tibshirani RJ, Foygel Barber R, Candes E, Ramdas A. 2019. Conformal prediction under covariate shift. *Adv Neural Inf Process Syst* **32**: 1–11.

Tokuriki N, Tawfik DS. 2009. Protein dynamism and evolvability. *Science* **324**: 203–207. doi:10.1126/science.1169375

Trinquier J, Uguzzoni G, Pagnani A, Zamponi F, Weigt M. 2021. Efficient generative modeling of protein sequences using simple autoregressive models. *Nat Commun* **12**: 5800. doi:10.1038/s41467-021-25756-4

Vovk V. 2012. Conditional validity of inductive conformal predictors. *PMLR* **25**: 475–490.

Vovk V, Gammerman A, Shafer G. 2005. *Algorithmic learning in a random world*. Springer, New York.

Wang J, Lisanza S, Juergens D, Tischer D, Watson JL, Castro KM, Ragotte R, Saragovi A, Milles LF, Baek M, et al. 2022. Scaffolding protein functional sites using deep learning. *Science* **377**: 387–394. doi:10.1126/science.abn2100

Watson JL, Juergens D, Bennett NR, Trippe BL, Yim J, Eisenach HE, Ahern W, Borst AJ, Ragotte RJ, Milles LF, et al. 2023. De novo design of protein structure and function with RFdiffusion. *Nature* **620**: 1089–1100.

Weinstein EN, Amin AN, Frazer J, Marks DS. 2022a. Non-identifiability and the blessings of misspecification in models of molecular fitness and phylogeny. In *Advances in neural information processing systems 35 (NeurIPS 2022)* (ed. Koyejo S, et al.), pp. 5484–5497. Curran Associates, Red Hook, NY.

Weinstein EN, Amin AN, Grathwohl WS, Kassler D, Disset J, Marks D. 2022b. Optimal design of stochastic DNA synthesis protocols based on generative sequence models. *PMLR* **151**: 7450–7482.

Wheelock LB, Malina S, Gerold J, Sinai S. 2022. Forecasting labels under distribution-shift for machine-guided sequence design. *PMLR* 166–180.

Wilson J, Hutter F, Deisenroth M. 2018. Maximizing acquisition functions for Bayesian optimization. *Adv Neural Inf Process Syst* **31**: 1–12.

Wittmann BJ, Yue Y, Arnold FH. 2021. Informed training set design enables efficient machine learning-assisted directed protein evolution. *Cell Syst* **12**: 1026–1045.e7. doi:10.1016/j.cels.2021.07.008

Wrenbeck EE, Faber MS, Whitehead TA. 2017. Deep sequencing methods for protein engineering and design. *Curr Opin Struct Biol* **45**: 36–44. doi:10.1016/j.sbi.2016.11.001

Wu J, Frazier P. 2016. The parallel knowledge gradient method for batch Bayesian optimization. *Adv Neural Inf Process Syst* **29**: 1–9.

Wu Z, Kan SBJ, Lewis RD, Wittmann BJ, Arnold FH. 2019. Machine learning-assisted directed protein evolution with combinatorial libraries. *Proc Natl Acad Sci* **116**: 8852–8858. doi:10.1073/pnas.1901979116

Wu Z, Yang KK, Liszka MJ, Lee A, Batzilla A, Wernick D, Weiner DP, Arnold FH. 2020. Signal peptides generated by attention-based neural networks. *ACS Synth Biol* **9**: 2154–2161. doi:10.1021/acssynbio.0c00219

Yang G, Anderson DW, Baier F, Dohmen E, Hong N, Carr PD, Kamerlin SCL, Jackson CJ, Bornberg-Bauer E, Tokuriki N. 2019a. Higher-order epistasis shapes the fitness landscape of a xenobiotic-degrading enzyme. *Nat Chem Biol* **15**: 1120–1128. doi:10.1038/s41589-019-0386-3

Yang KK, Chen Y, Lee A, Yue Y. 2019b. Batched stochastic Bayesian optimization via combinatorial constraints design. *PMLR* **89**: 3410–3419.

Yang J, Ducharme J, Johnston KE, Li FZ, Yue Y, Arnold FH. 2023. DeCOIL: optimization of degenerate codon libraries for machine learning–assisted protein engineering. *ACS Synth Biol* **12**: 2444–2454.

Zhang C, Butepage J, Kjellstrom H, Mandt S. 2019. Advances in variational inference. *IEEE Trans Pattern Anal Mach Intell* **41**: 2008–2026. doi:10.1109/TPAMI.2018.2889774

Zhu D, Brookes DH, Busia A, Carneiro A, Fannjiang C, Popova G, Shin D, Donohue KC, Chang EF, Nowakowski TJ, et al. 2022. Optimal trade-off control in machine learning-based library design, with application to adeno-associated virus (AAV) for gene therapy. bioRxiv doi:10.1101/2021.11.02.467003

Cite this article as *Cold Spring Harb Perspect Biol* doi: 10.1101/cshperspect.a041469

Exploring the Protein Sequence Space with Global Generative Models

Sergio Romero-Romero,[1,4] Sebastian Lindner,[2,4] and Noelia Ferruz[3]

[1]Department of Biochemistry, University of Bayreuth, 95447 Bayreuth, Germany

[2]University of Heidelberg, 69047 Heidelberg, Germany

[3]Barcelona Institute of Molecular Biology, 08028 Barcelona, Spain

Correspondence: noelia.ferruz@ibmb.csic.es

Recent advancements in specialized large-scale architectures for training images and language have profoundly impacted the field of computer vision and natural language processing (NLP). Language models, such as the recent ChatGPT and GPT-4, have demonstrated exceptional capabilities in processing, translating, and generating human language. These breakthroughs have also been reflected in protein research, leading to the rapid development of numerous new methods in a short time, with unprecedented performance. Several of these models have been developed with the goal of generating sequences in novel regions of the protein space. In this work, we provide an overview of the use of protein generative models, reviewing (1) language models for the design of novel artificial proteins, (2) works that use non-transformer architectures, and (3) applications in directed evolution approaches.

Proteins are highly attractive nanomaterials, capable of performing a wide range of functions under mild, nontoxic conditions. This has prompted significant research efforts in the field of protein design, with a particular emphasis on the development of functional proteins. In the last two decades, remarkable advances have been made in this area, including the design of novel de novo protein structures using traditional methods (Korendovych and DeGrado 2020; Pan and Kortemme 2021; Romero-Romero et al. 2021). The conventional approach to protein design involves providing a backbone scaffold as input and then using computational methods to identify optimal sequences that fold into the scaffold. This problem, often referred to as the "inverse folding problem," has been mathematically formulated as an optimization problem, where the goal is to find the global minimum of a high-dimensional physicochemical energy function (Fig. 1A). However, due to its computational complexity—with over 100^{20} possible sequences for a protein of 100 amino acids—approximations to both the algorithm and function are often employed (Huang et al. 2016), with the exception of a few algorithms that sample the energy space exhaustively and deterministically (Gainza et al. 2016).

---

[4]These authors contributed equally to this work.

Cite this article as *Cold Spring Harb Perspect Biol* doi: 10.1101/cshperspect.a041471

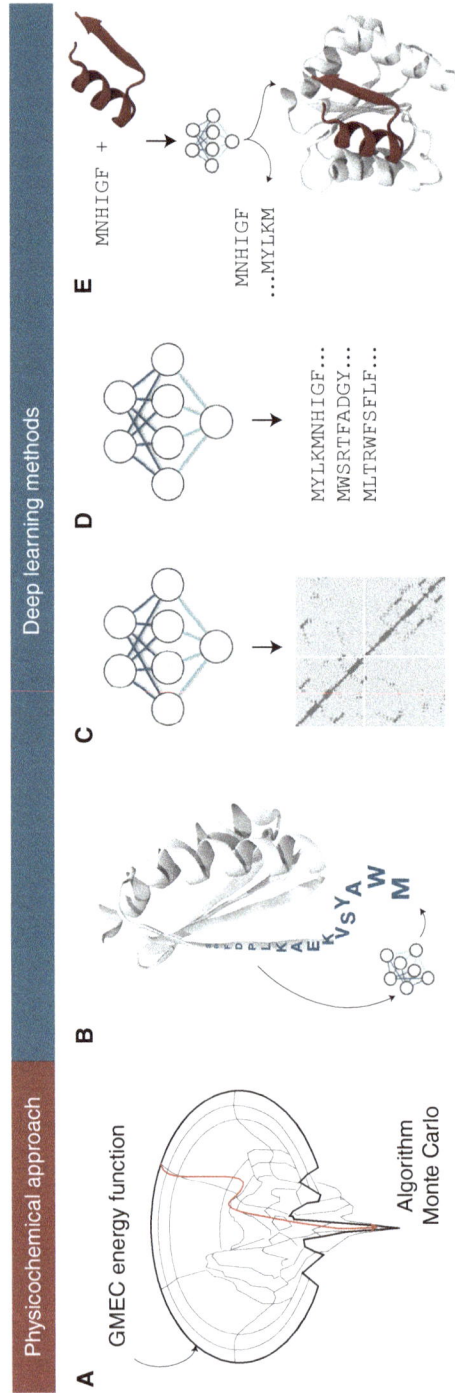

Figure 1. A paradigm shift in protein design. (*A*) The traditional protein design problem, where an approximated energy function, such as the global minimum energy conformation (GMEC), is searched with a heuristic algorithm. Deep learning techniques have enabled the design of fixed-backbone sequences (*B*), structure generation (*C*), sequence generation (*D*), and sequence and structure design around a scaffold (*E*).

Despite significant progress in the field of protein design in recent years (as detailed in Huang et al. 2016; Pan and Kortemme 2021), traditional approaches have two primary limitations. First, they typically require a predefined protein backbone as input, which may not be the optimal scaffold for a given function. Second, the integration of functional properties is typically performed in a separate step after protein design, a process that can be time-consuming, extending over several years. However, the integration of artificial intelligence (AI) methods in protein design has brought about a paradigm shift in the field. The rapid advancement of deep learning (DL) architectures and hardware has led to significant breakthroughs in various fields, such as the development of tools for image processing (DALLE-2, StableDiffusion), text generation (ChatGPT, GPT-3), audio assistants (Siri, Alexa), and even puts self-driving cars on the horizon. In the field of protein research, one notable example of this revolution is the structure prediction method Alphafold2, which has inspired the development of numerous DL-based protein design methods. In fact, more than 40 different methods have been developed in the past 3 years alone (Ferruz et al. 2022a). These methods provide not only new approaches to the traditional inverse folding problem (Fig. 1B) but also introduce novel ways of designing proteins, such as generating structures (Fig. 1C; Anand and Huang 2018), sequences (Fig. 1D; Ferruz et al. 2022b; Nijkamp et al. 2022), or concurrently designing both (Fig. 1E; Wang et al. 2022).

In this review, we delve into the potential of models that are capable of generating protein sequences across the entire protein space. We begin by reviewing transformer-based language models that can generate unconditioned protein sequences or sequences conditioned on a user-defined prompt. Then, we focus on generative models that employ DL architectures other than transformers, such as generative adversarial networks (GANs), variational autoencoders (VAEs), or long short-term memory (LSTM) networks. Finally, we examine the use of DL methods in the field of directed evolution and protein design. Our goal is to provide a comprehensive overview of the use of AI-based methods for sequence generation and to introduce readers to this emerging field of research.

TRANSFORMER-BASED LANGUAGE MODELS

The transformer has emerged as one of the most critical developments in AI in recent years (Vaswani et al. 2017), enabling the implementation of a myriad of language models. Its success is mainly attributable to the attention mechanism (Bahdanau et al. 2014), which originated as a solution to traditional sequence-to-sequence (seq2seq) models (Fig. 2). In seq2seq models, the input (a sentence) is stepwise processed in the encoder to produce a context vector passed to the decoder, an architecture that however exhibited degrading performance and increasing times with sequence length. The attention mechanism provided a solution to these problems since it allows the decoder to analyze the whole input and focus on specific parts, a notion similar to attention in the human mind. A simplified example of the attention mechanism is to focus on the input word "home" when outputting the word "maison" in an English-to-French translation (Fig. 2A). The transformer not only mediated the attention mechanism between the two modules but also throughout them, producing a much better performance in many tasks. Following these advances, researchers soon started exploring the modules' performance separately.

In this direction, Devlin et al. (2018) pretrained bidirectional encoder representations from transformers (BERT). BERT is also inspired by the transformer architecture. Still, given that in this case the interest lies in creating representations of text input, it only uses the encoder module (Fig. 2B). Models like BERT are trained by corrupting the input tokens in some way (e.g., by masking and trying to reconstruct the original sentence), such as in cloze tests. Soon after, OpenAI released GPT (generative pretrained transformer), the first of a series of highly performing generative models, the most recent being ChatGPT and GPT-4. GPT was pretrained on the classic language-modeling task, namely, predicting the next item of a sequence based on

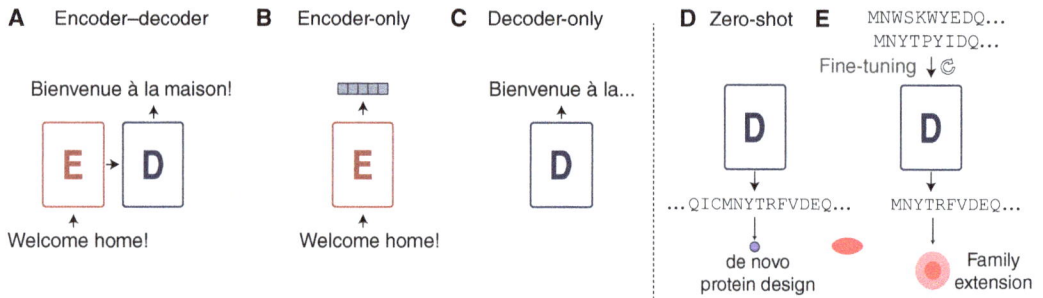

Figure 2. Architectural types of language models and their examples in natural language processing (NLP). (*A*) The original transformer has encoder (E) and decoder (D) modules. (*B*) The bidirectional encoder representations from transformers (BERT) model is based on the original transformer but contains only the encoder model. (*C*) The GPT-n models are based on the decoder-only part of the transformer and have generative capabilities. (*D*) In zero-shot, language models generate sequences unconditionally. (*E*) During fine-tuning, the model updates its weights and will sample new sequences with properties from the new training set.

the previous ones—a task that makes it particularly powerful for language generation. Models trained on this objective are termed "autoregressive" (AR), and their architecture corresponds to the decoder module (Fig. 2C).

Given this unprecedented success, researchers soon began to apply BERT and GPT-n architectures to the protein realm. Some of the earliest examples of encoder-only protein language models are ESM-1b (Rives et al. 2021) and ProtTrans (Elnaggar et al. 2022), which showed remarkable accuracy in a wide variety of tasks, such as contact prediction or functional annotation. In late 2022, META AI released a new ESM version, ESM2 (Lin et al. 2023). ESM2 contains 15 billion parameters and performs so extraordinarily that it has been used for protein structure prediction, such as in the more than 770 M (release v2023_02) predictions from the Mgnify database released in the ESMatlas (esmatlas.com). Despite not being a generative model, ESM2 has been used to design de novo proteins with remarkable success (Verkuil et al. 2022). Namely, in this work, the authors explore the design of sequences for a defined backbone by sampling the model via Markov chain Monte Carlo (MCMC) with simulated annealing and concomitantly designing sequence/structure generation by sampling from both distributions (Verkuil et al. 2022).

Another interesting example in this field is the multiple sequence alignment (MSA) trans-

former (Rao et al. 2021). This language model takes as input an MSA (see next section for a review of non-transformer methods using MSAs) and applies row and column attention through the input sequences using a modified masked language-modeling objective. The model has also been recently applied to the design of soluble, active malate dehydrogenases and superoxide dismutases (Johnson et al. 2023).

Decoder-only transformers have also emerged as very powerful architectures in protein research. Given their AR objective, they are particularly suitable for sequence generation, generating amino acids or k-mers in the amino- to carboxyl-terminus direction. Decoder-only models can be primarily categorized into two types, depending on their conditioning nature. Unconditional language models sample sequences from their learned probability distribution and, by doing so, they generate sequences in unpredictable regions of the protein space. In contrast, conditional models generate sequences conditioned on input data, which may specify a specific property, such as the protein function or target organism. The following section focuses on these types of models.

Unconditioned Generative Language Models

Several recent works have applied decoder-only architectures for unconditional protein design. In 2022, Moffat and colleagues (2022) implemented DARK, a 110-million-decoder-only transformer capable of designing novel structures, and Ferruz et al. (2022b) released Prot-GPT2, a 738 million transformer model based on the GPT-2 architecture. ProtGPT2 showed to generate de novo sequences in unexplored regions of the protein space while the sequences exhibited values akin to natural sequences in multiple properties such as disorder, dynamic properties, pLDDT values, or predicted stability.

RITA is a suite of generative decoder-only models for protein design ranging from 85M to 1.2B parameters, trained on the Uniref100 database for each sequence and its reverse. Besides releasing the models, the authors studied the relationship between model size and downstream performance and observed that performance at predicting protein fitness increased with model size (Hesslow et al. 2022). Nijkamp and colleagues (2022) released ProGen2, a family of models of up to 6.4B parameters trained on over a billion proteins from genomic, metagenomic, and immune repertoire databases. The models generate sequences predicted to adopt well-folded structures, despite being significantly distant from the current protein space, and are able to predict protein fitness without further training.

One particularly interesting property of protein language models is that they can be used in zero-shot or after fine-tuning (Fig. 2D,E; Radford and Narasimhan 2018). When used in zero-shot, models sample from their entire distribution and hence generate sequences unconditionally. In contrast, fine-tuning the model is a process that updates its parameters by learning from a new, narrower data set, like a protein family. In this case, they will generate new sequences from that group and can be used to augment protein family repertoires (Fig. 2E). Fine-tuning is possibly the best strategy in unconditional models to add some control over the generation process; alternatively, postgeneration filtering can also provide proteins with tailored properties given the fast inference times of these models. In this direction, Ferruz et al. (2022a,b) developed a pipeline that combines the synergistic nature of encoder and decoder-only architectures, where ProtGPT2 was used to generate

sequences in a high-throughput fashion while ProtT5 annotated their functions. A more sophisticated alternative is the use of conditional models (i.e., models that directly infer sequences given a user-defined set of conditions). The next section deals with these types of models.

Conditional Language Models: Tailored Protein Design

Conditional language models can be trained by coupling the training sequences with control tags, such as their annotated functions or properties. In this way, the model learns a joint sequence-function distribution. One of the most essential works in this direction was the development of the conditional transformer language (CTRL), an AR model including conditional tags capable of controllably generating text without relying on input sequences (Keskar et al. 2019). These tags, called control codes, allow users to influence genre, topic, or style more specifically —an enormous step toward goal-oriented text generation.

Shortly after CTRL implementation, some of the authors adapted this model to the protein realm by training on a data set of 281 million protein sequences (Madani et al. 2020). The model, named ProGen, contains as conditional tags UniProtKB Keywords, a vocabulary of ten categories including "biological process," "cellular component," or "molecular function." In total, the conditional tags comprised more than 1100 terms. ProGen's perplexities (a statistical measure of how confidently the model generates a certain sample) were observed to be very low, even on protein families not present in the training set (Madani et al. 2020). The generation of random sequences and their Rosetta energy evaluation revealed that the sequences had better scores than random ones. Then, the authors experimentally validated ProGen in generating lysozymes after fine-tuning on five different protein families (Madani et al. 2023). The results showed that the generated sequences possessed enzymatic activities in the range of natural lysozymes, even in cases with sequence identities as low as 40%–50%, and sometimes rivaled that of a natural hen egg white lysozyme. X-ray character-

Table 1. Summary of released language models ordered by inverse chronological order

Model name	Number of parameters	Date	Reference
Encoder-only (embedding)			
xTrimoPGLM	100B	Jul 2023	Chen et al. 2023
Ankh	1.15B	Jan 2023	Elnaggar et al. 2023
ESM2	15B	Oct 2022	Lin et al. 2023
DistilProtBert	230M	May 2022	Geffen et al. 2022
ProteinLM	200M–3B	Aug 2021	Xiao et al. 2021
ProteinBERT	16M	May 2021	Brandes et al. 2022
ESM1	43M–670M	April 2019	Rives et al. 2021
PRoBERTa	44M	Jun 2020	Nambiar et al. 2020
ProtTrans	420M–11B	Jul 2020	Elnaggar et al. 2021
TAPE	38M	Jun 2019	Rao et al. 2019
Decoder-only (generation)			
ZymCTRL	762M	Dec 2022	Munsamy et al. 2022
ProGEN2	151M–6.4B	Jun 2022	Nijkamp et al. 2022
RITA	85M–1.2B	May 2022	Hesslow et al. 2022
Tranception	700M	May 2022	Notin et al. 2022
ProtGPT2	762M	Mar 2022	Ferruz et al. 2022b
DARK	110M	Jan 2022	Moffat et al. 2022
ProGEN	1.2B	Mar 2020	Madani et al. 2020

ization of one of the variants showed that it recapitulated the native 3D structure.

In a similar direction, ZymCTRL is a conditional model trained on the *Corpora* of enzymes and their corresponding annotations (Munsamy et al. 2022). Enzymatic sequences are classified depending on their enzymatic commission (EC) numbers, a four-letter code that groups sequences catalyzing the same chemical reactions. By training a CTRL-like language model on a set of sequences and their EC classes, ZymCTRL learned a joint distribution of sequence properties and functional annotations and can effectively generate sequences that perform a user-defined enzymatic reaction. An overview of all released protein language models to date is summarized in Table 1.

NON-TRANSFORMER GENERATIVE MODELS

Although transformers have dominated the field of protein sequence generation in recent years, multiple other model architectures have been used for sequence generation. While most of these models had the general capability to generate sequences, only a few were developed with the specific task of global sequence generation in mind. Table 2 summarizes the latest non-transformer models with generative capabilities.

LSTMs were the precursors to transformers and were capable of detecting and learning from long-term dependencies as a type of recurrent neural network. UniRep (Alley et al. 2019) is an AR multiplicative LSTM (mLSTM) comprising 20.15 million parameters with nine layers that progressively decrease in dimensionality from 1900 to 64 dimensions (mLSTM strategy shown in Fig. 3B). UniRep's generative ability relies on a given input seed and results mostly in proteins with high identity (>50%) to natural proteins. In the directed evolution section, we will discuss the applicability of UniRep for protein engineering.

Convolutional neural networks (CNNs) have been an architecture that was mainly associated with image classification, but they also can be used analogously to encoder models in natural language processing (NLP). For this application, mainly diluting CNNs have been used, which amplify the feature extraction of the con-

Cite this article as *Cold Spring Harb Perspect Biol* doi: 10.1101/cshperspect.a041471

Table 2. A nonexhaustive list of recently released non-transformer generative models[a]

Model name	Architecture	Description	Reference
UniRep	LSTM	Learning of representation	Alley et al. 2019
Low-N Unirep	LSTM + linear model	Low-N protein fitness prediction	Biswas et al. 2021
ProteinGAN	GAN	MDH protein family (local)	Repecka et al. 2021
ProteoGAN	GAN	Conditional on GO-terms	Kucera et al. 2022
Antibody-GAN	GAN	Humanoid antibody design	Amimeur et al. 2020
FBGAN	GAN	Peptide design	Gupta and Zou 2019
MSA VAE/AR- VAE	VAE	Focus on protein family design (local)	Hawkins-Hooker et al. 2021
ProteinVAE	VAE	Generation of synthetic viral vector serotypes	Lyu et al. 2023
PepVAE	VAE	Focus on peptides paired with antimicrobial activity prediction	Dean et al. 2021
CARP	CNN	Pretrained on UniREF-50 to encode	Yang et al. 2022a
MIF/MIF-ST	GNN	Inverse folding with structural and sequence input	Yang et al. 2022b

[a]We indicate models that do not generate sequences in a global fashion.

volution to condense the information into a continuous numerical representation. Using convolutions has the major advantage that the necessary computer resources scale linearly with sequence length compared to quadratically scaling in transformer models. In this sense, Yang et al. (2022a) implemented convolutional autoencoding representations of proteins (CARP), a CNN pretrained with the semi-supervised masked language modeling objective of reconstructing sequences. They showed that their model exhibits state-of-the-art results in comparison to large language models in zero-shot as well as in downstream tasks after fine-tuning with specialized heads.

To tackle the inverse folding problem, Yang et al. (2022b) trained MIF (masked inverse folding), a language model trained on only 19k structures and sequences from the CATH database in a process termed "structure-conditioned masked language model (MLM) pre-training or masked inverse folding." As the model's core, they used a language model parameterized as a structured graph neural network. In the pre-training, the structure is fed to the model in the form of dihedral angles, planar angles, and Euclidean residue distances, and the model learns to reconstruct corrupted sequences conditioned on the input backbone structure. MIF outperformed other models in the sequence recovery objective. To extend the CATH data set

and improve the pre-training task performance, the authors developed a second version of MIF called MIF-ST, using CARP sequence embeddings. This sequence transfer led to a significantly higher sequence recovery rate than solely using the data from CATH.

ProteinGAN (Repecka et al. 2021) is a generative adversarial network with 45 layers spanning over 60 million trainable parameters. The general strategy of using GANs for sequence design is to first generate sequences in the local/global sequence space by a generator and then let a pre-trained discriminator decide whether the sequence is natural or generated (Fig. 3A). This enforces the model to generate sequences in the area of natural clusters. In a first work, the authors assessed the capability of ProteinGAN to generate catalytically active malate dehydrogenases (MDHs) locally, discovering fully functional MDHs even for identities as low as 66% to the natural space (Repecka et al. 2021). In subsequent work, by testing three distinct generative models in the generation of MDH and superoxide dismutases, the authors proposed computational metrics for predicting in vitro activity that could help in the selection process of active enzymes for experimental characterization (Johnson et al. 2023), demonstrating the potential of these models to generate highly diverse functional proteins with natural-like physical properties.

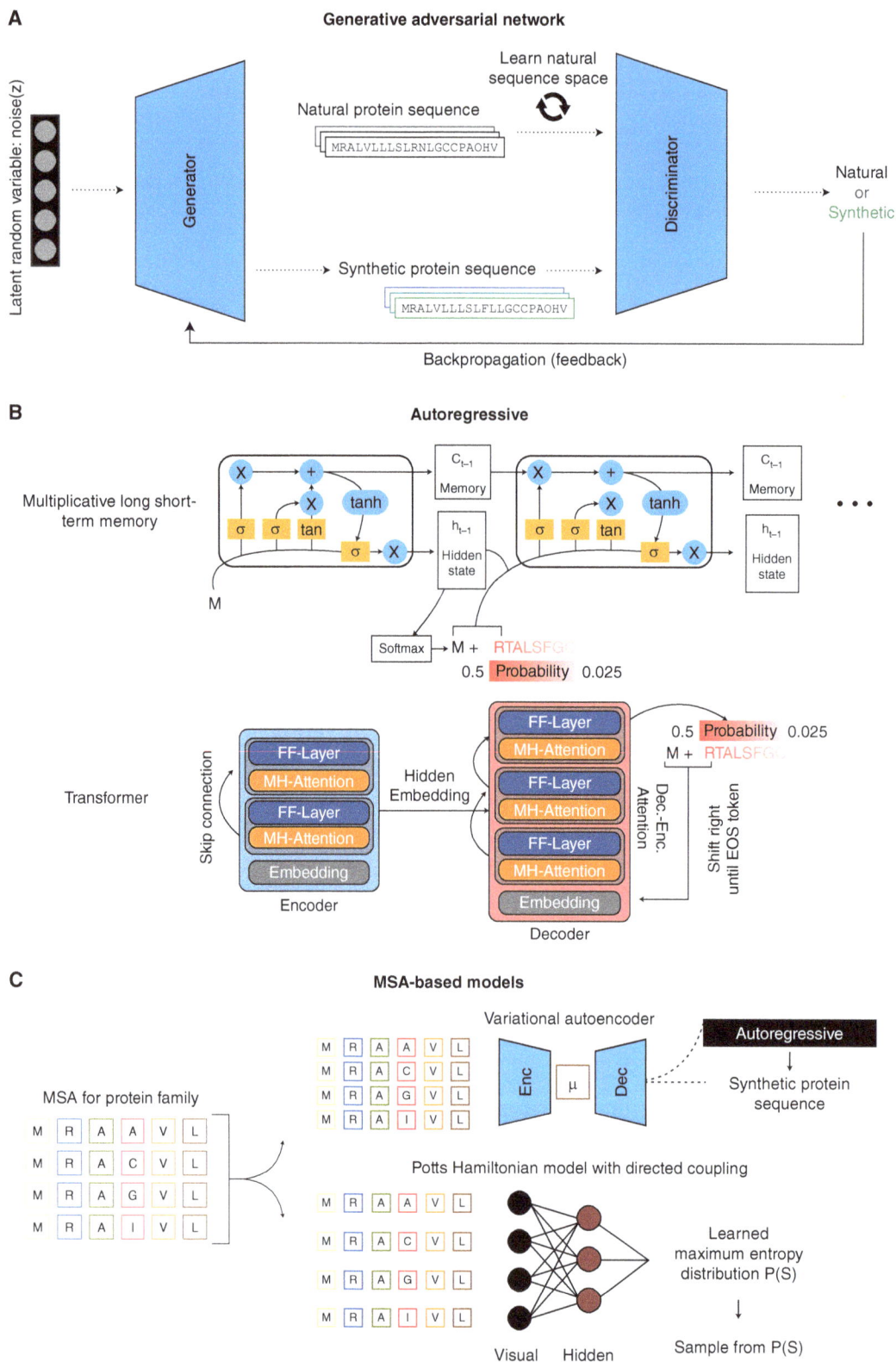

Figure 3. Approaches to global sequence design. (*A*) Generative adversarial networks (GANs) as generators from noise with a selection process through the discriminator. (*B*) Multiplicative long short-term memory (mLSTM) and transformer as autoregressive models predicting iteratively the next token based on the previous ones. (*C*) Models based on multiple sequence alignments of homologous proteins. (EOS) End of sentence.

This approach was recently adopted by ProteoGAN (Kucera et al. 2022), which conditions the generation of de novo proteins with labels from the hierarchically sorted gene ontology annotations. The method trains with conditional labels and focuses on a global data set allowing for a general approach to the problem of de novo protein design. Amimeur and colleagues presented Antibody-GAN as an approach capable of generating large and diverse libraries of novel antibodies mimicking the somatic response from humans. The method designs variants with improved stability and developability, allowing the control of properties to find suitable therapeutic antibodies (Amimeur et al. 2020). In another application of GANs, FBGAN is a feedback-loop mechanism that generates sequences with a prompted function of the gene product. It was optimized to produce α-helical antimicrobial peptides, showing the opportunity to go from global to local peptide sequence generation through fine-tuning (Gupta and Zou 2019).

Another approach used in the past involves using MSAs to generate sequences (Fig 3C). This method does not result in a complete de novo design of proteins but rather an interpolation between the variations in the MSA. The generated proteins tend to be more functional than those from a de novo design approach, as they have only minor deviations in the variable amino acid positions from natural proteins. Two different architectures have gained attention in this area: variational autoencoders (VAEs) and Potts Hamiltonian maximum entropy models, which are a type of restricted Boltzmann machines. VAEs can recognize higher-order epistasis that is often absent in rational design approaches, a crucial capability for navigating the protein fitness landscape.

Examples of VAEs with the general capability to draw new sequences from their learned distribution are MSA VAE (Hawkins-Hooker et al. 2021) and deep sequence (Riesselman et al. 2018). Although these two models were trained with the objective of local generation, they also exemplify the potential of this architecture for global sequence generation. Hawkins-Hooker et al. (2021) not only showed in their work that sequence generation with align-

ment-based VAEs is possible but also that it is possible without aligned input sequences by introducing an AR component to variational autoencoders. Both the MSA VAE and the AR-VAE models could capture and integrate learned physicochemical properties in their sequence generation. However, the authors emphasize that the models trained on raw sequence inputs miss more long-distance dependencies in the mutational patterns and 3D structure. To experimentally validate these architectures, they trained both models on sequences of the luciferase protein family. By this, they were in both cases able to generate novel proteins with an improved solubility compared to wild-type.

Also related to VAEs, but in applications on a more complex protein family with longer sequences, ProteinVAE was developed to generate synthetic viral vector serotypes without epitopes for preexisting antibodies (Lyu et al. 2023). This VAE used the knowledge learned by a previous model (ProtTrans; Elnaggar et al. 2022) but trained in a natural data set of adenovirus hexon sequences. ProteinVAE learns intrinsic relationships of long protein sequences and generates diverse proteins with patterns similar to the natural ones. The latent space exploited by ProteinVAE is structured and can be used to facilitate the selection of distant sequences to be tested in the laboratory.

VAEs also have been used to assist in the generation of bioactive peptides (Das et al. 2018; Dean and Walper 2020; Dean et al. 2021). From them, PepVAE (Dean et al. 2021) is the newest semi-supervised VAE model that designs active novel antimicrobial peptide sequences. By coupling it with antimicrobial activity prediction, PepVAE demonstrates the possibility of VAEs to expand the diversity and functionality of peptides.

Potts Hamiltonian models (Fig. 3C) seek to learn the covariations between any given position in a protein. This covariation matrix correlates and implies, in many cases, the structural contacts inside the given protein. Moreover, these types of models map sequence variations to their corresponding prevalence. They do this by transforming the statistical Potts energy prediction with the Boltzmann distribution. Here, the statistical Potts energy prediction is proportional to the Gibbs free energy (i.e., the model will output the likelihood that a certain sequence belongs to the distribution defined by an MSA). A recent study (McGee et al. 2021) developed a benchmark to examine the generative capacity of probabilistic protein sequence models trained on MSAs and found that Mi3, a Potts Hamiltonian model with only pairwise interaction terms (direct coupling), outperformed various VAEs as well as side-independent models.

Many frameworks using Potts models as efficiently as possible were developed and tested. For example, adabmDCA (Muntoni et al. 2021) uses a direct coupling analysis to capture amino acid propensities and other relevant properties, showing promising results in a local sequence-generation task that potentially would also be scalable to a global generative capacity. Further development of the algorithm proposed for the adabmDCA framework (Barrat-Charlaix et al. 2021) addresses an important aspect of generative modeling by achieving a parameter reduction to allow reducing the computational cost of generative modeling. In a related work, bmDCA is a model that generates sequences based on pairwise interactions among a large and diverse multiple sequence alignment (Russ et al. 2020). After being tested in the chorismate mutase, artificial enzymes recapitulate the catalytic parameters from natural proteins, demonstrating the potential of evolution-based models to find improved proteins.

A recent study (Shin et al. 2021) has shown that alignment-free AR models also have the ability to learn mutational effects and epistasis and subsequently use this knowledge to design sequences and predict mutational effects. In summary, MSA-based models are powerful tools to capture pairwise or higher-order mutational variation with high confidence of functional outcomes with comparatively few parameters, but the input MSAs are computationally and data intensive. Last, the realm of denoising diffusion probabilistic models for global sequence design has yet to be explored, but the rapid progress of this architecture especially in protein backbone design holds a promising future. Nevertheless, two papers have already explored and proposed the possibility to design sequences with this architecture tackling separately the inverse protein-folding problem and the pairwise design of amino acid sequences as well as structures from an initially random sequence (Lisanza et al. 2023; Yi et al. 2023).

ASSISTANCE IN DIRECTED EVOLUTION TECHNIQUES AND PROTEIN DESIGN

Directed evolution techniques have tremendously benefited from the advances in DL, including unsupervised learning. Given the growing literature in this field, we include several recent works in this area, even if, at times, they do not focus on global but local generation. Directed evolution has been a successful method for creating new biological molecules, such as scaffolds, enzymes, or antibodies, by mimicking the process of natural evolution. It involves creating a population of molecules with a desired property or function, selecting the best ones that perform that function, and then using genotypic and microbiological techniques to create new generations of molecules with even better performance (Fig. 4; Kuchner and Arnold 1997). Directed evolution, as well as generative models, is related to the general concept of "protein sequence space" or "fitness landscapes," a fundamental idea proposed in early 1970 by Maynard Smith to refer to the vast number of possible meaningful and nonmeaningful sequences that can form a protein based on the combination of amino acids, describing the relationship between protein sequences and their functions (Maynard Smith 1970; Ogbunugafor 2020).

This idea suggests that the possible sequences of a protein can be visualized as points on a landscape, with the height of each point representing the functional fitness of the corresponding sequence. Sequences with higher fitness would be located at higher points on the landscape, while less functional sequences would appear at the landscape valleys (Fig. 4B). However, it is known that many of the possible protein sequences and evolutionary trajectories will not yield folded and stable three-dimensional structures and, therefore, not any relevant function or fitness (Arnold et al. 2001;

Peisajovich and Tawfik 2007; Wagner 2008; Romero and Arnold 2009; Kondrashov and Kondrashov 2015; Starr and Thornton 2016; Wheeler et al. 2016). Consequently, only a reduced subset of sequences will form stable and functional proteins. This subset is known as the "protein sequence space of life" in the case of sequences found in nature and "post-evolution or dark matter sequence space" for sequences that can be successfully designed (Woolfson et al. 2015).

Meanwhile, although traditional directed evolution (TDE) approaches have yielded successful outcomes with high applicability in different fields (MacBeath et al. 1998; Shaner et al. 2004; Aharoni et al. 2005; Fasan et al. 2007; Giger et al. 2013; Hammer et al. 2017; Kan et al. 2017; Zhang et al. 2022), this method has some limitations and challenges (as described in Vidal et al. 2023), such as (1) it is a time-consuming and expensive process, requiring numerous rounds of mutagenesis-screening-selection; (2) the genetic diversity of a population may be limited, reducing the chance of finding a suitable variant; (3) the outcome can be unpredictable, making difficult to control the final variants; and (4) the evolved proteins may not have the desired specificity, leading to off-target effects or reduced efficacy. For these reasons, generative models can significantly help to improve protein sequence exploration and therefore get faster and more accurate sequence sampling, as has been extensively discussed in a previous review (Wittmann et al. 2021a). Several of these models, although not all of them, focus on a global sequence generation; in this section, we recapitulate some of the most relevant from the last few years.

Consecutive approaches by the Arnold's group have demonstrated the relevance of machine learning-assisted directed evolution (MLDE) to navigate protein epistatic fitness landscapes in silico. These approaches, in general, are implemented by training a machine learning (ML) model with a small-size experimentally characterized library (i.e., each combinatorial variant used in the training data set is labeled with a known experimental fitness (enzymatic activity, stability, function, etc.). Then

the model can be used to predict and explore an entire combinatorial space with commutated positions (Fig. 4), which enables exploring more sequences in subsequent analysis rounds (Wittmann et al. 2021b). Using this strategy, a human protein G domain (GB1-binding protein) was used to validate a model fitness landscape by exploring multiple positions simultaneously. Then, to determine if MLDE can find improved variants efficiently, a putative nitric oxide dioxygenase was engineered to evolve its enantiodivergent enzyme activity (Wu et al. 2019). Generated variants demonstrated that sampling in silico a larger sequence space of a specific protein family favors rapidly evolving parent enzymes to selectively perform the desired activity or function, considerably reducing the size of libraries that should be experimentally tested in the laboratory to obtain similar results. Details on the steps and requirements needed to build MLDE models for protein engineering applications are described in a previous review by Yang and colleagues (2019) using two case studies that exemplify relevant biological problems: the increase of thermostability of a protein (cytochrome P450) and maximize the productivity of an enzyme (halohydrin dehalogenase).

In a related work also training ML models directly on experimental data to explore a full diversity landscape, the adeno-associated virus 2 capsid protein was used as an example system to generate viable engineered capsids varying only in short sequence regions (Bryant et al. 2021). Employing complete, random, and additive sampling strategies to generate experimental libraries, these data sets were used to compare the performance of three model architectures. CNNs and recurrent neural networks were the more successful at deep diversification, generating functional variants that are viable for assembly and packing DNA payloads, suggesting that by using small, simple, and unbiased training sets, generative models can predict variable, robust, and functional variants.

Building on the concept of LSTMs, which preceded transformers as described in the previous section, Biswas and colleagues proposed a low-N approach that combines the global knowledge of fundamental functional features learned by the

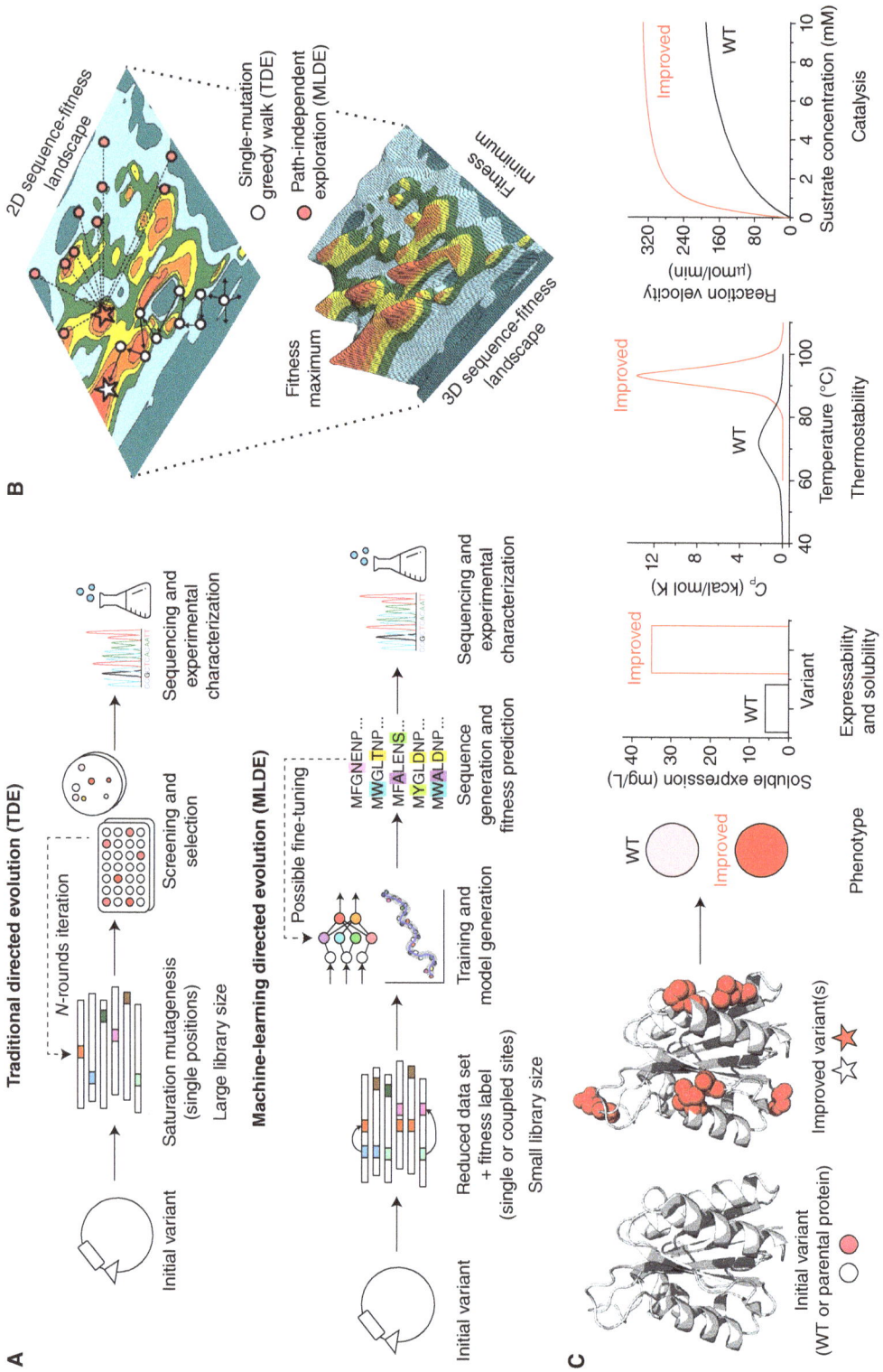

Figure 4. (See following page for legend.)

Cite this article as *Cold Spring Harb Perspect Biol* doi: 10.1101/cshperspect.a041471

language model UniRep with experimentally characterized proteins of a desired target. Applying this archetype idea, after fine-tuning the UniRep model with two target families, green fluorescent protein and TEM-1 β-lactamase, and characterizing a small number of random mutants of wild-type proteins, it was possible to perform MLDE and create a protein fitness landscape that was used to select candidates with an improved function. From the generated variants, they demonstrated that the candidates are fully functional and optimized proteins (Biswas et al. 2021). This work showed that supervised learning can simplify the sequence search by eliminating most of the nonfunctional variants and, simultaneously, the possibility of exploiting molecular epistasis and giving prominent applications to directed evolution.

Using a transformer-based architecture, Wu et al. generated a model capable of generating functional signal peptides (SPs) that can be used to perform specific protein secretion in living organisms (Wu et al. 2020). This model, trained and validated by using SP-protein pairs from all domains of life, generated novel and diverse SP sequences which, when used on in vivo expression of SP-protein pairs for four families, can produce not only functional enzymes but also proteins that exhibit activity comparable to natural SPs.

Tied to in silico analysis of sequence-fitness landscapes, using a latent variable model with nonlinear dependencies to capture higher-order and context-dependent constraints, DeepSequence is a probabilistic unsupervised model able to predict the effect of mutations for biological sequence families (Riesselman et al. 2018).

This model exhibits high accuracy in most of the data sets as well or better among those of site-independent or pairwise-interaction models. Using the β-lactamase family as an example, DeepSequence revealed the latent organization of this sequence family by learning interpretable structure for both macro variation and phylogeny; this information could be applied to other families to explore new regions of sequence space.

In another work to predict new sequences, Lu et al. (2022) used a three-dimensional self-supervised CNN (MutCompute) to identify stabilizing mutations of wild-type and previously engineered PETases. The best variant, FAST-PETase (functional, active, stable, and tolerant PETase), showed higher activity both at 40°C and 50°C and presented only five mutations compared to wild-type PETase. This new enzyme can degrade PET fragments embedded in textile fabrics, depolymerize pretreated bottle films, and consume nonphysically disrupted melted plastic pucks from entire bottles.

On the other hand, performing a model-guided sequence generation with two language models, ESM-1b (Rives et al. 2021) and ESM-1v (Meier et al. 2021), Hie et al. (2023) reported the affinity maturation of clinically relevant human antibodies. After experimentally screening only a small number of variants for each type, these new antibodies showed a maturation higher for both already highly mature wild-type sequences and unmatured antibodies, representing another example of how MLDE can generate quickly diverse and functional proteins.

Finally, Hsu and colleagues (2022) assessed previously published ML methods for protein fit-

---

Figure 4. Comparison between traditional directed evolution (TDE) and machine learning-assisted directed evolution (MLDE) approaches. (*A,B*) In TDE methods, a large library size is needed to explore broad sequence space and eventually reach variants close to the maximum performance (white star) in the sequence-fitness landscape, a process commonly known as single-mutation greedy walk (solid arrows in panel *B*). On the other hand, MLDE models use focused training experimental data libraries (small size) with multiple combinatorial positions (labeled or nonlabeled) following a path-independent exploration in the fitness landscape (dotted lines in panel *B*). Then, after training and model generation, it is possible to predict new improved variants (red star) and reach fitness maximums more easily (red zones in the landscape), reducing the necessity of a large experimental characterization. (*C*) Newly generated proteins (from both TDE and MLDE) are then analyzed to compare differences among the initial variant and the improved ones on biochemical and biophysical features such as phenotype (e.g., fluorescence, antibiotic resistance, using color-modifying reporter genes, etc.), expressability/solubility, thermostability, or catalytic activity, among others. (WT) Wild-type.

ness predictions (many of them discussed herein) and developed a combined approach using evolutionary and assay-labeled data. This simple baseline approach makes use of supervised data to enhance, with a linear regression model on site-specific amino acid features, the evolutionary density models such as a hidden Markov model (HMM) (Shihab et al. 2013), a Potts model (EV-mutation [Hopf et al. 2017]), a VAE (Deep-Sequence [Riesselman et al. 2018]), an LSTM (UniRep [Alley et al. 2019]), and a transformer (ESM-1b [Rives et al. 2021]). Augmented models generally performed better at ranking mutational effects overall, making the augmented Deep-Sequence VAE the most effective for this purpose.

All these examples demonstrate the versatility and potential of generative models to explore and access remote regions of protein sequence space, not only reaching functionalities similar to natural proteins but also unraveling the latent protein space toward sequences with new properties, efficiencies, and functions.

CONCLUDING REMARKS

Protein design is undergoing a rapid transformation due to impressive advances in the field of AI. In particular, the use of architectures that excel in other areas, such as computer vision and NLP, is proving to be highly successful in generating sequences in previously inaccessible regions of the protein space. However, the field still faces limitations and challenges. For example, only a few of the methods presented in this work are comprehensively evaluated experimentally. Often, the architectures are directly "plugged in" to the protein realm without specific modifications to this domain. Last, it is unknown whether training in the natural space may hamper the exploration of out-of-domain areas. In this work, however, we provide an overview of these advances in sequence generation and their potential applications in directed evolution. We also highlight the impressive pace they have achieved in the last couple of years. While there are limitations, we believe this progress marks an optimistic outlook, with the potential for designing à la carte protein functions and new-to-nature enzymes becoming realistic in the near future.

REFERENCES

Aharoni A, Gaidukov L, Khersonsky O, Gould SM, Roodveldt C, Tawfik DS. 2005. The "evolvability" of promiscuous protein functions. *Nat Genet* **37:** 73–76. doi:10.1038/ng1482

Alley EC, Khimulya G, Biswas S, AlQuraishi M, Church GM. 2019. Unified rational protein engineering with sequence-based deep representation learning. *Nat Methods* **16:** 1315–1322. doi:10.1038/s41592-019-0598-1

Amimeur T, Shaver JM, Ketchem RR, Taylor JA, Clark RH, Smith J, Van Citters D, Siska CC, Smidt P, Sprague M, et al. 2020. Designing feature-controlled humanoid antibody discovery libraries using generative adversarial networks. bioRxiv doi.org/10.1101/2020.04.12.024844

Anand N, Huang P. 2018. Generative modeling for protein structures. In *Advances in Neural Information Processing Systems 3. NeurIPS Proceedings*. 32nd Conference on Neural Information Processing Systems (NeurIPS 2018). Montréal, Canada.

Arnold FH, Wintrode PL, Miyazaki K, Gershenson A. 2001. How enzymes adapt: lessons from directed evolution. *Trends Biochem Sci* **26:** 100–106. doi:10.1016/s0968-0004(00)01755-2

Bahdanau D, Cho K, Bengio Y. 2014. Neural machine translation by jointly learning to align and translate. arXiv doi:10.48550/arXiv.1409.0473

Barrat-Charlaix P, Muntoni AP, Shimagaki K, Weigt M, Zamponi F. 2021. Sparse generative modeling via parameter reduction of Boltzmann machines: application to protein-sequence families. *Phys Rev E* **104:** 024407. doi:10.1103/PhysRevE.104.024407

Biswas S, Khimulya G, Alley EC, Esvelt KM, Church GM. 2021. Low-*N* protein engineering with data-efficient deep learning. *Nat Methods* **18:** 389–396. doi:10.1038/s41592-021-01100-y

Brandes N, Ofer D, Peleg Y, Rappoport N, Linial M. 2022. ProteinBERT: a universal deep-learning model of protein sequence and function. *Bioinformatics* **38:** 2102–2110. doi:10.1093/bioinformatics/btac020

Bryant DH, Bashir A, Sinai S, Jain NK, Ogden PJ, Riley PF, Church GM, Colwell LJ, Kelsic ED. 2021. Deep diversification of an AAV capsid protein by machine learning. *Nat Biotechnol* **39:** 691–696. doi:10.1038/s41587-020-00793-4

Chen B, Cheng X, Gengyang L, Li S, Zeng X, Wang B, Jing G, Liu C, Zeng A, Dong Y, et al. 2023. xTrimoPGLM: unified 100B-scale pre-trained transformer for deciphering the language of protein. bioRxiv doi:10.1101/2023.07.05.547496

Das P, Wadhawan K, Chang O, Sercu T, Santos CD, Riemer M, Chenthamarakshan V, Padhi I, Mojsilovic A. 2018. PepCVAE: semi-supervised targeted design of antimicrobial peptide sequences. arXiv doi:10.48550/arXiv.1810.07743

Dean SN, Walper SA. 2020. Variational autoencoder for generation of antimicrobial peptides. *ACS Omega* **5:** 20746–20754. doi:10.1021/acsomega.0c00442

Dean SN, Alvarez JAE, Zabetakis D, Walper SA, Malanoski AP. 2021. PepVAE: variational autoencoder framework for antimicrobial peptide generation and activity prediction. *Front Microbiol* **12:** 725727. doi:10.3389/fmicb.2021 .725727

Devlin J, Chang MW, Lee K, Toutanova K. 2018. BERT: pre-training of deep bidirectional transformers for language understanding. arXiv doi:10.48550/arXiv.1810.04805

Elnaggar A, Heinzinger M, Dallago C, Rehawi G, Wang Y, Jones L, Gibbs T, Feher T, Angerer C, Steinegger M, et al. 2022. Prottrans: toward understanding the language of life through self-supervised learning. *IEEE Trans Pattern Anal Mach Intell* **44:** 7112–7127. doi:10.1109/TPAMI .2021.3095381

Elnaggar A, Essam H, Salah-Eldin W, Moustafa W, Elker-dawy M, Rochereau C, Rost B. 2023. Ankh: optimized protein language model unlocks general-purpose model-ling. arXiv doi:10.48550/arXiv.2301.06568

Fasan R, Chen MM, Crook NC, Arnold FH. 2007. Engi-neered alkane-hydroxylating cytochrome P450BM3 ex-hibiting nativelike catalytic properties. *Angew Chem Int Ed* **46:** 8414–8418. doi:10.1002/anie.200702616

Ferruz N, Heinzinger M, Akdel M, Goncearenco A, Naef L, Dallago C. 2022a. From sequence to function through structure: deep learning for protein design. *Comput Struct Biotechnol J* **21:** 238–250. doi:10.1016/j.csbj.2022.11.014

Ferruz N, Schmidt S, Höcker B. 2022b. ProtGPT2 is a deep unsupervised language model for protein design. *Nat Commun* **13:** 4348. doi:10.1038/s41467-022-32007-7

Gainza P, Nisonoff HM, Donald BR. 2016. Algorithms for protein design. *Curr Opin Struct Biol* **39:** 16–26. doi:10 .1016/j.sbi.2016.03.006

Geffen Y, Ofran Y, Unger R. 2022. Distilprotbert: a distilled protein language model used to distinguish between real proteins and their randomly shuffled counterparts. *Bio-informatics* **38:** ii95–ii98. doi:10.1093/bioinformatics/ btac474

Giger L, Caner S, Obexer R, Kast P, Baker D, Ban N, Hilvert D. 2013. Evolution of a designed retro-aldolase leads to complete active site remodeling. *Nat Chem Biol* **9:** 494–498. doi:10.1038/nchembio.1276

Gupta A, Zou J. 2019. Feedback GAN for DNA optimizes protein functions. *Nat Mach Intell* **1:** 105–111. doi:10 .1038/s42256-019-0017-4

Hammer SC, Kubik S, Watkins E, Huang S, Minges H, Arnold FH. 2017. Anti-Markovnikov alkene oxidation by metal-oxo–mediated enzyme catalysis. *Science* **358:** 215–218. doi:10.1126/science.aao1482

Hawkins-Hooker A, Depardieu F, Baur S, Couairon G, Chen A, Bikard D. 2021. Generating functional protein variants with variational autoencoders. *PLOS Comput Biol* **17:** e1008736. doi:10.1371/journal.pcbi.1008736

Hesslow D, Zanichelli N, Notin P, Poli I, Marks D. 2022. RITA: a study on scaling up generative protein sequence models. arXiv doi:10.48550/arXiv.2205.05789

Hie BL, Shanker VR, Xu D, Bruun TUJ, Weidenbacher PA, Tang S, Wu W, Pak JE, Kim PS. 2023. Efficient evolution of human antibodies from general protein language mod-els. *Nat Biotechnol* doi:10.1038s41587-023-01763-2

Hopf TA, Ingraham JB, Poelwijk FJ, Schärfe CPI, Springer M, Sander C, Marks DS. 2017. Mutation effects predicted from sequence co-variation. *Nat Biotechnol* **35:** 128–135. doi:10.1038/nbt.3769

Hsu C, Nisonoff H, Fannjiang C, Listgarten J. 2022. Learning protein fitness models from evolutionary and assay-la-beled data. *Nat Biotechnol* **40:** 1114–1122. doi:10.1038/ s41587-021-01146-5

Huang PS, Boyken SE, Baker D. 2016. The coming of age of de novo protein design. *Nature* **537:** 320–327. doi:10 .1038/nature19946

Johnson SR, Fu X, Viknander S, Goldin C, Monaco S, Ze-lezniak A, Yang KK. 2023. Computational scoring and experimental evaluation of enzymes generated by neural networks. bioRxiv doi:10.1101/2023.03.04.531015

Kan SBJ, Huang X, Gumulya Y, Chen K, Arnold FH. 2017. Genetically programmed chiral organoborane synthesis. *Nature* **552:** 132–136. doi:10.1038/nature24996

Keskar NS, McCann B, Varshney LR, Xiong C, Socher R. 2019. CTRL: a conditional transformer language model for controllable generation. arXiv doi:10.48550/arXiv.1909 .05858

Kondrashov DA, Kondrashov FA. 2015. Topological fea-tures of rugged fitness landscapes in sequence space. *Trends Genet* **31:** 24–33. doi:10.1016/j.tig.2014.09.009

Korendovych IV, DeGrado WF. 2020. De novo protein de-sign, a retrospective. *Q Rev Biophys* **53:** e3. doi:10.1017/ S0033583519000131

Kucera T, Togninalli M, Meng-Papaxanthos L. 2022. Con-ditional generative modeling for de novo protein design with hierarchical functions. *Bioinformatics* **38:** 3454–3461. doi:10.1093/bioinformatics/btac353

Kuchner O, Arnold FH. 1997. Directed evolution of enzyme catalysts. *Trends Biotechnol* **15:** 523–530. doi:10.1016/ S0167-7799(97)01138-4

Lin Z, Akin H, Rao R, Hie B, Zhu Z, Lu W, Smetanin N, Verkuil R, Kabeli O, Shmueli Y, et al. 2023. Evolutionary-scale prediction of atomic-level protein structure with a language model. *Science* **379:** 1123–1130. doi:10.1126/sci ence.ade2574

Lisanza SL, Gershon JM, Tipps S, Arnoldt L, Hendel S, Sims JN, Li X, Baker D. 2023. Joint generation of protein sequence and structure with RoseTTAFold se-quence space diffusion. bioRxiv doi:10.1101/2023.05.08 .539766

Lu H, Diaz DJ, Czarnecki NJ, Zhu C, Kim W, Shroff R, Acosta DJ, Alexander BR, Cole HO, Zhang Y, et al. 2022. Machine learning-aided engineering of hydrolases for PET depolymerization. *Nature* **604:** 662–667. doi:10 .1038/s41586-022-04599-z

Lyu S, Sowlati-Hashjin S, Garton M. 2023. ProteinVAE: var-iational autoencoder for translational protein design. bioRxiv doi:10.1101/2023.03.04.531110

MacBeath G, Kast P, Hilvert D. 1998. Redesigning enzyme topology by directed evolution. *Science* **279:** 1958–1961. doi:10.1126/science.279.5358.1958

Madani A, McCann B, Naik N, Keskar NS, Anand N, Eguchi RR, Huang P-S, Socher R. 2020. Progen: language mod-eling for protein generation. arXiv doi:10.48550/arXiv .2004.03497

Madani A, Krause B, Greene ER, Subramanian S, Mohr BP, Holton JM, Olmos JL, Xiong C, Sun ZZ, Socher R, et al. 2023. Large language models generate functional protein sequences across diverse families. *Nat Biotechnol* doi:10 .1038/s41587-022-01618-2

Maynard Smith J. 1970. Natural selection and the concept of a protein space. *Nature* 225: 563–564. doi:10.1038/225563a0

McGee F, Hauri S, Novinger Q, Vucetic S, Levy RM, Carnevale V, Haldane A. 2021. The generative capacity of probabilistic protein sequence models. *Nat Commun* 12: 6302. doi:10.1038/s41467-021-26529-9

Meier J, Rao R, Verkuil R, Liu J, Sercu T, Rives A. 2021. Language models enable zero-shot prediction of the effects of mutations on protein function. bioRxiv doi:10.1101/2021.07.09.450648

Moffat L, Kandathil SM, Jones DT. 2022. Design in the DARK: learning deep generative models for de novo protein design. bioRxiv doi:10.1101/2022.01.27.478087

Munsamy G, Lindner S, Lorenz P, Ferruz N. 2022. ZymCTRL: a conditional language model for the controllable generation of artificial enzymes. In *Machine Learning for Structural Biology Workshop, NeurIPS 2022*. New Orleans, LA.

Muntoni AP, Pagnani A, Weigt M, Zamponi F. 2021. adabmDCA: adaptive Boltzmann machine learning for biological sequences. *BMC Bioinformatics* 22: 528. doi:10.1186/s12859-021-04441-9

Nambiar A, Heflin M, Liu S, Maslov S, Hopkins M, Ritz A. 2020. Transforming the language of life: transformer neural networks for protein prediction tasks. In *Proceedings of the 11th ACM International Conference on Bioinformatics, Computational Biology and Health Informatics, BCB '20*, pp. 1–8. doi:10.1145/3388440.3412467

Nijkamp E, Ruffolo J, Weinstein EN, Naik N, Madani A. 2022. Progen2: exploring the boundaries of protein language models. arXiv doi:10.48550/arXiv.2206.13517

Notin P, Dias M, Frazer J, Hurtado JM, Gomez AN, Marks D, Gal Y. 2022. Tranception: protein fitness prediction with autoregressive transformers and inference-time retrieval. In *Proceedings of the 39th International Conference on Machine Learning*, pp. 16990–17017.

Ogbunugafor CB. 2020. A reflection on 50 years of John Maynard Smith's "Protein Space." *Genetics* 214: 749–754. doi:10.1534/genetics.119.302764

Pan X, Kortemme T. 2021. Recent advances in de novo protein design: principles, methods, and applications. *J Biol Chem* 296: 100558. doi:10.1016/j.jbc.2021.100558

Peisajovich SG, Tawfik DS. 2007. Protein engineers turned evolutionists. *Nat Methods* 4: 991–994. doi:10.1038/nmeth1207-991

Radford A, Narasimhan K. 2018. Improving language understanding by generative pre-training. https://www.semanticscholar.org/paper/Improving-Language Understanding-by-Generative-Radford Narasimhan/cd18800a0fe0b668a1cc19f2ec95b5003d0a5035

Rao R, Bhattacharya N, Thomas N, Duan Y, Chen X, Canny J, Abbeel P, Song YS. 2019. Evaluating protein transfer learning with TAPE. arXiv doi:10.48550/arXiv.1906.08230

Rao RM, Liu J, Verkuil R, Meier J, Canny J, Abbeel P, Sercu T, Rives A. 2021. MSA transformer. In *Proceedings of the 38th International Conference on Machine Learning 139*, pp. 8844–8856.

Repecka D, Jauniskis V, Karpus L, Rembeza E, Rokaitis I, Zrimec J, Poviloniene S, Laurynenas A, Viknander S,

Abuajwa W, et al. 2021. Expanding functional protein sequence spaces using generative adversarial networks. *Nat Mach Intell* 3: 324–333. doi:10.1038/s42256-021-00310-5

Riesselman AJ, Ingraham JB, Marks DS. 2018. Deep generative models of genetic variation capture the effects of mutations. *Nat Methods* 15: 816–822. doi:10.1038/s41592-018-0138-4

Rives A, Meier J, Sercu T, Goyal S, Lin Z, Liu J, Guo D, Ott M, Zitnick CL, Ma J, et al. 2021. Biological structure and function emerge from scaling unsupervised learning to 250 million protein sequences. *Proc Natl Acad Sci* 118: e2016239118. doi:10.1073/pnas.2016239118

Romero PA, Arnold FH. 2009. Exploring protein fitness landscapes by directed evolution. *Nat Rev Mol Cell Biol* 10: 866–876. doi:10.1038/nrm2805

Romero-Romero S, Kordes S, Michel F, Höcker B. 2021. Evolution, folding, and design of TIM barrels and related proteins. *Curr Opin Struct Biol* 68: 94–104. doi:10.1016/j.sbi.2020.12.007

Russ WP, Figliuzzi M, Stocker C, Barrat-Charlaix P, Socolich M, Kast P, Hilvert D, Monasson R, Cocco S, Weigt M, et al. 2020. An evolution-based model for designing chorismate mutase enzymes. *Science* 369: 440–445. doi:10.1126/science.aba3304

Shaner NC, Campbell RE, Steinbach PA, Giepmans BNG, Palmer AE, Tsien RY. 2004. Improved monomeric red, orange and yellow fluorescent proteins derived from *Discosoma* sp. red fluorescent protein. *Nat Biotechnol* 22: 1567–1572. doi:10.1038/nbt1037

Shihab HA, Gough J, Cooper DN, Stenson PD, Barker GLA, Edwards KJ, Day INM, Gaunt TR. 2013. Predicting the functional, molecular, and phenotypic consequences of amino acid substitutions using hidden Markov models. *Hum Mutat* 34: 57–65. doi:10.1002/humu.22225

Shin J-E, Riesselman AJ, Kollasch AW, McMahon C, Simon E, Sander C, Manglik A, Kruse AC, Marks DS. 2021. Protein design and variant prediction using autoregressive generative models. *Nat Commun* 12: 2403. doi:10.1038/s41467-021-22732-w

Starr TN, Thornton JW. 2016. Epistasis in protein evolution. *Protein Sci* 25: 1204–1218. doi:10.1002/pro.2897

Vaswani A, Shazeer N, Parmar N, Uszkoreit J, Jones L, Gomez AN, Kaiser L, Polosukhin I. 2017. Attention is all you need. arXiv doi:10.48550/arXiv.1706.03762

Verkuil R, Kabeli O, Du Y, Wicky BIM, Milles LF, Dauparas J, Baker D, Ovchinnikov S, Sercu T, Rives A. 2022. Language models generalize beyond natural proteins. bioRxiv doi:10.1101/2022.12.21.521521

Vidal LS, Isalan M, Heap JT, Ledesma-Amaro R. 2023. A primer to directed evolution: current methodologies and future directions. *RSC Chem Biol* 4: 271–291. doi:10.1039/d2cb00231k

Wagner A. 2008. Neutralism and selectionism: a network-based reconciliation. *Nat Rev Genet* 9: 965–974. doi:10.1038/nrg2473

Wang J, Lisanza S, Juergens D, Tischer D, Watson JL, Castro KM, Ragotte R, Saragovi A, Milles LF, Baek M, et al. 2022. Scaffolding protein functional sites using deep learning. *Science* 377: 387–394. doi:10.1126/science.abn2100

Cite this article as *Cold Spring Harb Perspect Biol* doi: 10.1101/cshperspect.a041471

Wheeler LC, Lim SA, Marqusee S, Harms MJ. 2016. The thermostability and specificity of ancient proteins. *Curr Opin Struct Biol* **38**: 37–43. doi:10.1016/j.sbi.2016.05.015

Wittmann BJ, Johnston KE, Wu Z, Arnold FH. 2021a. Advances in machine learning for directed evolution. *Curr Opin Struct Biol* **69**: 11–18. doi:10.1016/j.sbi.2021.01.008

Wittmann BJ, Yue Y, Arnold FH. 2021b. Informed training set design enables efficient machine learning-assisted directed protein evolution. *Cell Syst* **12**: 1026–1045.e7. doi:10.1016/j.cels.2021.07.008

Woolfson DN, Bartlett GJ, Burton AJ, Heal JW, Niitsu A, Thomson AR, Wood CW. 2015. De novo protein design: how do we expand into the universe of possible protein structures? *Curr Opin Struct Biol* **33**: 16–26. doi:10.1016/j.sbi.2015.05.009

Wu Z, Kan SBJ, Lewis RD, Wittmann BJ, Arnold FH. 2019. Machine learning-assisted directed protein evolution with combinatorial libraries. *Proc Natl Acad Sci* **116**: 8852–8858. doi:10.1073/pnas.1901979116

Wu Z, Yang KK, Liszka MJ, Lee A, Batzilla A, Wernick D, Weiner DP, Arnold FH. 2020. Signal peptides generated by attention-based neural networks. *ACS Synth Biol* **9**: 2154–2161. doi:10.1021/acssynbio.0c00219

Xiao Y, Qiu J, Li Z, Hsieh C-Y, Tang J. 2021. Modeling protein using large-scale pretrain language model. arXiv doi:10.48550/arXiv.2108.07435

Yang KK, Wu Z, Arnold FH. 2019. Machine-learning-guided directed evolution for protein engineering. *Nat Methods* **16**: 687–694. doi:10.1038/s41592-019-0496-6

Yang KK, Fusi N, Lu AX. 2022a. Convolutions are competitive with transformers for protein sequence pretraining. bioRxiv doi:10.1101/2022.05.19.492714

Yang KK, Yeh H, Zanichelli N. 2022b. Masked inverse folding with sequence transfer for protein representation learning. bioRxiv doi:10.1101/2022.05.25.493516

Yi K, Zhou B, Shen Y, Liò P, Wang YG. 2023. Graph denoising diffusion for inverse protein folding. arXiv doi:10.48550/arXiv.2306.16819

Zhang L, King E, Black WB, Heckmann CM, Wolder A, Cui Y, Nicklen F, Siegel JB, Luo R, Paul CE, et al. 2022. Directed evolution of phosphite dehydrogenase to cycle noncanonical redox cofactors via universal growth selection platform. *Nat Commun* **13**: 5021. doi:10.1038/s41467-022-32727-w

Protein Design Using Structure-Prediction Networks: AlphaFold and RoseTTAFold as Protein Structure Foundation Models

Jue Wang,[1,2,3,4] Joseph L. Watson,[1,2] and Sidney L. Lisanza[1,2,3]

[1]Department of Biochemistry, University of Washington, Seattle, Washington 98195, USA

[2]Institute for Protein Design, University of Washington, Seattle, Washington 98195, USA

[3]Graduate Program in Biological Physics, Structure and Design, University of Washington, Seattle, Washington 98195, USA

[4]DeepMind, London EC4A 3BF, United Kingdom

Correspondence: juewang@post.harvard.edu

Designing proteins with tailored structures and functions is a long-standing goal in bioengineering. Recently, deep learning advances have enabled protein structure prediction at near-experimental accuracy, which has catalyzed progress in protein design as well. We review recent studies that use structure-prediction neural networks to design proteins, via approaches such as activation maximization, inpainting, or denoising diffusion. These methods have led to major improvements over previous methods in wet-lab success rates for designing protein binders, metalloproteins, enzymes, and oligomeric assemblies. These results show that structure-prediction models are a powerful foundation for developing protein-design tools and suggest that continued improvement of their accuracy and generality will be key to unlocking the full potential of protein design.

Proteins are composed of linear chains of 20 possible amino acids, yet are able to perform an amazing array of natural functions such as harvesting energy from the sun (Nelson and Yocum 2006), producing chemicals for self-defense (Christianson 2017) and performing logical computations inside the cell (Antebi et al. 2017). Drawing upon these functions and molding them through directed evolution, protein engineers have developed countless useful medicines and catalysts to address human problems (Arnold 2018; Wang et al. 2021). However, natural proteins occupy only a small part of the combinatorial vastness of sequence space (Huang et al. 2016). Therefore, a long-standing goal in protein science is to design "de novo" proteins unrelated to any natural starting point, whose functions go beyond what has been explored by evolution (Kuhlman and Bradley 2019).

Traditionally, de novo protein design aimed to generate proteins that minimize an energy function based on physics and natural protein statistics (Alford et al. 2017). This approach has successfully created a variety of structural scaffolds and functional proteins but suffers from low success rates due to the inaccuracy of the energy function (Huang et al. 2016). Recently, neural networks trained on the Protein Data Bank (PDB) have made it possible to predict protein structure from sequence at near-experimental accuracy (Baek et al. 2021; Jumper et al. 2021). This was not possible with energy-function-based methods, which raised the possibility that structure-prediction networks could advance the state-of-the-art in protein design as well.

Here, we review recent studies where models such as AlphaFold2 (AF) and RoseTTAFold (RF) are used directly or fine-tuned to design proteins. In particular, the approach of fine-tuning a structure-prediction network has led to one of the most powerful current methods for de novo protein design (Watson et al. 2023). This suggests that structure-prediction models may serve as "foundation models" for protein-structure-centric tasks, analogous to large language models for natural language understanding and generation (Bommasani et al. 2021; OpenAI 2023). Where relevant, we will mention machine-learning protein design methods that do not explicitly train on structure prediction, but we defer to other sources for comprehensive reviews of traditional or machine-learning-based protein design (Huang et al. 2016; Kuhlman and Bradley 2019; Gao et al. 2020; Ding et al. 2022; Strokach and Kim 2022; Ferruz et al. 2023; Winnifrith et al. 2023; Yang 2023).

SINGLE-SEQUENCE STRUCTURE PREDICTION

To achieve high accuracy, AF and RF require not only an input sequence but also a multiple-sequence alignment (MSA) of homologs for inferring evolutionary covariation between residue pairs (Fig. 1B). Covarying residues tend to be near each other in 3D space. This knowledge drastically reduces the search space for structure prediction (Marks et al. 2012). Accordingly, AF and RF perform poorly when predicting natural protein structures from only a single input sequence (Dauparas et al. 2022; Wang et al. 2022a). Recent methods based on protein language models such as ESMFold (Lin et al.

2023), OmegaFold (Wu et al. 2022), and others (Chowdhury et al. 2022; Wang et al. 2022b) do not require an alignment at prediction time, but their accuracy on natural proteins is nevertheless correlated to the number of homologs in the training set (Lin et al. 2023). This suggests that evolutionary covariation information has been "stored" in the parameters of the language model. There are no current methods capable of predicting natural protein structures to high accuracy from single sequences without using some form of homolog information. Achieving this ability would likely require advances in incorporating physical principles into structure prediction and is an important open research problem.

It is surprising then that AF, RF, ESMFold, and other models can predict the structures of designed proteins—proteins unrelated to natural sequences, generated by an energy-function or machine-learning-based method—with very high accuracy from only a single sequence (Yang et al. 2020; Wang et al. 2022a; Watson et al. 2023). This is fortunate, as these proteins lack evolutionary homologs by construction. This empirical observation is the original motivation (and justification) for using structure-prediction networks in the protein-design methods described below. It is not clear why these networks have this ability, but one possible explanation comes from the observation that natural proteins are marginally stable, or only a few mutations away from misfolding (Dill 1990; Thomas et al. 2010; Goldenzweig and Fleishman 2018). Therefore, natural proteins probably do not have the lowest-energy (or most probable) sequence for their structures, obscuring the sequence–structure relationship and necessitating homolog information to increase "signal-to-noise" during structure prediction. On the other hand, de novo protein sequences have been designed to have the most probable amino acids for their structure and a large energy gap between unfolded and folded states (Fleishman and Baker 2012; Baker 2019), making their structure prediction easier.

STRUCTURE PREDICTION AS DESIGN FILTER

Given that structure-prediction networks can accurately predict de novo proteins, the simplest

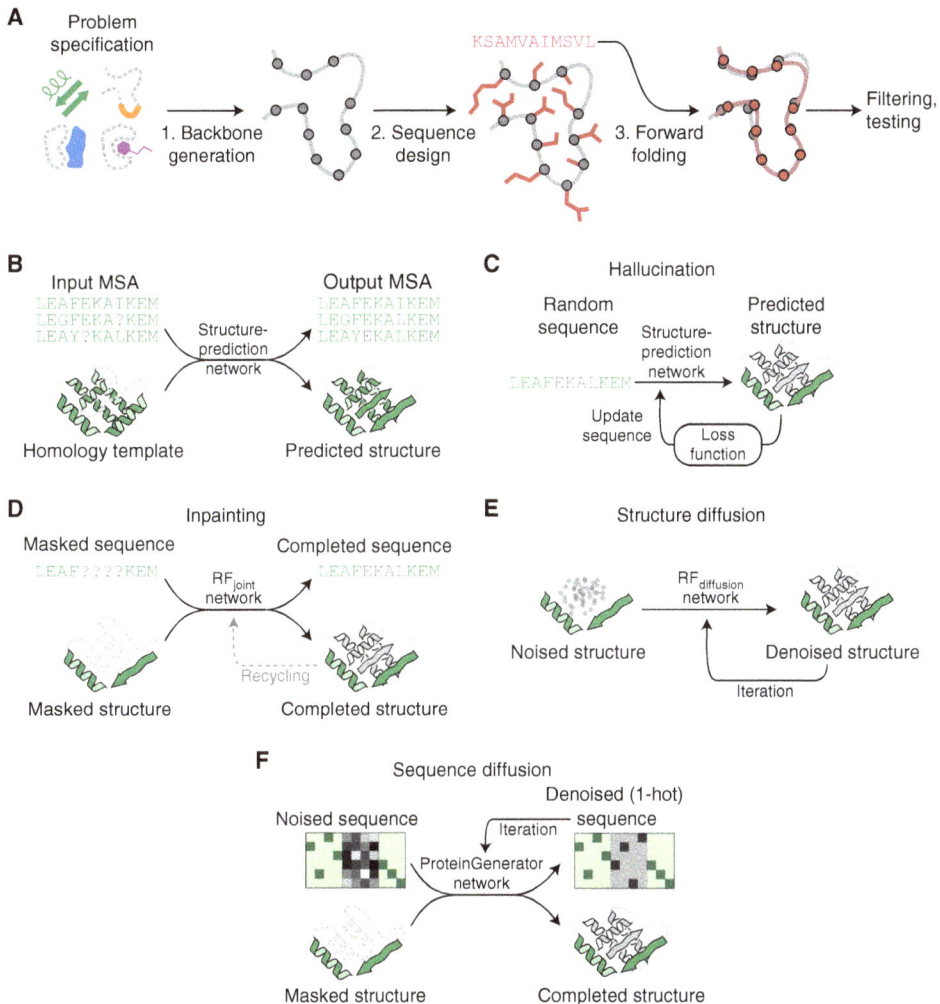

Figure 1. Schematic of de novo protein design methods. (*A*) Typical pipeline for de novo protein design. Initially, the design problem is specified by choosing, for example, a desired set of secondary structures (green), a motif to be scaffolded (orange), or a protein (blue) or ligand (purple) target to bind. Then, step 1 is to generate a protein backbone that is compatible with the design goals. Step 2 is to design a sequence that is predicted to fold into the backbone (sequence and sidechains shown in red). Step 3 is to verify that the designed sequence indeed folds into the desired backbone, using an independent structure-prediction method (predicted structure shown in red). Finally, designs are selected based on various in silico metrics and tested in the wet lab. (*B*) Input and outputs of structure-prediction networks such as AlphaFold and RoseTTAFold. Random residues in the input multiple-sequence alignment (MSA) are masked (replaced with a special token) and re-predicted during training, thereby enabling sequence information to be output by the network for fine-tuning tasks. Structures of "template" proteins homologous to the query sequence are used to aid structure prediction, thereby allowing structure information to be input during fine-tuning. (*C–E*) Protein design methods based on structure-prediction networks: (*C*) hallucination, (*D*) inpainting with RF$_{joint}$, (*E*) structure diffusion with RF$_{diffusion}$, and (*F*) sequence diffusion with ProteinGenerator. An example of a motif-scaffolding task is shown, where green represents a prespecified structural motif and gray represents what is being generated by the design method. In principle, hallucination and diffusion are inherently iterative while RF$_{joint}$ inpainting can perform one-shot design. However, iteration or "recycling" of outputs is also needed to obtain good design in practice (gray dotted arrow in *D*). In *F*, sequence is represented schematically as a matrix of logits, where the region to be generated is initialized to random real-numbered values and denoised until it becomes 1-hot.

way to use them in protein design is for in silico validation. The pipeline for structure-based protein design typically follows three steps (Fig. 1A; Huang et al. 2016). First, a protein backbone is designed that is thought to be capable of carrying out a particular function, such as scaffolding a functional motif or forming a binding interface. Second, a sequence is generated that is designed to fold to the desired structure, and therefore carry out the desired function. This second step is known as "fixed backbone sequence design," "inverse folding," or simply "sequence design." Finally, the designed sequence is input to an independent structure-prediction calculation to assess whether it will fold—and therefore function—as intended. This step, called "forward folding," is required because methods for sequence design aim only to find the lowest-energy sequence for a given structure, which does not guarantee that there is not another structure for which the chosen sequence is even lower energy (Norn et al. 2021). In other words, once a sequence has been designed, it is necessary to predict whether it will fold to the designed backbone structure. Forward folding was traditionally done using energy-function methods (Shortle et al. 1998), which are substantially less accuracy than deep learning. Replacing these low-accuracy methods with AF structure prediction has provided, for example, a 10-fold increase in experimental success rates for de novo binder design (Bennett et al. 2023). Similar increases in success rate have been observed in most other design problems to which AF filtering has been applied.

INVERTING STRUCTURE-PREDICTION NETWORKS

Traditional protein design tools make heavy use of the Markov Chain Monte Carlo (MCMC) algorithm: Starting from an initial structure and/or sequence, a small change is proposed by a sampling method and then accepted or rejected based on its energy according to an energy function (Kuhlman and Baker 2000). This is repeated for thousands of iterations until a low-energy protein is found. In an analogous approach called "hallucination," a structure-pre-

diction network is used instead of an energy function to evaluate incremental designs (Fig. 1C). This is similar to the "activation maximization" method in machine learning (Erhan et al. 2009) (but unrelated to the "hallucination problem" of language models generating nonfactual outputs).

Many different hallucination methods have been developed, but they all share the same basic principle (Anishchenko et al. 2021; Jendrusch et al. 2021; Moffat et al. 2021; Norn et al. 2021; Hie et al. 2022; Verkuil et al. 2022; Wang et al. 2022a; Wicky et al. 2022; An et al. 2023a; Frank et al. 2023; Goverde et al. 2023; Jeliazkov et al. 2023). A random starting sequence is given to a structure-prediction network to predict an initial, usually poorly folded, structure. Using this structure, as well as other outputs from the network such as prediction confidence, a user-defined "loss" function (analogous to the traditional energy function) is computed that represents the design objective. Subsequently, the sequence is mutated, given to the network again to predict a new structure, and the loss function recomputed. The new loss value is compared to the previous loss and some rule (typically the Metropolis–Hastings condition) is used to decide whether to "accept" or "reject" the mutated sequence. This process is repeated until the loss has converged, and the final sequence and its predicted structure are taken as outputs.

Hallucination's biggest advantage is simplicity and flexibility: novel tasks can be addressed by formulating new loss functions on top of otherwise unmodified structure-prediction models (Hie et al. 2022; Wang et al. 2022a). This can be a practical benefit for users who do not have the resources or expertise to train models. Early hallucination studies used a simple loss function based on model confidence—for example, the predicted error from the network (Jendrusch et al. 2021) or the entropy of predicted distributions of residue–residue distances (Anishchenko et al. 2021). This type of loss can be seen as promoting "free hallucination" of random but well-structured proteins, also known as "unconditional structure generation." Early studies demonstrated this idea in silico (Jendrusch et al. 2021), and Anishchenko et al. (2021) showed

 Cite this article as *Cold Spring Harb Perspect Biol* doi: 10.1101/cshperspect.a041472

in wet-lab experiments that free hallucinations with diverse secondary structures fold to their designed structures (Fig. 2B). Elaborating on this idea, Wicky et al. (2022) performed free hallucination with a chain break input feature to create assemblies of multiple protein chains (i.e., oligomers), which could be homo- or hetero-oligomeric (composed of identical or different subunits) depending on whether different chains were constrained to have the same sequences. Adding a cyclic symmetry loss term allowed the design of C33 assemblies with up to 1550 total residues (Fig. 2F). Using symmetry losses, sequence repetition, but no chain break, An et al. (2023a) hallucinated pseudo symmetric proteins (i.e., a single chain consisting of identical repeating units) with pockets for ligand binding or catalysis (Fig. 2D). Both studies experimentally tested their designs and found them to fold to the intended structures. These design problems would have been extremely difficult to address using traditional methods.

Although free hallucination can create useful building block proteins, practical applications usually require directly specifying the structure (or function) of a protein. To achieve this, a "structure recapitulation" loss can be defined as the deviation between the predicted structure and a prespecified structure, either in terms of 3D coordinates (e.g., frame-aligned point error or FAPE; Goverde et al. 2023) or residue–residue distance and orientation distributions (e.g., cross-entropy; Norn et al. 2021; Wang et al. 2022a). When this loss is applied over an entire protein, hallucination becomes a method for fixed-backbone sequence design (Fig. 2A; Norn et al. 2021; Verkuil et al. 2022; Goverde et al. 2023). If this loss is combined with a confidence-based loss—i.e., freely hallucinating part of a protein but constraining another part to recapitulate a predefined structural motif—then hallucination can create de novo proteins that scaffold functional motifs such as binding interfaces and catalytic sites (Tischer et al. 2020). Using this approach, Wang et al. (2022a) designed and experimentally validated metal- and protein-binding proteins, including scaffolding an epitope from the respiratory syncytial virus (Fig. 2H) that could not be tackled with previous

methods. The method also created promising de novo enzymes in silico by scaffolding catalytic sites. However, these were not tested experimentally because AF and RF cannot model small molecules and therefore cannot hallucinate substrate-binding pockets (see below). To overcome this, Yeh et al. (2023) used hallucination to design a library of pocket-containing proteins constrained to the natural NTF2 fold family. Then they used Rosetta, an energy-function-based method, to install de novo luciferase active sites into the hallucinated scaffolds and design pocket residues for binding nonnatural luciferin analogs (Fig. 2L). This pipeline, combined with three rounds of directed evolution, resulted in de novo luciferases with near-native activity. Key to success was the ability of hallucination to create pocket geometries better suited to the nonnatural substrates than those of native proteins from the PDB.

In principle, hallucination can jointly design sequence and structure (i.e., perform backbone and sequence design steps simultaneously) (Anishchenko et al. 2021; Wang et al. 2022a). Unfortunately, sequences from standard hallucination are poorly expressed and insoluble (Dauparas et al. 2022; Verkuil et al. 2022; Wicky et al. 2022). This is likely due to the phenomenon of adversarial examples, where applying activation maximization to deep neural networks leads to sampling from outside the training distribution (Goodfellow et al. 2015; Stern et al. 2023). The most practical solution, which dramatically increases wet-lab success rates, is to input hallucinated backbones to a dedicated sequence design method to generate new sequences. Traditionally, sequence design was done using Rosetta; however, recently, the ProteinMPNN neural network, which is trained to predict the amino acid at each residue position given the local structural context, has become a state-of-the-art widely adopted method for this step (Dauparas et al. 2022; Wicky et al. 2022; An et al. 2023a; Frank et al. 2023; Glasscock et al. 2023; Goverde et al. 2023; Kim et al. 2023; Krishna et al. 2023; Sumida et al. 2023; Torres et al. 2023; Yeh et al. 2023). However, because joint sequence-structure design has the potential to improve overall design quality, various strategies

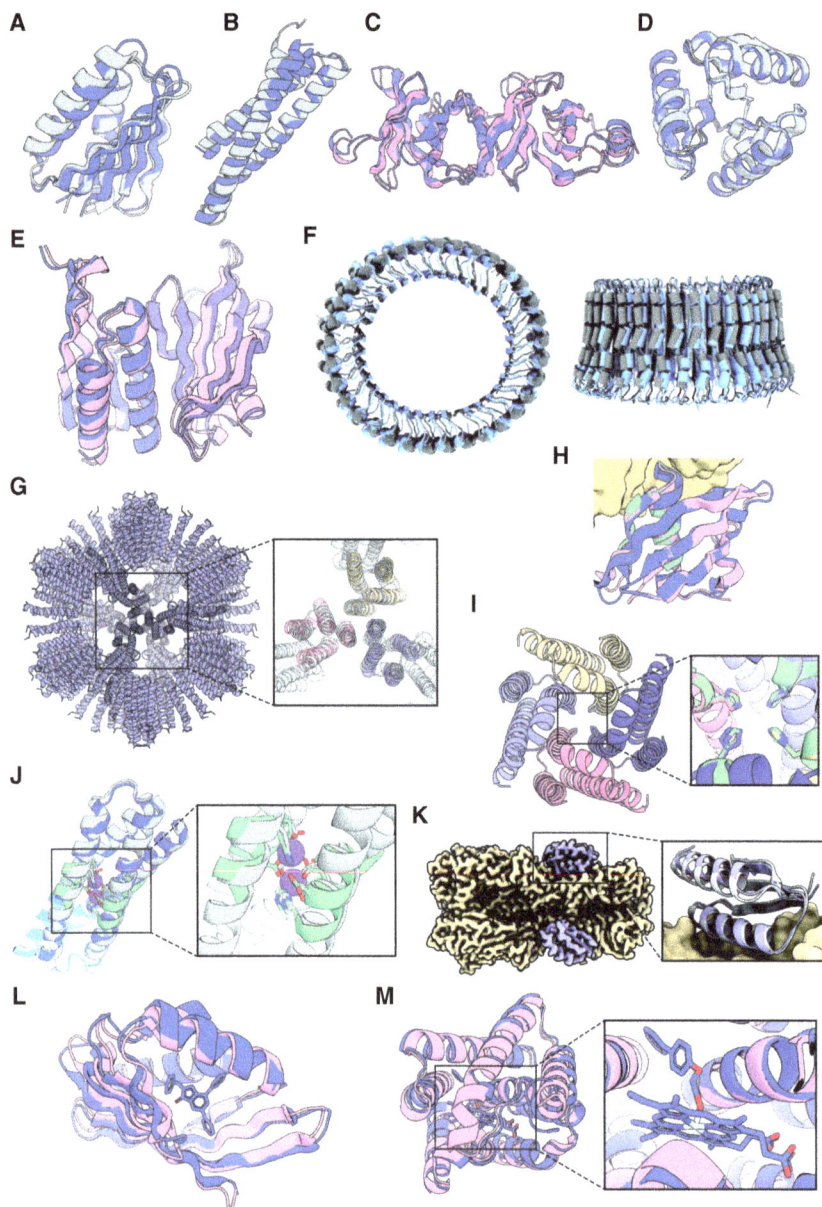

Figure 2. Experimentally validated proteins designed by structure-prediction networks. In all panels, experimentally determined structures (crystal or cryo-EM) are shown in gray, design models are shown in blue or other colors, and AlphaFold2 (AF) models are shown in pink. (*A*) A fixed-backbone sequence design to the Top7 protein backbone using AF hallucination (Goverde et al. 2023). (*B*) Unconditionally generated monomer from Markov Chain Monte Carlo (MCMC) trRosetta hallucination (Anishchenko et al. 2021). (*C,D*) Repeat protein monomers (An et al. 2023a; Lisanza et al. 2023). (*C*) Repeat protein with secondary structure (all-β) conditioning from ProteinGenerator sequence diffusion (An et al. 2023a). (*D*) Repeat cyclic protein with pocket for potential ligand binding, from AF hallucination (An et al. 2023a). (*E–G*) Homo- and hetero-oligomers. (*E*) Hetero-oligomer from continuous-sequence AF hallucination (Frank et al. 2023). (*F*) Large C33-symmetric oligomer from MCMC AF hallucination (Wicky et al. 2022). (*G*) Icosahedral cage from RF$_{diffusion}$ structure diffusion (Watson et al. 2023). (*H–J*) Motif-scaffolding. (*H*) RF hallucination presenting site V epitope from respiratory syncytial virus (Wang et al. 2022a). (*I*) C4-symmetric oligomer scaffolding a Ni^{2+}-binding motif from RF$_{diffusion}$ (Watson et al. 2023). (*J*) Di-iron binding protein from RF$_{joint}$ inpainting, scaffolding iron-coordinating motif (Wang et al. 2022a). (*K*) Protein binder against influenza H1 hemagglutinin from RF$_{diffusion}$ (Watson et al. 2023). (*L*) De novo luciferase, from fold-constrained trRosetta hallucination and Rosetta (Yeh et al. 2023). (*M*) De novo heme protein, from RF$_{diffusion}$ all-atom (Krishna et al. 2023).

Cite this article as *Cold Spring Harb Perspect Biol* doi: 10.1101/cshperspect.a041472

have been pursued to improve hallucinated sequences. One idea is to simply add loss terms to control the amino acid composition and properties such as solvent-exposed hydrophobic residues (a cause of poor expression), but this has only modest benefits (Anishchenko et al. 2021; Wang et al. 2022a; Goverde et al. 2023). A more principled strategy is to constrain sequence updates by another model that has learned the distribution of plausible sequences (Moffat et al. 2021; Verkuil et al. 2022; Jeliazkov et al. 2023). One example used the ESM-2 protein language model to propose MCMC mutations and achieved protein expression success rates qualitatively similar to redesign by Protein-MPNN (Verkuil et al. 2022). However, this method is impractical, requiring more than 100,000 MCMC steps per trajectory. Given the above, hallucination is most useful as a backbone-generation tool and hallucinated sequences are usually discarded.

The biggest practical limitation of conventional hallucination is its high computational cost, often requiring minutes to hours of GPU time per design and sampling of 10^3-10^4 designs to obtain experimental success (Anishchenko et al. 2021; Wang et al. 2022a). This is partly because MCMC requires many steps to converge, and partly because discrete amino acids form a rugged, difficult-to-optimize loss landscape. To avoid MCMC, one can take advantage of the differentiability of structure-prediction networks and use backpropagation to calculate the derivative of the loss function with respect to the input sequence (Norn et al. 2021; Wang et al. 2022a; Frank et al. 2023). This allows mutational steps to be made in the direction of most improved loss, greatly accelerating convergence. However, this comes at the cost of increased GPU memory usage and leads to worse losses than MCMC, likely due to the need for gradient approximations when backpropagating through discrete amino acids (Anishchenko et al. 2021; Linder and Seelig 2021; Wang et al. 2022a). To address this last issue, Frank et al. (2023) hallucinated proteins with a continuous (real-valued) sequence representation, completely dispensing with the goal of generating a usable sequence. This avoids the need to approximate gradients

and the resulting loss landscape is smoother, leading to faster convergence. This also allows structures to be realized more precisely, improving the success rate and reducing the number of trajectories needed. Using this approach for backbone generation and ProteinMPNN for sequence design, the authors achieved the highest in silico design quality per compute time out of all current hallucination methods, in some cases rivaling more sophisticated generative models (see below). They created a number of oligomeric assemblies that successfully expressed in the wet lab and validated the structures of a subset of designs by cryo-EM (Fig. 2E).

STRUCTURE PREDICTION AS PRETRAINING

The most direct solution to the drawbacks of hallucination is to train a model explicitly for protein design. For fixed-backbone sequence design, a number of effective methods already exist (Alford et al. 2017), with performance improving even more as modern machine learning has been applied to the problem (Dauparas et al. 2022). By contrast, most early neural networks for backbone generation gave poor results (Anand and Huang 2018; Anand et al. 2019; Guo et al. 2020; Lin et al. 2021; Harteveld et al. 2022; Lai et al. 2022) unless the sampling space was constrained to a specific protein fold (Eguchi et al. 2020). As these networks typically used convolutional layers and were trained as variational autoencoders (VAEs) or generative adversarial networks (GANs), they probably lacked the inductive bias and expressive capacity to fully model the distribution of high-quality protein structures.

AF and RF showed that equivariant, transformer-based models excel at reasoning about protein structures, inspiring a new wave of generative methods. Wang et al. (2022a) noticed that RF itself could represent arbitrary design tasks because it already had sequence and structure as both inputs and outputs (Fig. 1B). This is because, in addition to the usual sequence input and structure output used for structure prediction, the network also receives homology template structures as input and outputs sequence logits to perform auxiliary masked-token prediction. Therefore, motif scaffolding or unconditional structure generation could be performed simply by masking out portions of the input sequence and structure and asking the network to "inpaint" or predict the missing regions (Fig. 1D). (For simplicity, we use "inpainting" even when the majority of the input is masked, which is sometimes called "outpainting.") By starting with RF weights pretrained for structure prediction and fine-tuning them on inpainting and sequence design, Wang et al. obtained a model, RF$_{joint}$, that could design metalloproteins (Fig. 2J) and protein binders with higher wet-lab success rates than hallucination. Success rates increased even further as subsequent versions of RF$_{joint}$ were fine-tuned on top of improved variants of the base structure-prediction model.

Compared to hallucination, inpainting scaffolds motifs more accurately and requires less compute at design time (Wang et al. 2022a; Watson et al. 2023). This comes at the cost of an initial fine-tuning run, but it becomes negligible when amortized over all downstream design work. Like hallucination, inpainting can in theory design sequence and structure jointly, and, in fact, inpainted sequences have reasonably high in silico and wet-lab success rates (Wang et al. 2022a). However, success rates are still higher when sequences are redesigned using ProteinMPNN (Sumida et al. 2023; Watson et al. 2023). Inpainting excels at "idealizing" local backbone structure, or replacing loopy regions with α helices or β sheets (Glasscock et al. 2023; Sumida et al. 2023; Torres et al. 2023). This is a common "polishing" step applied to designs generated by any method, because it improves the quality of sequences from ProteinMPNN and wet-lab success rates. On the other hand, inpainting's weaknesses are low design diversity and poor performance on unconditional generation of full structures. This is not surprising given that the training task defines one specific solution, rather than a distribution of solutions, for any given input. A better formulation would pose protein design as a probabilistic modeling problem, but traditional approaches like VAEs and GANs have shown limited success in generating proteins with high quality and diversity.

COMBINING STRUCTURE PREDICTION AND DIFFUSION GENERATIVE MODELS

Recently, denoising diffusion probabilistic models (DDPMs or "diffusion models") (Sohl-Dickstein et al. 2015; Ho et al. 2020; Song and Ermon 2020; Yang et al. 2023) have achieved state-of-the-art in image generation (Dhariwal and Nichol 2021; Ramesh et al. 2022; Rombach et al. 2022). These models are trained to take a data sample to which noise has been added and predict either the noise or the original un-noised sample. At inference time, the model is used to transform a sample of pure noise into an example from the data distribution by many steps of iterative denoising. In the limit of infinitesimally small step sizes, a correctly trained diffusion model can theoretically sample from the true data distribution. This framework is also highly amenable to adding conditioning information to control properties of the generated samples (Ramesh et al. 2022; Rombach et al. 2022).

For proteins, the key challenge is to represent structure and sequence as quantities that can be diffused. To date, the majority of protein diffusion models have focused on diffusing protein structure (Anand and Achim 2022; Trippe et al. 2022; Wu et al. 2022; Chu et al. 2023; Ingraham et al. 2023; Lin and AlQruaishi 2023; Watson et al. 2023; Yim et al. 2023). Protein structures are inherently three dimensional, and their biological function is determined by the position of atoms relative to one another, rather than their absolute coordinates in some arbitrary reference frame. As such, protein structure diffusion models are all SE(3)-equivariant, and the majority have reused architectural components directly from structure-prediction networks. The challenges in formulating the diffusion process in three dimensions has commonly led to practical compromises, such as suboptimal protein representations or unprincipled "diffusion-like" processes that are not theoretically guaranteed to learn the true data distribution.

ProtDiff, one of the earliest protein diffusion models, simplified the problem of backbone generation to C-α coordinate denoising, which allowed a principled formulation of diffusion with a mean-squared error loss (Trippe et al. 2022).

Cite this article as *Cold Spring Harb Perspect Biol* doi: 10.1101/cshperspect.a041472

However, the simplistic backbone representation and an architecture invariant to reflections led to incorrect chirality in some of the designed backbones. FoldingDiff performed diffusion over backbone torsion angles, but this representation is known to be vulnerable to error propagation through lever-arm effects (Wu et al. 2022). Anand and Achim (2022) trained a protein diffusion model that adopted the frame-based representation of protein backbones used in AF and RF (treating backbones as elements of the $SE(3)^N$ Lie group). Like ProtDiff, they also diffused C-α atoms, but further used spherical interpolation to define a diffusion-like process over residue backbone orientations. This permitted the use of AlphaFold's FAPE loss to jointly train denoising of the backbone position and orientation, enforcing the chirality of natural proteins. This model generated quite realistic protein structures when the samples were conditioned on fold information (secondary structure blocks and their relative adjacency), but unconditional sampling was not explored. The above studies demonstrated the promise of diffusion for protein design, but used relatively small neural networks and did not pursue wet-lab testing of designs.

Currently, the state of the art for protein diffusion models (and perhaps de novo protein design in general) is $RF_{diffusion}$, a fine-tuned version of the RF structure-prediction network (Fig. 1E). Watson et al. (2023) noticed that inpainting using RF_{joint} was almost identical to one step of denoising, and therefore proposed that structure-prediction weights could serve as the foundation for a diffusion model. Diffusion fine-tuning was done with noised inputs, an input feature representing the denoising step (to specify the expected level of noise), mean-squared error loss on coordinates as well as frame orientations, and self-conditioning, where the network is given its own outputs from a previous step. To perform specific design tasks such as motif scaffolding or specifying the overall fold, additional fine-tuning was performed with task-specific conditioning inputs. The resulting $RF_{diffusion}$ model proved to be extremely capable on a wide variety of protein design tasks. Simply by adding a symmetrization step to inference and leveraging the $SE(3)$ equivariance of the RF architecture, $RF_{diffusion}$

was able to generate large, homo-oligomeric assemblies with cyclic and point group symmetry that adopted the intended quaternary structure by cryo-EM (Fig. 2G). Additionally, $RF_{diffusion}$ was able to scaffold symmetric motifs (Fig. 2I) and very small motifs such as single residues from an enzyme active site that were almost impossible with hallucination or RF_{joint}. Finally, the model designed protein binders conditioned only on the target structure with wet-lab success rates of up to 50% on some targets (Fig. 2K). For comparison, this is around an order of magnitude higher success rate than binder design with energy-based methods (combined with improved AF-based filtering) (Bennett et al. 2023; Watson et al. 2023). For two different peptide-binder design problems, one out of 96 $RF_{diffusion}$ designs tested experimentally had picomolar affinity for the target without any experimental optimization (Torres et al. 2023). These are the highest-affinity de novo protein binders ever designed.

Alongside and after $RF_{diffusion}$, several other general-purpose protein diffusion models have been reported (as well as application-specific diffusion models such as for antibody design (Luo et al. 2022), which will not be detailed here). Chroma, a model based on a "polymer diffusion" approach, is trained only unconditionally, relying on classifier-based guidance at inference time to condition on design objectives such as functional motifs, text descriptions, and arbitrary 3D shapes (Ingraham et al. 2023). Genie uses C-α diffusion and an AF-like architecture to perform unconditional generation (Lin and AlQuraishi 2023). FrameDiff uses a similar frame representation as $RF_{diffusion}$ but within a simpler, lighter-weight network architecture and a more-exact score-matching objective (Yim et al. 2023). Finally, Protpardelle diffuses coordinates of all atoms corresponding to all possible sidechains at each residue, and then uses these "superposed atom" positions to predict amino acid identities (Chu et al. 2023). This avoids the complexity of residue orientation diffusion and provides a framework for joint sequence-structure design. All of these models are trained "from scratch" rather than fine-tuned from a structure-prediction network. Currently, however, in silico performance of such models lags behind RF$_{diffusion}$, and only one has reported results from wet-lab testing.

Interestingly, RF$_{joint}$ or RF$_{diffusion}$ performance is substantially worse when protein design fine-tuning (i.e., a relatively short training run) is performed from randomly initialized weights instead of pretrained structure-prediction weights (Wang et al. 2022a; Watson et al. 2023). Given this, it is tempting to conclude that something special about structure-prediction pretraining makes it *necessary* for generating high-quality designs. However, since RF$_{diffusion}$ fine-tuning typically uses <2% of the GPU time used to train RF structure prediction, it is not surprising that omitting the pretraining step reduces design quality (a full RF$_{diffusion}$ training run "from scratch" with the same amount of compute as structure-prediction pretraining was not attempted). Similarly, the lower performance of other diffusion models could simply be due to lower model capacity. For example, FrameDiff, which has a similar diffusion scheme as RF$_{diffusion}$, only has 1/4 as many parameters, yet already has reasonable performance (Yim et al. 2023). There is no obvious reason in principle why the performance of RF$_{diffusion}$ cannot be matched or exceeded by training on protein design from scratch, given sufficient model capacity and training time. However, the fact that structure-prediction pretraining provides such a strong starting point for design has important practical benefits, allowing rapid experimentation with different fine-tuning strategies at relatively low computational expense.

An interesting alternative to protein structure diffusion is to diffuse protein sequences instead. Diffusion of discrete variables (such as amino acids) currently has worse performance than autoregressive or masked-token models for natural language modeling (Austin et al. 2021; Hoogeboom et al. 2021; Dieleman et al. 2022). However, for proteins, sequence diffusion can be much simpler than structure diffusion, and a wealth of protein sequence-function data (binding or enzymatic activity) exists that can potentially be used to train oracles to perform classifier guidance on sequences. This is the motivation

behind ProteinGenerator, a diffusion model fine-tuned from RF that performs diffusion in sequence space (Fig. 1F; Lisanza et al. 2023). Amino acids are represented as one-hot encodings which have Gaussian noise added, and the network is trained to denoise them via a cross-entropy loss. The model was used to design proteins conditioned on sequence-centric design objectives such as enrichment for certain amino acids, masked bioactive peptide scaffolding, and repeat protein design (Fig. 2C). Sequences directly from ProteinGenerator (without additional sequence design) expressed well and had the expected secondary structure content in the wet lab. Because ProteinGenerator is based on RF, it can also predict structure (which was maintained during fine-tuning by continuing to apply a FAPE loss), and therefore can perform structure-conditioned tasks such as motif scaffolding and multistate design. However, overall, in silico metrics and wet-lab success rates were lower than in the RF$_{diffusion}$-ProteinMPNN pipeline. Other sequence diffusion models (Alamdari et al. 2023) have also recently been reported. The true potential of this approach will be realized when it can design functional proteins guided by a sequence-to-function oracle.

PROTEIN DESIGN IN THE CONTEXT OF NONPROTEIN MOLECULES

Many important protein-design problems require modeling the interactions between proteins and DNA, RNA, and small molecules (Tinberg et al. 2013; Huang et al. 2016; Dou et al. 2018; Glasscock et al. 2023; Kalvet et al. 2023). Classical methods handle these problems naturally because their force fields are defined on explicitly represented atoms (Alford et al. 2017). However, these methods do not have the ability of hallucination and diffusion to generate full proteins from scratch, and so require complex, multistep design pipelines with an initial input of a library of predefined protein "scaffold" structures for docking against the intended binding target (Tinberg et al. 2013; Dou et al. 2018; An et al. 2023b; Lee et al. 2023; Lu et al. 2023). Machine-learning-based design methods are simpler and more flexible, but AF and RF cannot model nonprotein molecules. Therefore, until recently, protein-design tools based on these models could only design protein monomers and oligomers, with any interactions to nonprotein molecules being modeled in a separate step by classical methods (Glasscock et al. 2023; Kalvet et al. 2023; Yeh et al. 2023) or machine learning (Dauparas et al. 2023). This means that overall protein shape (and therefore the binding pocket) cannot be defined in a ligand- or nucleic-acid-aware manner, which limits the degree of achievable shape and chemical complementarity between the protein and the nonprotein molecule (Wang et al. 2022a; An et al. 2023a; Watson et al. 2023). This is likely a key reason why existing methods have struggled to obtain high binding affinity "out of the computer," before doing any experimental optimization via directed evolution (Tinberg et al. 2013; Dou et al. 2018; Yeh et al. 2023).

Recently, RF was generalized to model protein-nucleic-acid complexes (RF nucleic acid [RFNA]; Baek et al. 2022) as well as protein-small-molecule complexes and covalently modified amino acids (RF all-atom [RFAA]; Krishna et al. 2023). The latter model, RFAA, can model arbitrary biomolecular assemblies and has been trained on all protein and nonprotein structures in the PDB. Unlike traditional docking methods that take as input a known protein and ligand structure and predict how they will bind as a complex (Trott and Olson 2010; Corso et al. 2023), RFAA predicts the structure of a protein along with any binding partners, effectively performing folding and docking simultaneously. For structure modeling, this potentially allows the prediction of binding-induced conformational changes. For design, this potentially enables designing proteins with highly tailored binding interfaces to small molecules and nucleic acids.

To test this idea, Krishna et al. fine-tuned their RFAA model to create $RF_{diffusion}$ AA (Krishna et al. 2023), a diffusion model that can generate proteins conditioned on small molecule ligands (Fig. 1A, purple input). This was used to design protein binders for a number of biological and nonbiological small molecules, and designs against three targets—digoxigenin,

heme (Fig. 2M), and bilin—were tested and confirmed to have binding activity in wet-lab experiments. Notably, one of the tested digoxigenin binders had a K_d of 10 nM, comparable to a previously designed binder that had been optimized by directed evolution to have three mutations relative to its original design (Tinberg et al. 2013). This demonstrates that the "folding and docking" capability of RFAA translates, after fine-tuning, to an ability to generate binding pockets highly tailored to specific binding partners, thus eliminating (and improving on) the traditional step of docking the target against a library of scaffolds. These results open up several directions for future work. DNA- (or RNA)-binding proteins could be designed by generating proteins in the presence of a nucleic acid, potentially improving on similar recent efforts using more traditional tools (Glasscock et al. 2023). Enzymes could be designed by combining ligand-conditioned protein generation with motif scaffolding of catalytic residues (Huang et al. 2016; Yeh et al. 2023). These problems would pose a number of additional scientific challenges, but the key technical obstacle of generating novel proteins with highly complementary binding pockets now seems to have been overcome.

AlphaFold AND RoseTTAFold AS PROTEIN STRUCTURE FOUNDATION MODELS

In the domains of natural language and computer vision, a common practice is to train high-capacity models on large data sets using a simple self-supervised objective, and then use these "foundation" models, either directly or with fine-tuning, on a variety of downstream tasks (Bommasani et al. 2021; OpenAI 2023). Although AF and RF were originally trained to solve the supervised task of protein structure prediction, they have proven useful in unanticipated applications such as model accuracy estimation (Roney and Ovchinnikov 2022) and mutation effect prediction (Mansoor et al. 2023). Given the above and the success of hallucination, RF_{joint} and $RF_{diffusion}$, ProteinGenerator, it is clear that AF and RF function practically as foundation models for protein structure analysis and generation.

Therefore, to improve design success rates and expand the range of addressable problems, a key overarching challenge is improving the structural accuracy of foundation models. Applications such as enzyme design and binder design against polar targets (whether protein, DNA, or RNA) require subangstrom positioning of side-chain atoms. Although tools like RF all-atom have made it possible to model arbitrary biomolecular assemblies, their accuracy on protein-small-molecule and protein-nucleic-acid complexes lags behind that of protein monomers and oligomers (Linder and Seelig 2021; Krishna et al. 2023). Addressing this will require improving machine learning methods as well as increasing the size and diversity of training data sets.

The field of de novo protein design has only recently begun to succeed in practically important problems, partly due to the advances reviewed here. A key theme for future work will be identifying novel design problems and devising new hallucination losses or diffusion conditioning mechanisms to address them. At a technical level, even current applications often require designing degrees of freedom for which optimal values are not known a priori. A simple example of this is the length of a designed protein and those of linker regions between discontinuous motifs (Tischer et al. 2020; Wang et al. 2022a; Watson et al. 2023). A more complex example is the fact that infinitely many motifs can be consistent with the optimal geometric constraints between residues in a given enzyme active site (as computed by quantum chemistry) (Tantillo et al. 1998; Zanghellini et al. 2006). For both problems, there are currently no methods better than exhaustive enumeration, blind sampling, or human intuition (Wang et al. 2022a). If these variables can be chosen "automatically" by neural networks based on optimal patterns learned during training, design quality could improve substantially.

Joint sequence-structure design is a long-sought goal in protein design because it could potentially improve design quality by letting backbone coordinates depend on the sequence and vice versa (Huang et al. 2011; Loshbaugh and Kortemme 2020; Maguire et al. 2021; Bennett et al. 2023). The most widely used traditional method for this is "flexible backbone" sequence design using Rosetta, where multiple rounds of sequence design are alternated with relaxation of the backbone conformation (Huang et al. 2011; Maguire et al. 2021; Bennett et al. 2023). However, this requires an already-designed backbone, or a coarse-grained description of one, as an input. The few non-deep-learning methods that could design a sequence and structure from scratch, given a minimal description of the design problem (such as a motif to be scaffolded), have had poor success rates (Bonet et al. 2018; Sesterhenn et al. 2020; Yang et al. 2021). Hallucination, inpainting, and diffusion represent the state-of-the-art in solving this problem. However, the fact that a separate ProteinMPNN sequence-design step increases wet-lab success rates suggests that current generative models still cannot perfectly sample from the joint distribution of sequences and structures. Given that sequence-only (e.g., ProGen; Madani et al. 2023), structure-only (e.g., $RF_{diffusion}$), and structure-conditioned sequence generation (e.g., ProteinMPNN) all perform quite well, the key unsolved problem seems to be the way in which simultaneous sequence- and structure-generative processes are coupled. This will be an important near-term research direction and a likely source of additional gains in success rate.

Finally, a major underexplored area in de novo design is the modeling of protein function as part of the design process. The field of directed evolution has long used machine learning models trained on protein activity data to guide in silico design efforts (Yang et al. 2019; Johnston et al. 2023). To date, there are relatively few applications of these ideas to de novo design (Rocklin et al. 2017; Gligorijević et al. 2021). However, many practically important protein characteristics, such as immunogenicity or catalytic activity, are not understood well enough to be easily formulated as structure or sequence conditioning. Therefore, learning from and conditioning on direct functional measurements may be the most reliable way to design these functions. In the short term, functional measurements can be most easily incorporated during fine-tuning or inference and applied to design problems with a narrow sequence space (such as a specific fold type or enzyme class) (Madani et al. 2023; Yeh et al. 2023). However, a potentially even more powerful approach could be training a general foundation model on all existing function measurements, provided that the data can be standardized across different sources. The representations and capabilities of sequence- and structure-centric protein models are already converging (Yang et al. 2022; Heinzinger et al. 2023; Lin et al. 2023; Mansoor et al. 2023; Su et al. 2023). We may soon begin to see the application of multimodal sequence-structure-function models to the most difficult protein design problems.

ACKNOWLEDGMENTS

We thank Christopher Frank, Caper Goverde, Alexis Courbet, Linna An, Indrek Kalvet, and Ivan Anishchenko for sharing design structures and models for Figure 2; David Baker and other members of the Institute for Protein Design for valuable discussions informing the perspectives shared here; and Rob Fergus for feedback on the manuscript.

REFERENCES

Alamdari S, Thakkar N, van den Berg R, Lu AX, Fusi N, Amini AP, Yang KK. 2023. Protein generation with evolutionary diffusion: sequence is all you need. bioRxiv doi:10.1101/2023.09.11.556673

Alford RF, Leaver-Fay A, Jeliazkov JR, O'Meara MJ, DiMaio FP, Park H, Shapovalov MV, Renfrew PD, Mulligan VK, Kappel K, et al. 2017. The Rosetta all-atom energy function for macromolecular modeling and design. *J Chem Theory Comput* 13: 3031–3048. doi:10.1021/acs.jctc.7b00125

An L, Hicks DR, Zorine D, Dauparas J, Wicky BIM, Milles LF, Courbet A, Bera AK, Nguyen H, Kang A, et al. 2023a. Hallucination of closed repeat proteins containing central pockets. *Nat Struct Mol Biol* doi:10.1038/s41594-023-01112-6

An L, Meerit S, Long T, Sagardip M, Goreshnik I, Lee GR, Juergens D, Dauparas J, Anishchenko I, Coventry B, et al. 2023b. De novo design of diverse small molecule binders and sensors using Shape Complementary Pseudocycles. bioRxiv doi:10.1101/2023.12.20.572602

Anand N, Achim T. 2022. Protein structure and sequence generation with equivariant denoising diffusion probabilistic models. arXiv doi:10.48550/arXiv.2205.15019

Anand N, Huang P. 2018. Generative modeling for protein structures. In *Advances in neural information processing systems 31* (ed. Bengio S, Wallach H, Larochelle H, Grau-

man K, Cesa-Bianchi N, Garnett R), pp. 7494–7505. Curran Associates, Red Hook, NY.

Anand N, Eguchi R, Huang PS. 2019. Fully differentiable full-atom protein backbone generation. https://openreview.net/pdf?id=SJxnVL8YOV

Anishchenko I, Pellock SJ, Chidyausiku TM, Ramelot TA, Ovchinnikov S, Hao J, Bafna K, Norn C, Kang A, Bera AK, et al. 2021. De novo protein design by deep network hallucination. *Nature* 600: 547–552. doi:10.1038/s41586-021-04184-w

Antebi YE, Linton JM, Klumpe H, Bintu B, Gong M, Su C, McCardell R, Elowitz MB. 2017. Combinatorial signal perception in the BMP pathway. *Cell* 170: 1184–1196.e24. doi:10.1016/j.cell.2017.08.015

Arnold FH. 2018. Directed evolution: bringing new chemistry to life. *Angew Chem Int Ed Engl* 57: 4143–4148. doi:10.1002/anie.201708408

Austin J, Johnson DD, Ho J, Tarlow D, van den Berg R. 2021. Structured denoising diffusion models in discrete state-spaces. arXiv doi:10.48550/arXiv.2107.03006

Baek M, DiMaio F, Anishchenko I, Dauparas J, Ovchinnikov S, Lee GR, Wang J, Cong Q, Kinch LN, Schaeffer RD, et al. 2021. Accurate prediction of protein structures and interactions using a three-track neural network. *Science* 373: 871–876. doi:10.1126/science.abj8754

Baek M, McHugh R, Anishchenko I, Baker D, DiMaio F. 2022. Accurate prediction of nucleic acid and protein-nucleic acid complexes using RoseTTAFoldNA. bioRxiv doi:10.1101/2022.09.09.507333

Baker DA. 2019. What has de novo protein design taught us about protein folding and biophysics? What has de novo protein design taught us? *Protein Sci* 28: 678–683. doi:10.1002/pro.3588

Bennett NR, Coventry B, Goreshnik I, Huang B, Allen A, Vafeados D, Peng YP, Dauparas J, Baek M, Stewart L, et al. 2023. Improving de novo protein binder design with deep learning. *Nat Commun* 14: 2625. doi:10.1038/s41467-023-38328-5

Bommasani R, Hudson DA, Adeli E, Altman R, Arora S, von Arx S, Bernstein MS, Bohg J, Bosselut A, Brunskill E, et al. 2021. On the opportunities and risks of foundation models. arXiv https://arxiv.org/pdf/2108.07258.pdf%C3%82%C2%A0

Bonet J, Wehrle S, Schriever K, Yang C, Billet A, Sesterhenn F, Scheck A, Sverrisson F, Veselkova B, Vollers S, et al. 2018. Rosetta FunFolDes—a general framework for the computational design of functional proteins. *PLoS Comput Biol* 14: e1006623. doi:10.1371/journal.pcbi.1006623

Chowdhury R, Bouatta N, Biswas S, Floristean C, Kharkar A, Roy K, Rochereau C, Ahdritz G, Zhang J, Church GM, et al. 2022. Single-sequence protein structure prediction using a language model and deep learning. *Nat Biotechnol* 40: 1617–1623. doi:10.1038/s41587-022-01432-w

Christianson DW. 2017. Structural and chemical biology of terpenoid cyclases. *Chem Rev* 117: 11570–11648. doi:10.1021/acs.chemrev.7b00287

Chu AE, Cheng L, Nesr GE, Xu M, Huang PS. 2023. An all-atom protein generative model. bioRxiv doi:10.1101/2023.05.24.542194

Corso G, Stärk H, Jing B, Barzilay R, Jaakkola T. 2023. Diff-dock: diffusion steps, twists, and turns for molecular docking. arXiv doi:10.48550/arXiv.2210.01776

Dauparas J, Anishchenko I, Bennett N, Bai H, Ragotte RJ, Milles LF, Wicky BIM, Courbet A, de Haas RJ, Bethel N, et al. 2022. Robust deep learning-based protein sequence design using ProteinMPNN. *Science* **378:** 49–56. doi:10.1126/science.add2187

Dauparas J, Lee GR, Pecoraro R, An L, Anishchenko I, Glasscock C, Baker D. 2023. Atomic context-conditioned protein sequence design using LigandMPNN. bioRxiv doi:10.1101/2023.12.22.573103

Dhariwal P, Nichol A. 2021. Diffusion models beat GANs on image synthesis. arXiv doi:10.48550/arXiv.2105.05233

Dieleman S, Sartran L, Roshannai A, Savinov N, Ganin Y, Richemond PH, Doucet A, Strudel R, Dyer C, Durkan C, et al. 2022. Continuous diffusion for categorical data. arXiv doi:10.48550/arXiv.2211.15089

Dill KA. 1990. Dominant forces in protein folding. *Biochemistry* **29:** 7133–7155. doi:10.1021/bi00483a001

Ding W, Nakai K, Gong H. 2022. Protein design via deep learning. *Brief Bioinform* **23:** bbac102. doi:10.1093/bib/bbac102

Dou J, Vorobieva AA, Sheffler W, Doyle LA, Park H, Bick MJ, Mao B, Foight GW, Lee MY, Gagnon LA, et al. 2018. De novo design of a fluorescence-activating β-barrel. *Nature* **561:** 485–491. doi:10.1038/s41586-018-0509-0

Eguchi RR, Anand N, Choe CA, Huang PS. 2020. Ig-VAE: generative modeling of immunoglobulin proteins by direct 3D coordinate generation. bioRxiv doi:10.1101/2020.08.07.242347

Erhan D, Bengio Y, Courville A, Vincent P, Box PO. 2009. Visualizing higher-layer features of a deep network. https://www.researchgate.net/publication/265022827_Visualizing_Higher-Layer_Features_of_a_Deep_Network

Ferruz N, Heinzinger M, Akdel M, Goncearenco A, Naef L, Dallago C. 2023. From sequence to function through structure: deep learning for protein design. *Comput Struct Biotechnol J* **21:** 238–250. doi:10.1016/j.csbj.2022.11.014

Fleishman SJ, Baker D. 2012. Role of the biomolecular energy gap in protein design, structure, and evolution. *Cell* **149:** 262–273. doi:10.1016/j.cell.2012.03.016

Frank C, Khoshouei A, de Stigter Y, Schiewitz D, Feng S, Ovchinnikov S, Dietz H. 2023. Efficient and scalable de novo protein design using a relaxed sequence space. bioRxiv doi:10.1101/2023.02.24.529906

Gao W, Mahajan SP, Sulam J, Gray JJ. 2020. Deep learning in protein structural modeling and design. *Patterns* **1:** 100142. doi:10.1016/j.patter.2020.100142

Glasscock CJ, Pecoraro R, McHugh R, Doyle LA, Chen W, Boivin O, Lonnquist B, Na E, Politanska Y, Haddox HK, et al. 2023. Computational design of sequence-specific DNA-binding proteins. bioRxiv doi:10.1101/2023.09.20.558720

Gligorijević V, Berenberg D, Ra S, Watkins S, Kelow S, Cho K, Bonneau R. 2021. Function-guided protein design by deep manifold sampling. bioRxiv doi:10.1101/2021.12.22.473759

Goldenzweig A, Fleishman S. 2018. Principles of protein stability and their application in computational design.

Annu Rev Biochem **87:** 105–129. doi:10.1146/annurev-biochem-062917-012102

Goodfellow IJ, Shlens J, Szegedy C. 2015. Explaining and harnessing adversarial examples. arXiv doi:10.48550/arXiv.1412.6572

Goverde CA, Wolf B, Khakzad H, Rosset S, Correia BE. 2023. De novo protein design by inversion of the AlphaFold structure prediction network. *Protein Sci* **32:** e4653. doi:10.1002/pro.4653

Guo X, Tadepalli S, Zhao L, Shehu A. 2020. Generating tertiary protein structures via an interpretative variational autoencoder. arXiv doi:10.48550/arXiv.2004.07119

Harteveld Z, Southern J, Defferrard M, Loukas A, Vandergheynst P, Bronstein M, Correia BE. 2022. Deep sharpening of topological features for de novo protein design. https://openreview.net/pdf?id=DwN81YIXGQP

Heinzinger M, Weissenow K, Sanchez JG, Henkel A, Steinegger M, Rost B. 2023. Prostt5: bilingual language model for protein sequence and structure. bioRxiv doi:10.1101/2023.07.23.550085

Hie B, Candido S, Lin Z, Kabeli O, Rao R, Smetanin N, Sercu T, Rives A. 2022. A high-level programming language for generative protein design. bioRxiv doi:10.1101/2022.12.21.521526

Ho J, Jain A, Abbeel P. 2020. Denoising diffusion probabilistic models. arXiv doi:10.48550/arXiv.2006.11239

Hoogeboom E, Nielsen D, Jaini P, Forré P, Welling M. 2021. Argmax flows and multinomial diffusion: learning categorical distributions. arXiv doi:10.48550/arXiv.2102.05379

Huang PS, Ban YE, Richter F, Andre I, Vernon R, Schief WR, Baker D. 2011. Rosettaremodel: a generalized framework for flexible backbone protein design. *PLoS ONE* **6:** e24109. doi:10.1371/journal.pone.0024109

Huang PS, Boyken SE, Baker D. 2016. The coming of age of de novo protein design. *Nature* **537:** 320–327. doi:10.1038/nature19946

Ingraham JB, Baranov M, Costello Z, Barber KW, Wang W, Ismail A, Frappier V, Lord DM, Ng-Thow-Hing C, Van Vlack ER, et al. 2023. Illuminating protein space with a programmable generative model. *Nature* **623:** 1070–1078. doi:10.1038/s41586-023-06728-8

Jeliazkov JR, Del Alamo D, Karpiak JD. 2023. ESMFold hallucinates native-like protein sequences. bioRxiv doi:10.1101/2023.05.23.541774

Jendrusch M, Korbel JO, Sadiq SK. 2021. Alphadesign: a de novo protein design framework based on AlphaFold. bioRxiv doi:10.1101/2021.10.11.463937

Johnston KE, Fannjiang C, Wittmann BJ, Hie BL, Yang KK, Wu Z. 2023. Machine learning for protein engineering. arXiv doi:10.48550/arXiv.2305.16634

Jumper J, Evans R, Pritzel A, Green T, Figurnov M, Ronneberger O, Tunyasuvunakool K, Bates R, Žídek A, Potapenko A, et al. 2021. Highly accurate protein structure prediction with AlphaFold. *Nature* **596:** 583–589. doi:10.1038/s41586-021-03819-2

Kalvet I, Ortmayer M, Zhao J, Crawshaw R, Ennist NM, Levy C, Roy A, Green AP, Baker D. 2023. Design of heme enzymes with a tunable substrate binding pocket adjacent to an open metal coordination site. *J Am Chem Soc* **145:** 14307–14315. doi:10.1021/jacs.3c02742

Kim DE, Jensen DR, Feldman D, Tischer D, Saleem A, Chow CM, Li X, Carter L, Milles L, Nguyen H, et al. 2023. De novo design of small beta barrel proteins. *Proc Natl Acad Sci* **120**: e2207974120. doi:10.1073/pnas.2207974120

Krishna R, Wang J, Ahern W, Sturmfels P, Venkatesh P, Kalvet I, Lee GR, Morey-Burrows FS, Anishchenko I, Humphreys IR, et al. 2023. Generalized biomolecular modeling and design with RoseTTAFold all-atom. bioRxiv doi:10.1101/2023.10.09.561603

Kuhlman B, Baker D. 2000. Native protein sequences are close to optimal for their structures. *Proc Natl Acad Sci* **97**: 10383–10388. doi:10.1073/pnas.97.19.10383

Kuhlman B, Bradley P. 2019. Advances in protein structure prediction and design. *Nat Rev Mol Cell Biol* **20**: 681–697. doi:10.1038/s41580-019-0163-x

Lai B, McPartlon M, Xu J. 2022. End-to-end deep structure generative model for protein design. bioRxiv doi:10.1101/2022.07.09.499440

Lee GR, Pellock SJ, Norn C, Tischer D, Dauparas J, Anishchenko I, Mercer JAM, Kang A, Bera A, Nguyen H, et al. 2023. Small-molecule binding and sensing with a designed protein family. bioRxiv doi:10.1101/2023.11.01.565201

Lin Y, AlQuraishi M. 2023. Generating novel, designable, and diverse protein structures by equivariantly diffusing oriented residue clouds. arXiv doi:10.48550/arxiv.org/abs/2301.12485

Lin Z, Sercu T, LeCun Y, Rives A. 2021. Deep generative models create new and diverse protein structures. In *Machine learning for structural biology workshop NeurIPS2021*. https://www.mlsb.io/papers_2021/MLSB2021_Deep_generative_models_create.pdf

Lin Z, Akin H, Rao R, Hie B, Zhu Z, Lu W, Smetanin N, Verkuil R, Kabeli O, Shmueli Y, et al. 2023. Evolutionary-scale prediction of atomic-level protein structure with a language model. *Science* **379**: 1123–1130. doi:10.1126/science.ade2574

Linder J, Seelig G. 2021. Fast activation maximization for molecular sequence design. *BMC Bioinformatics* **22**: 510. doi:10.1186/s12859-021-04437-5

Lisanza SL, Gershon JM, Tipps S, Arnoldt L, Hendel S, Sims JN, Li X, Baker D. 2023. Joint generation of protein sequence and structure with RoseTTAFold sequence space diffusion. bioRxiv doi:10.1101/2023.05.08.539766

Loshbaugh AL, Kortemme T. 2020. Comparison of Rosetta flexible-backbone computational protein design methods on binding interactions. *Proteins Struct Funct Bioinforma* **88**: 206–226. doi:10.1002/prot.25790

Lu L, Gou X, Tan SK, Mann SI, Yang H, Zhong X, Gazgalis D, Valdiviezo J, Jo H, Wu Y, et al. 2023. De novo design of drug-binding proteins with predictable binding energy and specificity. bioRxiv doi:10.1101/2023.12.23.573178

Luo S, Su Y, Peng X, Wang S, Peng J, Ma J. 2022. Antigen-Specific antibody design and optimization with diffusion-based generative models. bioRxiv doi:10.1101/2022.07.10.499510

Madani A, Krause B, Greene ER, Subramanian S, Mohr BP, Holton JM, Olmos JL, Xiong C, Sun ZZ, Socher R, et al. 2023. Large language models generate functional protein sequences across diverse families. *Nat Biotechnol* **41**: 1099–1106. doi:10.1038/s41587-022-01618-2

Maguire JB, Haddox HK, Strickland D, Halabiya SF, Coventry B, Griffin JR, Pulavarti SVSRK, Cummins M, Thieker DF, Klavins E, et al. 2021. Perturbing the energy landscape for improved packing during computational protein design. *Proteins Struct Funct Bioinforma* **89**: 436–449. doi:10.1002/prot.26030

Mansoor S, Baek M, Juergens D, Watson JL, Baker D. 2023. Zero-shot mutation effect prediction on protein stability and function using RoseTTAFold. *Protein Sci* **32**: e4780. doi:10.1002/pro.4780

Marks DS, Hopf TA, Chris S, Sander C. 2012. Protein structure prediction from sequence variation. *Nat Biotechnol* **30**: 1072–1080. doi:10.1038/nbt.2419

Moffat L, Greener JG, Jones DT. 2021. Using AlphaFold for rapid and accurate fixed backbone protein design. bioRxiv doi:10.1101/2021.08.24.457549

Nelson N, Yocum CF. 2006. Structure and function of photosystems I and II. *Annu Rev Plant Biol* **57**: 521–565. doi:10.1146/annurev.arplant.57.032905.105350

Norn C, Wicky BIM, Juergens D, Liu S, Kim D, Tischer D, Koepnick B, Anishchenko I, Players F, Baker D, et al. 2021. Protein sequence design by conformational landscape optimization. *Proc Natl Acad Sci* **118**: e2017228118. doi:10.1073/pnas.2017228118

OpenAI. 2023. GPT-4 technical report. arXiv doi:10.48550/arXiv.2303.08774

Ramesh A, Dhariwal P, Nichol A, Chu C, Chen M. 2022. Hierarchical text-conditional image generation with CLIP latents. arXiv doi:10.48550/arXiv.2204.06125

Rocklin GJ, Chidyausiku TM, Goreshnik I, Ford A, Houliston S, Lemak A, Carter L, Ravichandran R, Mulligan VK, Chevalier A, et al. 2017. Global analysis of protein folding using massively parallel design, synthesis, and testing. *Science* **357**: 168–175. doi:10.1126/science.aan0693

Rombach R, Blattmann A, Lorenz D, Esser P, Ommer B. 2022. High-resolution image synthesis with latent diffusion models. arXiv doi:10.48550/arXiv.2112.10752

Roney JP, Ovchinnikov S. 2022. State-of-the-art estimation of protein model accuracy using AlphaFold. bioRxiv doi:10.1101/2022.03.11.484043

Sesterhenn F, Yang C, Bonet J, Cramer JT, Wen X, Wang Y, Chiang CI, Abriata LA, Kucharska I, Castoro G, et al. 2020. De novo protein design enables the precise induction of RSV-neutralizing antibodies. *Science* **368**: eaay5051. doi:10.1126/science.aay5051

Shortle D, Simons KT, Baker D. 1998. Clustering of low-energy conformations near the native structures of small proteins. *Proc Natl Acad Sci* **95**: 11158–11162. doi:10.1073/pnas.95.19.11158

Sohl-Dickstein J, Weiss EA, Maheswaranathan N, Ganguli S. 2015. Deep unsupervised learning using nonequilibrium thermodynamics. arXiv doi:10.48550/arXiv.1503.03585

Song Y, Ermon S. 2020. Generative modeling by estimating gradients of the data distribution. arXiv doi:10.48550/arXiv.1907.05600

Stern JA, Free TJ, Stern KL, Gardiner S, Dalley NA, Bundy BC, Price JL, Wingate D, Corte DD. 2023. A probabilistic view of protein stability, conformational specificity, and design. *Sci Rep* **13**: 15493. doi:10.1038/s41598-023-42032-1

Strokach A, Kim PM. 2022. Deep generative modeling for protein design. *Curr Opin Struct Biol* **72:** 226–236. doi:10.1016/j.sbi.2021.11.008

Su J, Han C, Zhou Y, Shan J, Zhoouu X, Yuan F. 2023. Saprot: protein language modeling with structure-aware vocabulary. bioRxiv doi:10.1101/2023.10.01.560349

Sumida KH, Núñez-Franco R, Kalvet I, Pellock SJ, Wicky BIM, Milles LF, Dauparas J, Wang J, Kipnis Y, Jameson N, et al. 2023. Improving protein expression, stability, and function with ProteinMPNN. bioRxiv doi:10.1101/2023.10.03.560713

Tantillo DJ, Jiangang C, Houk KN. 1998. Theozymes and compuzymes: theoretical models for biological catalysis. *Curr Opin Chem Biol* **2:** 743–750. doi:10.1016/S1367-5931(98)80112-9

Thomas VL, McReynolds AC, Shoichet BK. 2010. Structural bases for stability–function tradeoffs in antibiotic resistance. *J Mol Biol* **396:** 47–59. doi:10.1016/j.jmb.2009.11.005

Tinberg CE, Khare SD, Dou J, Doyle L, Nelson JW, Schena A, Jankowski W, Kalodimos CG, Johnsson K, Stoddard BL, et al. 2013. Computational design of ligand-binding proteins with high affinity and selectivity. *Nature* **501:** 212–216. doi:10.1038/nature12443

Tischer D, Lisanza S, Wang J, Dong R, Anishchenko I, Milles LF, Ovchinnikov S, Baker D. 2020. Design of proteins presenting discontinuous functional sites using deep learning. bioRxiv doi:10.1101/2020.11.29.402743

Torres SV, Leung PJY, Venkatesh P, Lutz ID, Hink F, Huynh HH, Becker J, Yeh AHW, Juergens D, Bennett NR, et al. 2023. De novo design of high-affinity binders of bioactive helical peptides. *Nature* **626:** 435–442. doi:10.1038/s41586-023-06953-1

Trippe BL, Yim J, Tischer D, Baker D, Broderick T, Barzilay R, Jaakola T. 2022. Diffusion probabilistic modeling of protein backbones in 3D for the motif-scaffolding problem. arXiv doi:10.48550/arXiv.2206.04119

Trott O, Olson AJ. 2010. Autodock vina: improving the speed and accuracy of docking with a new scoring function, efficient optimization, and multithreading. *J Comput Chem* **31:** 455–461. doi:10.1002/jcc.21334

Verkuil R, Kabeli O, Du Y, Wicky BIM, Milles LF, Dauparas J, Baker D, Ovchinnikov S, Sercu T, Rives A. 2022. Language models generalize beyond natural proteins. bioRxiv doi:10.1101/2022.12.21.521521

Wang Y, Xue P, Cao M, Yu T, Lane ST, Zhao H. 2021. Directed evolution: methodologies and applications. *Chem Rev* **121:** 12384–12444. doi:10.1021/acs.chemrev.1c00260

Wang J, Lisanza S, Juergens D, Tischer D, Watson JL, Castro KM, Ragotte R, Saragovi A, Milles LF, Baek M, et al. 2022a. Scaffolding protein functional sites using deep learning. *Science* **377:** 387–394. doi:10.1126/science.abn2100

Wang W, Peng Z, Yang J. 2022b. Single-sequence protein structure prediction using supervised transformer protein language models. *Nat Comput Sci* **2:** 804–814. doi:10.1038/s43588-022-00373-3

Watson JL, Juergens D, Bennett NR, Trippe BL, Yim J, Eisenach HE, Ahern W, Borst AJ, Ragotte RJ, Milles LF, et al. 2023. De novo design of protein structure and function with RF$_{diffusion}$. *Nature* **620:** 1089–1100. doi:10.1038/s41586-023-06415-8

Wicky BIM, Milles LF, Courbet A, Ragotte RJ, Dauparas J, Kinfu E, Tipps S, Kibler RD, Baek M, DiMaio F, et al. 2022. Hallucinating symmetric protein assemblies. *Science* **378:** 56–61. doi:10.1126/science.add1964

Winnifrith A, Outeiral C, Hie B. 2023. Generative artificial intelligence for de novo protein design. arXiv doi:10.48550/arXiv.2310.09685

Wu KE, Yang KK, van den Berg R, Zou JY, Lu AX, Amini AP. 2022a. Protein structure generation via folding diffusion. arXiv doi:10.48550/arXiv.2209.15611

Wu R, Ding F, Wang R, Shen R, Zhang X, Luo S, Su C, Wu Z, Xie Q, Berger B, et al. 2022b. High-resolution de novo structure prediction from primary sequence. bioRxiv doi:10.1101/2022.07.21.500999

Yang KK. 2023. List of papers about proteins design using deep learning. Yangkky/machine-learning-for-proteins. https://github.com/yangkky/Machine-learning-for-proteins

Yang KK, Wu Z, Arnold FH. 2019. Machine-learning-guided directed evolution for protein engineering. *Nat Methods* **16:** 687–694. doi:10.1038/s41592-019-0496-6

Yang J, Anishchenko I, Park H, Peng Z, Ovchinnikov S, Baker D. 2020. Improved protein structure prediction using predicted interresidue orientations. *Proc Natl Acad Sci* **117:** 1496–1503. doi:10.1073/pnas.1914677117

Yang C, Sesterhenn F, Bonet J, van Aalen EA, Scheller L, Abriata LA, Cramer JT, Wen X, Rosset S, Georgeon S, et al. 2021. Bottom-up de novo design of functional proteins with complex structural features. *Nat Chem Biol* **17:** 492–500. doi:10.1038/s41589-020-00699-x

Yang KK, Zanichelli N, Yeh H. 2022. Masked inverse folding with sequence transfer for protein representation learning. bioRxiv doi:10.1101/2022.05.25.493516

Yang L, Zhang Z, Song Y, Hong S, Xu R, Zhao Y, Zhang W, Cui B, Yang MH. 2023. Diffusion models: a comprehensive survey of methods and applications. arXiv doi:10.48550/arXiv.2209.00796

Yeh AH, Norn C, Kipnis Y, Tischer D, Pellock SJ, Evans D, Ma P, Lee GR, Zhang JZ, Anishchenko I, et al. 2023. De novo design of luciferases using deep learning. *Nature* **614:** 774–780. doi:10.1038/s41586-023-05696-3

Yim J, Trippe BL, De Bortoli V, Mathieu E, Doucet A, Barzilay R, Jaakola T. 2023. SE(3) diffusion model with application to protein backbone generation. arXiv doi:10.48550/arXiv.2302.02277

Zanghellini A, Jiang L, Wollacott AM, Cheng G, Meiler J, Althoff EA, Röthlisberger D, Baker D. 2006. New algorithms and an in silico benchmark for computational enzyme design. *Protein Sci* **15:** 2785–2794. doi:10.1110/ps.062353106

Backbone Conditional Protein Sequence Design

Justas Dauparas

Institute for Protein Design, University of Washington, Seattle, Washington 98195, USA

Correspondence: justas@uw.edu

A protein is defined by its amino acid sequence. This sequence and environmental factors shape a protein's 3D structural landscape, which is crucial for the protein's function and activity. Protein design aims to develop novel protein sequences or modify existing ones to perform specific functions or have desired protein properties. The protein sequence space is exponentially large, making protein sequence design a tough problem. This problem can be simplified by considering a backbone conditional protein sequence design that factorizes the design problem into two parts: protein backbone design and backbone-dependent sequence design. This allows for a more efficient search over the sequence space for desired structural features. In this review, we discuss when backbone conditional sequence design is possible and how to assess the performance of different design methods, training data, symmetric design, and the combination of unconditional and conditional sequence models.

The primary objective of computational protein design involves the identification of a protein sequence that can display a specific function. The exponentially vast sequence space can often be narrowed down by knowing something about the approximate protein structure that would favor the desired function. This means that the protein design problem could be split into two stages. The first one is generating or reusing existing protein backbone structures. Previously, the backbone generative methods like SEWING (Jacobs et al. 2016) and the Rosetta blueprint (Huang et al. 2011) have been used; but, more recently, hallucination and diffusion-based methods have been very successful in coming up with new backbone structures (Ingraham et al. 2019; Anishchenko et al. 2021; Anand and Achim 2022; Hie et al. 2022; Wang et al. 2022; Wicky et al. 2022; Watson et al. 2023). The second one is coming up with sequences that would approximately adopt desired backbone structures and have some other features needed for the desired function. Traditionally, the Rosetta FastDesign protocol (Maguire et al. 2021) has been successful in designing backbone conditional protein sequences (Dou et al. 2018; Silva et al. 2019; Cao et al. 2022). This protocol iterates Markov chain Monte Carlo (MCMC) optimization of fixed backbone side chain rotamer and amino acid identities and gradient-based Rosetta energy minimization of side chain and backbone torsion angles. Each round of MCMC optimization involves hundreds of thousands of substitutions, making this method quite slow on larger proteins. More recently, deep learning–based protein sequence design methods have been developed and successfully used to design new proteins (Anand and Achim 2022; Dauparas et al. 2022; Wicky et al. 2022; Lutz et al. 2023).

These methods usually use high-resolution protein crystal structures to learn a mapping from the backbone geometry to categorical distributions for amino acids. To capture a joint amino acid distribution, autoregressive (Jacobs et al. 2016; Ingraham et al. 2019; Jing et al. 2020), masked language (Qi and Zhang 2020; Anand and Achim 2022), or Potts (Ingraham et al. 2023; Li et al. 2023) models can be used. Some methods split 3D space into cubes with, for example, edge sizes of 20°A centered at the residue of interest to capture local interactions and then use convolutional neural networks with 1°A voxels to predict amino acid identities (Qi and Zhang 2020; Anand and Achim 2022). Some other methods represent proteins as graphs with residues as nodes and interactions between residues as edges (Ingraham et al. 2019; Jing et al. 2020; Strokach et al. 2020; Dauparas et al. 2022; Hsu et al. 2022). Because of the locality of the problem, these protein graphs can be sparse, meaning that it is enough to consider only 30 nearest-neighbor interactions in 3D space. The full structure-based protein design pipeline consists of backbone generation, sequence design, structure prediction to validate sequences, and applying other filters (Marcos and Silva 2018; Ovchinnikov and Huang 2021; Ferruz et al. 2023). In this work, we are interested in reviewing different sequence design models and discussing how models are evaluated and used in practice (Fig. 1).

WHAT IS A BACKBONE CONDITIONAL SEQUENCE DESIGN?

Atoms forming protein structures are usually divided into two groups. The first one is called the protein backbone. It consists of N, $C\alpha$, C, and O atoms. These atoms are found in every canonical amino acid and form the foundation of the protein structures (Branden and Tooze 2012). The second group is called protein side chains. Different side chain atom types and their bonds determine the type of amino acid. Since every amino acid shares the same backbone atom types, one might want to try to guess, depending on the geometry of the backbone, what side chain atoms (or amino acid type) should fit for every backbone residue. This is what we call a backbone conditional sequence design. It would be quite easy to learn this mapping if some local structural features would allow one to know the answer (e.g., the bond length between N and $C\alpha$ atoms for the same residue is smaller for glycine and larger for proline or preferred backbone dihedral angles for different residues). Indeed, this can be the case; there are small differences in bond lengths and angles between canonical amino acids (Jaskolski et al. 2007; Engh and Huber 2012). These standard values are used as parameters for modeling the crystal content because the diffraction data obtained from protein crystals only provide limited resolution coordinates. The trivial local bond length/angle solution can be removed either by adding noise to backbone coordinates (Dauparas et al. 2022) or discretizing 3D space (Qi and Zhang 2020; Anand and Achim 2022). We are interested in conditional sequence design because it allows us to split the protein sequence design problem into two simpler problems: backbone coordinate design and backbone conditional sequence design (see Fig. 1). Designing protein backbones is denoted as modeling p(backbone) distribution, and a backbone conditional sequence design refers to the models that are learning p(sequence|backbone) distribution. The reverse structure sequence factorization is to model sequences unconditionally, that is, p(sequence), and to fit a structure prediction model capturing p(backbone|sequence) (see Fig. 1). This idea of inverting structure prediction models for backbone sequence design is called hallucination (Anishchenko et al. 2021; Jendrusch et al. 2021; Moffat et al. 2021, 2022; Norn et al. 2021; Verkuil et al. 2022; Wang et al. 2022; Wicky et al. 2022). Only optimizing the p(backbone|sequence) term over sequences can lead to adversarial solutions (Verkuil et al. 2022; Wicky et al. 2022). This can be fixed either by using the p(sequence) model as a prior (Verkuil et al. 2022) or redesigning sequences using a conditional model p(sequence|backbone) (Wicky et al. 2022). Often, models from both factorizations are used in practice. For example, a backbone diffusion model can be used to generate protein backbones, a graph-based backbone conditional sequence design can be

Cite this article as *Cold Spring Harb Perspect Biol* doi: 10.1101/cshperspect.a041517

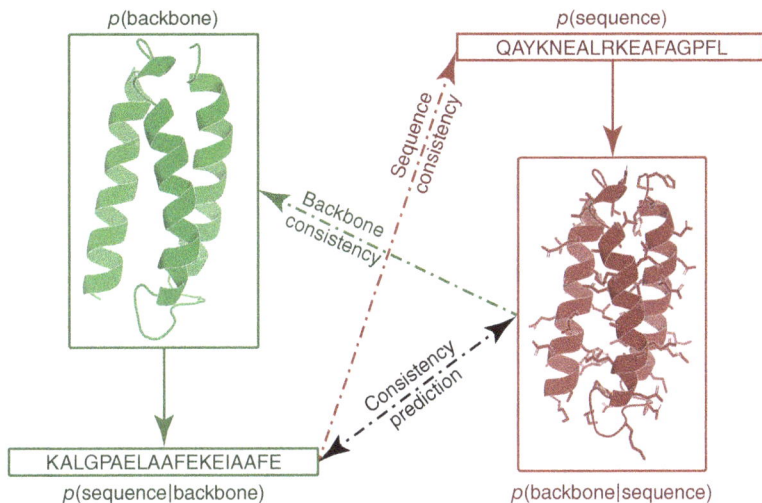

Figure 1. Graphical models illustrating protein backbone and sequence models. The green model with solid lines on the *left* represents p(sequence, backbone) = p(backbone) · p(sequence|backbone) factorization. The red model on the *right* shows p(sequence, backbone) = p(sequence) · p(backbone|sequence) factorization. Once the sequence is generated using the p(sequence|backbone) model, the output sequence can be used together with the p(backbone|sequence) model to get the backbone consistency score depicted in a green dash–dotted line. Equivalently, the sequence consistency score is shown in a red dash–dotted line.

used to generate many sequences per backbone, and then, finally, structure prediction models can be used to rank sequences according to the designed and predicted backbone match.

HOW TO EVALUATE THE PERFORMANCE OF SEQUENCE DESIGN METHODS?

Native Sequence Recovery

The most common metric reported when evaluating the performance of backbone conditional sequence design methods is the native sequence recovery (NSR). It measures the average percentage agreement between the native sequence for a given experimentally determined backbone and the sequence generated by the design method. This measure does not account for the similarity between amino acids. It is generally true that the NSR is higher for the protein core residues and lower for the surface residues (Kuhlman and Baker 2000; Anand and Achim 2022; Dauparas et al. 2022; Hsu et al. 2022). The sequence recovery is almost a monotonic function of the number of geometric constraints presented

for the residue of interest (e.g., these geometric constraints can be measured by calculating the average virtual $C\beta-C\beta$ distance of eight closest neighbors) (Dauparas et al. 2022). It should be noted that sequence recovery can be very sensitive to the protein backbone quality—lower backbone quality means lower sequence recovery (Dauparas et al. 2022; Castorina et al. 2023). To avoid backbone quality sensitivity, a random Gaussian noise can be added to the input coordinates, deleting the smallest scale structural features. For example, a ProteinMPNN model trained without any backbone noise gets an average NSR of 55%, but adding Gaussian noise with a standard deviation (SD) of 0.02 Å during training reduces the NSR to 51%. Further adding noise with SD equal to 0.20 Å leads to an NSR of 45%. We want to have a high sequence recovery, but at the same time, we want our model to pay attention to the global features of the folds instead of picking up on local fine details. Finally, it should be noted that even a single residue mutation of the native sequence can drastically change the protein backbone structure, and hence the high NSR does not necessarily lead to successful protein sequences (Petukh et al. 2015). The often-reported NSR only reflects the average performance over single- or multichain residues. This number can be greatly different for specific protein regions. For example, sequence recovery over protein surface residues is ~25% using Rosetta (average is 33%) or 40% using ProteinMPNN (SD = 0.02 Å, average is 52%). These numbers can be even lower for surface residues interacting with DNA/RNA or other molecules. The average NSR would also not reflect the recapitulation of metal-binding residues, transmembrane residues, and other subgroups of residues or folds.

Bias and Confusion Matrix

First-order average statistics of generated sequences are often compared with native sequences. This can reveal the model's bias to generate sequences with a larger fraction of specific residues. It is often the case that the deep learning models with the objective of maximizing the likelihood of the native sequences fit these first-order statistics very well (Dauparas et al. 2022; Hsu et al. 2022; Castorina et al. 2023). However, the amino acid biases are present for sequences sampled at low temperatures. For example, the ProteinMPNN (Dauparas et al. 2022) generates more charged amino acids at the expense of the polar ones when used at a temperature of 0.1. Similarly, Rosetta-designed sequences have an overrepresentation of alanines in the core and boundary (Maguire et al. 2021). It is also common to report confusion matrices averaged over many proteins to show what mistakes the models are making (Anand and Achim 2022; Hsu et al. 2022). These matrices resemble the amino acid substitution matrices (e.g., BLOSUM62) (Henikoff and Henikoff 1992), suggesting that these models are capturing natural amino acid pairwise similarities. It should be noted that amino acid bias for Rosetta design can be corrected by adding various constraints on allowed amino acids for different protein regions. Residues can also be assigned to one of three layers (core, boundary, or surface, depending on their environment) and designed with amino acids compatible with only the assigned

layer (Marcos and Silva 2018). Furthermore, one can avoid amino acids with low individual propensities for forming either α-helices or β-strands. Evolutionary and consensus-based constraints can be added for loop residues. The same domain knowledge restrictions can be applied to deep learning–based methods.

Sequence–Structure Compatibility

One of the most important metrics used to evaluate or filter designed sequences is predicting the structures of sequences at different levels of complexity. At the most coarse level, the protein secondary structure could be predicted and checked against the intended backbone geometry (Jones 1999). Another way to check sequence and structure compatibility is to collect fragments for each nine-residue frame in the designed sequence from a set of experimentally determined structures and rank sequences according to the fragment and backbone structural matches (Rohl et al. 2004). One can also use the PDB-derived statistics representing relationships between sequence and structure via local tertiary structure motifs (TERMs) (Zheng et al. 2015).

With the advance of the protein structure prediction methods powered by deep learning (Baek et al. 2021; Jumper et al. 2021; Lin et al. 2023), it turned out that computationally designed sequences can be more easily predicted compared to the native sequences using a single sequence prediction model (Yang et al. 2020; Dauparas et al. 2022) (i.e., not using templates, or searching for related sequences in sequence databases). This suggests that sequence design and structure prediction methods are picking up on the same idealized sequence–structure motifs. The structural match between the desired and predicted backbones can be quantified in terms of local distance difference test-Cα (Mariani et al. 2013), template modeling score (Zhang and Skolnick 2004), or other structural metrics. Using these metrics to filter designed sequences can increase design success rates nearly 10-fold for protein binder design problems (Bennett et al. 2023). Interestingly, NSR and structural metrics predicted by AlphaFold

Cite this article as *Cold Spring Harb Perspect Biol* doi: 10.1101/cshperspect.a041517

(Jumper et al. 2021) can be anticorrelated. Sequence design models trained with noisy input backbones have lower sequence recovery. However, the structural match metrics predicted by AlphaFold (Jumper et al. 2021) increase until the ~0.3 Å Gaussian noise level (Dauparas et al. 2022). This mismatch in two metrics might be coming from the sequence design models picking up on local small-scale features to correctly guess native sequences instead of optimizing for more global residue–residue interactions. It is also likely that a more blurry backbone structure leads to more averaged/consensus protein sequences that are easier to predict (Dauparas et al. 2022). Furthermore, the structural prediction success rate is a function of the structure model's capability to correctly predict structures using a single sequence as an input. For example, increasing the number of recycling steps for the AlphaFold models (up to 12 recycles or more) increases the match between the desired and predicted backbones (Dauparas et al. 2022). Furthermore, having a structural match does not guarantee that the sequence will succeed experimentally. It has been shown (Li et al. 2023) that feeding native sequences with a fraction of residues randomly mutated to nonnative residues can still lead to correct backbone predictions coming from AlphaFold. Also, directly optimizing protein sequences using MCMC methods to obtain the desired backbone without any sequence prior can lead to poorly behaved sequences (Anishchenko et al. 2021; Verkuil et al. 2022; Wicky et al. 2022). These either partially random or adversarial sequences for structure prediction models can be identified using p(sequence) or p(sequence|backbone) model likelihoods (Verkuil et al. 2022).

Sequence Diversity and Pairwise Statistics

Another desired characteristic of sequence design models is the ability to generate diverse, consistent solutions. The diversity and sequence quality for the log-likelihood models can be controlled by rescaling output logits, and this parameter is called temperature. Low-temperature sampling leads to higher sequence recovery and better structural prediction metrics but provides lower sequence diversity (Dauparas et al. 2022). Most of the sequence diversity is coming from structurally less-constrained regions like protein surfaces. Furthermore, it has been suggested that model's inference speed can be increased using a one-shot prediction of all amino acids (Gao et al. 2022) without the need to either use autoregressive or resampling techniques. However, single-shot decoding would not capture any pairwise (second order) statistics between amino acids for the same backbone. Sequences sampled from autoregressive models like Protein MPNN show coevolution-like correlations between residues that are close in 3D space. Capturing a joint sequence distribution is more important for cases where the backbone quality is lower; in the limit of the bad-quality backbones, conditional sequence design models are modeling unconditional sequence probability where capturing pairwise statistics is very important (Rao et al. 2020). To explicitly capture single and pair residue interactions, the Potts model (Wu 1982) can be used to approximate p(sequence|backbone) = exp($-E$[sequence, backbone]/Z[backbone]), where E denotes energy for a sequence–backbone pair, and Z is a partition function for that backbone (Ingraham et al. 2023; Li et al. 2023). Energy is chosen to be of the form $E_{ij} = B_i X_i + W_{ij} X_i X_j$, where $X_i \in \mathbb{R}^{20}$ is one hot-encoded amino acid at position i, and $B_i \in \mathbb{R}^{20}$ and $W_{ij} \in \mathbb{R}^{20 \times 20}$ are learnable Potts model parameters for single and pair residue terms. Pair terms can be used for predicting binding energies for protein complexes (Li et al. 2023).

TRAINING DATA AND OVERFITTING

It is common to generate training data by splitting all proteins into individual chains and then partitioning these chains into training, validation, and test sets such that there is no CATH (a hierarchical domain classification of protein structures in the Protein Data Bank) overlap (Ingraham et al. 2019; Anand and Achim 2022; Hsu et al. 2022). Models trained with only single protein chains might not learn about protein-binding interfaces explicitly. To train models that are more suited for binder design problems, multichain data sets can be used for this task (Dauparas et al. 2022; Li et al. 2023). Similarly to the protein structure prediction data splitting (Evans et al. 2021), single-chain sequences can be clustered at 30%–40% sequence identity cutoff using mmseqs2 (Steinegger and Söding 2017). Training, validation, and test data sets are split, ensuring that none of the chains from the protein biounits (or asymmetric units) would be in the other two groups. Even though the data splitting for the multichain data set is less rigorous, one can still significantly reduce training loss while keeping the same or increasing validation loss by simply training the model with more parameters (i.e., it is still easy to overfit). Most of the neural networks trained for the conditional sequence design are small (1–10 million trainable parameters) because of the overfitting issue. This has been addressed by using additional sequence–structure data generated using the AlphaFold model (Hsu et al. 2022). Training with additional AlphaFold data and evaluating native sequence recoveries on crystal structure can hurt the model's performance if the model is quite small. However, scaling models to 21–142 million parameters and training with extra AlphaFold data significantly improves the model's performance (Hsu et al. 2022) in terms of the NSR. On one hand, having more training data, even if it is noisy, might help with the generalization of the task; but on the other hand, this extra data is likely to be biased toward single-chain structures because it is much harder to confidently predict protein complexes. This bias might result in a worse protein–protein interaction design performance.

SYMMETRIES AND MULTISTATE DESIGN

Many soluble and membrane-bound proteins in cells are symmetric oligomeric complexes with two or more subunits (Goodsell and Olson 2000; Garcia-Seisdedos et al. 2017). Proteins can exhibit cyclic (e.g., porin [Weiss and Schulz 1992]), dihedral (e.g., phosphofructokinase [Evans et al. 1981]), octahedral (e.g., ferritin [Lawson et al. 1991]), and other symmetries

and pseudosymmetries. Therefore, sequence design methods need to be able to design required symmetric sequences. This can be done using masked language models, and during sampling, averaging out predicted logits (Anand and Achim 2022). Imposing sequence symmetries can also work with autoregressive sequence methods if the model is trained with arbitrary decoding orders. The decoding order can be chosen in a way such that all the symmetric residues are decoded at the same time (Dauparas et al. 2022). Experimental successes have been reported in designing TIM-barrels (Anand and Achim 2022), hallucinating dimers, trimers, and larger rings (Wicky et al. 2022) and more recently using reinforcement learning to come up with 60-subunit icosahedra (Lutz et al. 2023). Furthermore, symmetry constraints have been used to propose a high-level programming language for generative protein design (Hie et al. 2022).

Multistate backbone design can be thought of as symmetric sequence design if all the desired and unwanted backbone states are known (Dauparas et al. 2022; Hsu et al. 2022). For example, the predicted logits for different states can be linearly combined to form final logits from which the amino acids are sampled. Stimulus-responsive two-state hinge proteins have recently been designed using this idea, and adjusting the weights in which predicted logits are combined can help in designing sequences with higher or lower preference for a particular state (Praetorius et al. 2023).

MISSING BACKBONE INFORMATION AND LARGE LANGUAGE MODELS

Some proteins lack a well-structured 3D fold. Disordered protein regions can fold upon binding to their targets or make flexible linkers that play a role in the assembly of macromolecular arrays (Dyson and Wright 2005). It might be much harder to design sequences for these proteins using a fixed backbone approach. One way to deal with partially missing structural information is to train a model that predicts sequence and/or structure for the missing regions (Hsu et al. 2022; Wang et al. 2022). For example,

RoseTTAFold (Baek et al. 2021) has been fine-tuned to inpaint missing structure and sequence regions (Wang et al. 2022). Also, the ESM-IF1 model (Hsu et al. 2022) has been tasked with predicting sequences for missing structural regions. The authors found that masking backbones during training did not significantly change test performance on unmasked backbones but allowed running the model with partially missing backbone information. The residues that are missing structural information can no longer rely on 3D local graph structure, and, hence, they might need to be connected to all other residues like in the large language models (Rives et al. 2021). In the limit when all backbone information is not available during the sequence design, the problem changes from modeling $p(\text{sequence}|\text{backbone})$ to $p(\text{sequence})$, which is an unconditional sequence modeling. Guessing missing amino acids without or with partial backbone information is a much harder task. For example, the ESM-IF1 model's uncertainty almost doubles in regions with 30 masked backbone residues (Hsu et al. 2022). Structure-informed language models have been proposed to combine backbone conditional and unconditional sequence design models (Yang et al. 2023; Zheng et al. 2023). In the case of the masked inverse folding with sequence transfer model (Yang et al. 2023), both conditional and unconditional sequence models are trained with masked language model objectives. The convolutional neural network–based protein language model called CARP (Yang et al. 2024) was pretrained on sequence data first, and then it was used to generate inputs for a structure-based sequence design model. Using a pretrained language model improves the NSR from 45.4% to 55.6% (Yang et al. 2023). Similarly, Zheng et al. (2023) added a structural encoder adaptor module (Houlsby et al. 2019) to an already pretrained protein language model like ESM-1b (Rives et al. 2021) and improved sequence recovery from 54.4% to 59.4%. In both works, authors observed increased NSR because a conditional sequence design model can use evolutionary conservation and coevolution information via unconditional model weights in addition to the structural in-formation. For example, a sequence for a highly evolutionarily conserved kinase activation loop can be better recovered using a model that is using unconditional language model features (Zheng et al. 2023). One can imagine adding similar single and pair evolutionary statistics from multiple sequence alignments to any protein sequence design methods to preserve specific sequence regions or functions.

CONCLUDING REMARKS

Backbone conditional sequence design models are very useful in practice for various protein design problems. As for any design methods, it can be quite tricky to compare different models and their performances since there is no ground truth for new designs, and even a single incorrect amino acid can lead to misfolded and/or nonfunctional proteins. A single number like the NSR, which is easy to calculate and is often reported, does not fully capture the quality of the design method. It has been experimentally shown that protein sequences designed using machine learning–based methods have high expression levels, are thermostable, and fold into their intended structures (Anand and Achim 2022; Dauparas et al. 2022; Wicky et al. 2022; Lutz et al. 2023). The big advantage of the recent deep learning–based sequence design methods over more traditional Rosetta-based protein sequence design is that it reduces the amount of expert knowledge needed to design proteins and speeds up the computation. Furthermore, symmetric, multistate, and evolutionary information can be incorporated into current sequence design methods. The performance of deep learning–based sequence design models is heavily dependent on the training data available for specific protein subgroups (e.g., transmembrane proteins). Hence, either more data or domain knowledge is needed to correctly model these cases.

REFERENCES

Anand N, Achim T. 2022. Protein structure and sequence generation with equivariant denoising diffusion probabilistic models. arXiv doi:10.48550/arXiv.2205.15019

Anishchenko I, Pellock SJ, Chidyausiku TM, Ramelot TA, Ovchinnikov S, Hao J, Bafna K, Norn C, Kang A, Bera AK, et al. 2021. De novo protein design by deep network hallucination. *Nature* **600**: 547–552. doi:10.1038/s41586-021-04184-w

Baek M, DiMaio F, Anishchenko I, Dauparas J, Ovchinnikov S, Lee GR, Wang J, Cong Q, Kinch LN, Schaeffer RD, et al. 2021. Accurate prediction of protein structures and interactions using a three-track neural network. *Science* **373**: 871–876. doi:10.1126/science.abj8754

Bennett N, Coventry B, Goreshnik I, Huang B, Allen A, Vafeados D, Peng YP, Dauparas J, Baek M, Stewart L, et al. 2023. Improving de novo protein binder design with deep learning. *Nat Commun* **14**: 2625. doi:10.1038/s41467-023-38328-5

Branden CI, Tooze J. 2012. *Introduction to protein structure*. Garland Science, New York.

Cao L, Coventry B, Goreshnik I, Huang B, Sheffler W, Park JS, Jude KM, Marković I, Kadam RU, Verschueren KHG, et al. 2022. Design of protein-binding proteins from the target structure alone. *Nature* **605**: 551–560. doi:10.1038/s41586-022-04654-9

Castorina LV, Petrenas R, Subr K, Wood CW. 2023. PDBench: evaluating computational methods for protein-sequence design. *Bioinformatics* **39**: btad027. doi:10.1093/bioinformatics/btad027

Dauparas J, Anishchenko I, Bennett N, Bai H, Ragotte RJ, Milles LF, Wicky BIM, Courbet A, de Haas RJ, Bethel N, et al. 2022. Robust deep learning–based protein sequence design using protein mpnn. *Science* **378**: 49–56. doi:10.1126/science.add2187

Dou J, Vorobieva AA, Sheffler W, Doyle LA, Park H, Bick MJ, Mao B, Foight GW, Lee MY, Gagnon LA, et al. 2018. De novo design of a fluorescence-activating β-barrel. *Nature* **561**: 485–491. doi:10.1038/s41586-018-0509-0

Dyson HJ, Wright PE. 2005. Intrinsically unstructured proteins and their functions. *Nat Rev Mol Cell Biol* **6**: 197–208. doi:10.1038/nrm1589

Engh RA, Huber R. 2012. Structure quality and target parameters. Structure quality and target parameters. In *International tables for crystallography* (ed. Brock CP, et al.). Wiley Online Library, Hoboken, NJ.

Evans PR, Farrants GWT, Hudson PJ. 1981. Phosphofructokinase: structure and control. *Philos Trans R Soc Lond B Biol Sci* **293**: 53–62.

Evans R, O'Neill M, Pritzel A, Antropova N, Senior A, Green T, Žídek A, Bates R, Blackwell S, Yim J, et al. 2021. Protein complex prediction with AlphaFold-multimer. bioRxiv doi:10.1101/2021.10.04.463034

Ferruz N, Heinzinger M, Akdel M, Goncearenco A, Naef L, Dallago C. 2023. From sequence to function through structure: deep learning for protein design. *Comput Struct Biotechnol J* **21**: 238–250. doi:10.1016/j.csbj.2022.11.014

Gao Z, Tan C, Li SZ. 2022. PiFold: toward effective and efficient protein inverse folding. arXiv doi:10.48550/arXiv.2209.12643

Garcia-Seisdedos H, Empereur-Mot C, Elad N, Levy ED. 2017. Proteins evolve on the edge of supramolecular self-assembly. *Nature* **548**: 244–247. doi:10.1038/nature23320

Goodsell DS, Olson AJ. 2000. Structural symmetry and protein function. *Annu Rev Biophys Biomol Struct* **29**: 105–153. doi:10.1146/annurev.biophys.29.1.105

Henikoff S, Henikoff JG. 1992. Amino acid substitution matrices from protein blocks. *Proc Natl Acad Sci* **89**: 10915–10919. doi:10.1073/pnas.89.22.10915

Hie B, Candido S, Lin Z, Kabeli O, Rao R, Smetanin N, Sercu T, Rives A. 2022. A high-level programming language for generative protein design. bioRxiv doi:10.1101/2022.12.21.521526

Houlsby N, Giurgiu A, Jastrzebski S, Morrone B, De Laroussilhe Q, Gesmundo A, Attariyan M, Gelly S. 2019. Parameter-efficient transfer learning for NLP. In *International Conference on Machine Learning*, pp. 2790–2799. PMLR, New York.

Hsu C, Verkuil R, Liu J, Lin Z, Hie B, Sercu T, Lerer A, Rives A. 2022. Learning inverse folding from millions of predicted structures. In *International Conference on Machine Learning*, pp. 8946–8970. PMLR, New York.

Huang PS, Ban YEA, Richter F, Andre I, Vernon R, Schief WR, Baker D. 2011. RosettaRemodel: a generalized framework for flexible backbone protein design. *PLoS One* **6**: e24109. doi:10.1371/journal.pone.0024109

Ingraham J, Garg V, Barzilay R, Jaakkola T. 2019. Generative models for graph-based protein design. In *Advances in Neural Information Processing Systems 32. Proceedings of the 33rd International Conference on Neural Information Processing Systems*. Curran Associates, Red Hook, NY.

Ingraham J, Baranov M, Costello Z, Barber KW, Wang W, Ismail A, Frappier V, Lord DM, Ng-Thow-Hing C, Van Vlack ER, et al. 2023. Illuminating protein space with a programmable generative model. *Nature* **623**: 1070–1078. doi:10.1038/s41586-023-06728-8

Jacobs TM, Williams B, Williams T, Xu X, Eletsky A, Federizon JF, Szyperski T, Kuhlman B. 2016. Design of structurally distinct proteins using strategies inspired by evolution. *Science* **352**: 687–690. doi:10.1126/science.aad8036

Jaskolski M, Gilski M, Dauter Z, Wlodawer A. 2007. Stereochemical restraints revisited: how accurate are refinement targets and how much should protein structures be allowed to deviate from them? *Acta Crystallogr D Biol Crystallogr* **63**: 611–620. doi:10.1107/S090744490700978X

Jendrusch M, Korbel JO, Sadiq SK. 2021. AlphaDesign: a de novo protein design framework based on AlphaFold. bioRxiv doi:10.1101/2021.10.11.463937

Jing B, Eismann S, Suriana P, Townshend RJL, Dror R. 2020. Learning from protein structure with geometric vector perceptrons. arXiv doi:10.48550/arXiv.2009.01411

Jones DT. 1999. Protein secondary structure prediction based on position-specific scoring matrices. *J Mol Biol* **292**: 195–202. doi:10.1006/jmbi.1999.3091

Jumper J, Evans R, Pritzel A, Green T, Figurnov M, Ronneberger O, Tunyasuvunakool K, Bates R, Žídek A, Potapenko A, et al. 2021. Highly accurate protein structure prediction with AlphaFold. *Nature* **596**: 583–589. doi:10.1038/s41586-021-03819-2

Kuhlman B, Baker D. 2000. Native protein sequences are close to optimal for their structures. *Proc Natl Acad Sci* **97**: 10383–10388. doi:10.1073/pnas.97.19.10383

Cite this article as *Cold Spring Harb Perspect Biol* doi: 10.1101/cshperspect.a041517

Lawson DM, Artymiuk PJ, Yewdall SJ, Smith JMA, Livingstone JC, Treffry A, Luzzago A, Levi S, Arosio P, Cesareni G, et al. 1991. Solving the structure of human H ferritin by genetically engineering intermolecular crystal contacts. *Nature* **349:** 541–544. doi:10.1038/349541a0

Li AJ, Lu M, Desta I, Sundar V, Grigoryan G, Keating AE. 2023. Neural network-derived Potts models for structure-based protein design using backbone atomic coordinates and tertiary motifs. *Protein Sci* **32:** e4554. doi:10.1002/pro.4554

Lin Z, Akin H, Rao R, Hie B, Zhu Z, Lu W, Smetanin N, Verkuil R, Kabeli O, Shmueli Y, et al. 2023. Evolutionary-scale prediction of atomic-level protein structure with a language model. *Science* **379:** 1123–1130. doi:10.1126/science.ade2574

Lutz ID, Wang S, Norn C, Courbet A, Borst AJ, Zhao YT, Dosey A, Cao L, Xu J, Leaf EM, et al. 2023. Top-down design of protein architectures with reinforcement learning. *Science* **380:** 266–273. doi:10.1126/science.adf6591

Maguire JB, Haddox HK, Strickland D, Halabiya SF, Coventry B, Griffin JR, Pulavarti SVSRK, Cummins M, Thieker DF, Klavins E, et al. 2021. Perturbing the energy landscape for improved packing during computational protein design. *Proteins* **89:** 436–449. doi:10.1002/prot.26030

Marcos E, Silva DA. 2018. Essentials of de novo protein design: methods and applications. *Wiley Interdiscip Rev Comput Mol Sci* **8:** e1374. doi:10.1002/wcms.1374

Mariani V, Biasini M, Barbato A, Schwede T. 2013. lDDT: a local superposition-free score for comparing protein structures and models using distance difference tests. *Bioinformatics* **29:** 2722–2728. doi:10.1093/bioinformatics/btt473

Moffat L, Greener JG, Jones DT. 2021. Using AlphaFold for rapid and accurate fixed backbone protein design. bioRxiv doi:10.1101/2021.08.24.457549

Moffat L, Kandathil SM, Jones DT. 2022. Design in the dark: learning deep generative models for de novo protein design. bioRxiv doi:10.1101/2022.01.27.478087

Norn C, Wicky BIM, Juergens D, Liu S, Kim D, Tischer D, Koepnick B, Anishchenko I, Players F, Baker D, et al. 2021. Protein sequence design by conformational landscape optimization. *Proc Natl Acad Sci* **118:** e2017228118. doi:10.1073/pnas.2017228118

Ovchinnikov S, Huang PS. 2021. Structure-based protein design with deep learning. *Curr Opin Chem Biol* **65:** 136–144. doi:10.1016/j.cbpa.2021.08.004

Petukh M, Kucukkal TG, Alexov E. 2015. On human disease-causing amino acid variants: statistical study of sequence and structural patterns. *Hum Mutat* **36:** 524–534. doi:10.1002/humu.22770

Praetorius F, Leung PJY, Tessmer MH, Broerman A, Demakis C, Dishman AF, Pillai A, Idris A, Juergens D, Dauparas J, et al. 2023. Design of stimulus-responsive two-state hinge proteins. bioRxiv doi:10.1101/2023.01.27.525968

Qi Y, Zhang JZH. 2020. DenseCPD: improving the accuracy of neural-network-based computational protein sequence design with DenseNet. *J Chem Inf Model* **60:** 1245–1252. doi:10.1021/acs.jcim.0c00043

Rao R, Meier J, Sercu T, Ovchinnikov S, Rives A. 2020. Transformer protein language models are unsupervised structure learners. bioRxiv doi:10.1101/2020.12.15.422761

Rives A, Meier J, Sercu T, Goyal S, Lin Z, Liu J, Guo D, Ott M, Zitnick CL, Ma J, et al. 2021. Biological structure and function emerge from scaling unsupervised learning to 250 million protein sequences. *Proc Natl Acad Sci* **118:** e2016239118. doi:10.1073/pnas.2016239118

Rohl CA, Strauss CEM, Misura KMS, Baker D. 2004. Protein structure prediction using Rosetta. In *Methods in enzymology*, Vol. 383, pp. 66–93. Elsevier, New York.

Silva DA, Yu S, Ulge UY, Spangler JB, Jude KM, Labão-Almeida J, Ali LR, Quijano-Rubio A, Ruterbusch M, Leung I, et al. 2019. De novo design of potent and selective mimics of il-2 and il-15. *Nature* **565:** 186–191. doi:10.1038/s41586-018-0830-7

Steinegger M, Söding J. 2017. MMseqs2 enables sensitive protein sequence searching for the analysis of massive data sets. *Nat Biotechnol* **35:** 1026–1028. doi:10.1038/nbt.3988

Strokach A, Becerra D, Corbi-Verge C, Perez-Riba A, Kim PM. 2020. Fast and flexible protein design using deep graph neural networks. *Cell Syst* **11:** 402–411.

Verkuil R, Kabeli O, Du Y, Wicky BIM, Milles LF, Dauparas J, Baker D, Ovchinnikov S, Sercu T, Rives A. 2022. Language models generalize beyond natural proteins. bioRxiv doi:10.1101/2022.12.21.521521

Wang J, Lisanza S, Juergens D, Tischer D, Watson JL, Castro KM, Ragotte R, Saragovi A, Milles LF, Baek M, et al. 2022. Scaffolding protein functional sites using deep learning. *Science* **377:** 387–394.

Watson JL, Juergens D, Bennett NR, Trippe BL, Yim J, Eisenach HE, Ahern W, Borst AJ, Ragotte RJ, Milles LF, et al. 2023. De novo design of protein structure and function with RFdiffusion. *Nature* **620:** 1089–1100.

Weiss MS, Schulz GE. 1992. Structure of porin refined at 1.8 Å resolution. *J Mol Biol* **227:** 493–509.

Wicky BIM, Milles LF, Courbet A, Ragotte RJ, Dauparas J, Kinfu E, Tipps S, Kibler RD, Baek M, DiMaio F, et al. 2022. Hallucinating symmetric protein assemblies. *Science* **378:** 56–61.

Wu FY. 1982. The Potts model. *Rev Mod Phys* **54:** 235.

Yang J, Anishchenko I, Park H, Peng Z, Ovchinnikov S, Baker D. 2020. Improved protein structure prediction using predicted interresidue orientations. *Proc Natl Acad Sci* **117:** 1496–1503.

Yang KK, Zanichelli N, Yeh H. 2023. Masked inverse folding with sequence transfer for protein representation learning. *Protein Eng Des Sel* **36:** gzad015. doi:10.1093/protein/gzad015

Yang KK, Fusi N, Lu AX. 2024. Convolutions are competitive with transformers for protein sequence pretraining. *Cell Syst* **15:** 286–294.e2.

Zhang Y, Skolnick J. 2004. Scoring function for automated assessment of protein structure template quality. *Proteins* **57:** 702–710.

Zheng F, Zhang J, Grigoryan G. 2015. Tertiary structural propensities reveal fundamental sequence/structure relationships. *Structure* **23:** 961–971.

Zheng Z, Deng Y, Xue D, Zhou Y, Ye F, Gu Q. 2023. Structure-informed language models are protein designers. arXiv doi:10.48550/arXiv.2302.01649

Environmental Impacts of Machine Learning Applications in Protein Science

Loïc Lannelongue[1,2,3,4] and Michael Inouye[1,2,3,4,5,6]

[1]Cambridge Baker Systems Genomics Initiative, Department of Public Health and Primary Care, University of Cambridge, Cambridge CB2 0SR, United Kingdom

[2]British Heart Foundation Cardiovascular Epidemiology Unit, Department of Public Health and Primary Care, University of Cambridge, Cambridge CB2 0SR, United Kingdom

[3]Victor Phillip Dahdaleh Heart and Lung Research Institute, University of Cambridge, Cambridge CB2 0BB, United Kingdom

[4]Health Data Research UK Cambridge, Wellcome Genome Campus and University of Cambridge, Cambridge, United Kingdom

[5]Cambridge Baker Systems Genomics Initiative, Baker Heart and Diabetes Institute, Melbourne 3004, Victoria, Australia

[6]British Heart Foundation Centre of Research Excellence, University of Cambridge, Cambridge CB2 0BB, United Kingdom

Correspondence: ll582@medschl.cam.ac.uk

Computing tools and machine learning models play an increasingly important role in biology and are now an essential part of discoveries in protein science. The growing energy needs of modern algorithms have raised concerns in the computational science community in light of the climate emergency. In this work, we summarize the different ways in which protein science can negatively impact the environment and we present the carbon footprint of some popular protein algorithms: molecular simulations, inference of protein–protein interactions, and protein structure prediction. We show that large deep learning models such as AlphaFold and ESMFold can have carbon footprints reaching over 100 tonnes of CO_2e in some cases. The magnitude of these impacts highlights the importance of monitoring and mitigating them, and we list actions scientists can take to achieve more sustainable protein computational science.

Algorithms, computer simulations, and machine learning models are now an essential part of protein science. Statistical modeling was already used in the early 2000s to predict protein–protein interactions in yeast (Jansen et al. 2003), and the following 20 years saw algorithmic complexity increase in line with hardware and software abilities. AlphaFold (Jumper et al. 2021) is one of the recent examples; released in 2021 with millions of trainable parameters, it marked a leap forward for predicted protein structures. As a consequence of increased complexity, it is not uncommon for models to run for hours using hundreds, if not thousands, of processing cores.

While there is no doubt that such technological developments have enabled impressive discoveries, the contribution of these models to the climate emergency has raised concerns in recent years (Schwartz et al. 2020; Bender et al. 2021; Lannelongue et al. 2021a, 2023). High-performance computing has tangible impacts on the environment—mostly energy consumption but also water usage and ecological consequences—which raise ethical dilemmas for computational biologists: how to balance trying to cure diseases with the health impacts of climate change, partly fueled by large data centers?

Data centers' global greenhouse gas (GHG) emissions are estimated to be $\sim100 \times 10^6$ tonnes of CO_2e per year, similar to American commercial aviation (Lannelongue et al. 2021a). Estimates for the yearly carbon footprint of a scientist range between 4 and 37 tCO_2e (Stevens et al. 2020; ALLEA 2022; Knödlseder et al. 2022), far greater than the upper bound of 2 tCO_2e per person set by the International Panel on Climate Change (IPCC) to keep global warming under 1.5°C (Arias et al. 2021). Statistics from XSEDE (a now-concluded network of American research institutes) show the scale of computing in biology: in 2020, 586 million compute hours were dedicated to biochemistry or molecular structure and function (XSEDE Impact—usage statistics, portal .xsede.org/#/gallery).

The GHG emissions arising from computations need acknowledging and mitigating when possible. In this work, we highlight the main environmental impacts of computations used in protein science, and we explore the carbon footprint of some popular algorithms from molecular simulation, protein–protein interactions, and protein structure prediction.

THE ENVIRONMENTAL IMPACTS OF COMPUTATIONS

The standardized metric for carbon footprint is a quantity (usually in grams) of CO_2-equivalent (gCO_2e), which summarizes the environmental impact of GHG emissions with just one number. One of the challenges is that each gas has different impacts on climate change and different lifetimes; for example, it is estimated that over 100 years, 1 kg of methane will have the same impact on global warming as 28 kg of carbon dioxide. This is accounted for by giving methane a global warming potential (GWP$_{100}$) of 28 (Myhre et al. 2013) (the GWP of carbon dioxide is 1 by definition). The final carbon footprint is calculated by weighting each GHG by its GWP$_{100}$. For example, a mix of 4 kg of carbon dioxide and 2 kg of methane will have a carbon footprint of $4 \times 1 + 2 \times 28 = 60$ kgCO_2e. Generally, the GHGs considered are the ones included in the Kyoto basket, namely, carbon dioxide (CO_2), methane (CH_4), and nitrous oxide (N_2O) (Hill et al. 2020), which constitute 97.9% of total GHG emissions (Our World in Data 2017). Notably, this definition of carbon footprint does not consider all environmental impacts, such as water usage, impact on wildlife, etc., and the values of the GWPs are debated as they may misestimate the impact of short-lived climate pollutants such as methane (Allen et al. 2018).

Each stage of the hardware's life cycle results in some environmental impacts. Starting with manufacturing, when the mining of raw materials, assembly, and shipping alone can account for 70% to 90% of the total footprint in the case of smartphones and laptops (Clément et al. 2020; Apple Environmental Reports 2023; Dell Technologies 2023). Although a lower percentage for data center hardware, with between 15% and 40% of the total impacts due to manufacturing (Dell Technologies 2023), the environmental costs remain substantial. Disposing of technological wastes (e-waste) is also responsible for considerable environmental impacts—water, air, and soil pollution—notwithstanding consequences for the health of waste workers. A recent report by the World Health Organization (2021) found that more than 82% of the 53.6 million tonnes of e-waste were not processed formally or recycled. Informal waste processing involves between 12 and 56 million people worldwide, including millions of children. In dump sites, often in low- and middle-income countries, informal waste workers are exposed to a range of toxic by-products and hazardous com-

Cite this article as *Cold Spring Harb Perspect Biol* doi: 10.1101/cshperspect.a041473

pounds, such as mercury, lead, cadmium, and other heavy metals.

The rest of the environmental impacts of computing comes from usage, mainly through electricity consumption, but also water use and the ecological impact of the facilities. The carbon footprint from energy usage (C in gCO_2e) is the easiest aspect to quantify at the scale of one analysis or project. It depends on how much energy is needed (E in kWh) and how this energy is produced, called carbon intensity (CI in gCO_2e/kWh) (Lannelongue et al. 2021a):

$$C = E \times CI,$$

The energy needed can be estimated by focusing on runtime (t), power draw from processors (P_p), and memory (P_m) and the efficiency of the data center (PUE). PUE, which stands for power usage effectiveness, is the ratio between the total power delivered to the facility and the power used by the servers and is a measure of overheads, mostly cooling. Lannelongue et al. (2021a) can

be consulted for more details on this equation.

$$E = t \times (P_p + P_m) \times PUE,$$

As most data centers are powered by the general power grid, the carbon intensity depends on the energy production methods where the hardware is located. Because of differences between production methods, there is a wide variation between countries, with up to 3 orders of magnitude between Iceland (0.10 gCO_2e/kWh) and Australia (770 gCO_2e/kWh) (Fig. 1).

Data storage tends to be considered separately from computations, as it is typically low power but constant over multiple years. The order of magnitude of the carbon footprint of storing one terabyte of data for one year is ~10 $kgCO_2e$ (Nguyen et al. 2020; Seagate 2023). However, there can be great variations between hardware options. Just looking at Seagate's hard drives, cradle-to-grave carbon footprints range from 2 $kgCO_2e$/TB/year to 66 $kgCO_2e$/TB/year (Seagate 2023).

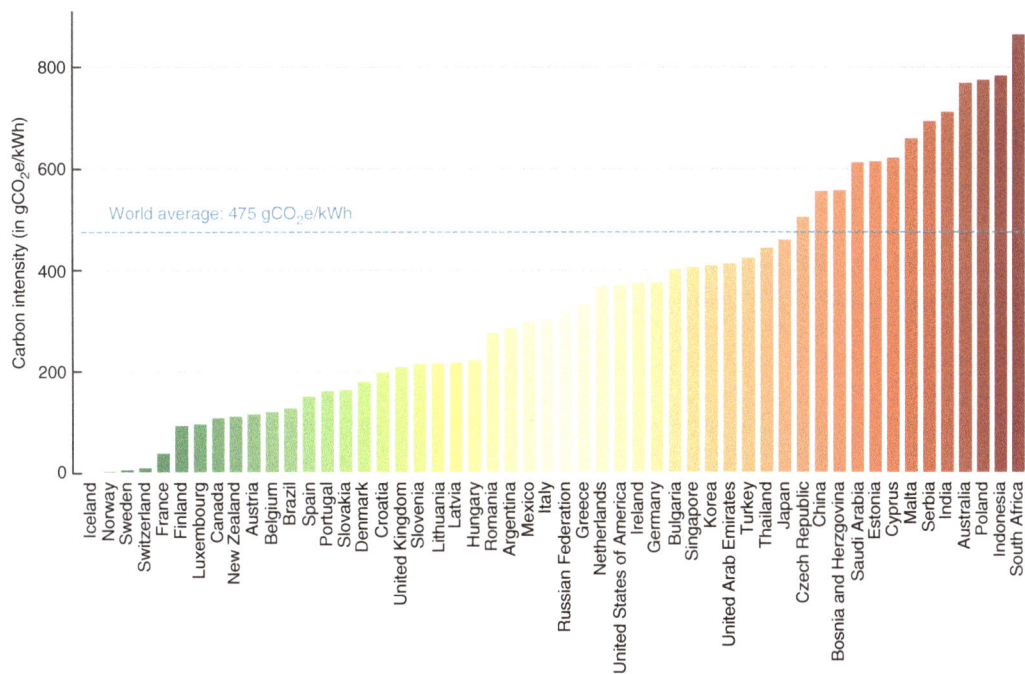

Figure 1. 2022 carbon intensity of electricity usage by country. Created using data from Carbon Footprint (2023). The world average value is available in the Global Energy & CO_2 Status Report 2019 (IEA 2019).

Moreover, different phases in computational projects may need to be assessed separately. Machine learning pipelines generally involve a research phase to identify and fine-tune the best model, followed by a training phase to build the final product. Then the model can be used to make predictions at scale (inference). Depending on how heavily used the final model is, most energy needs may come from training or prediction.

Estimating Carbon Footprints in Practice

Different tools exist to estimate the carbon footprint of computations: online calculators (e.g., Green Algorithms [Lannelongue et al. 2021a] or MLCO2 [Lacoste et al. 2019]), dedicated packages in Python to track energy usage (e.g., Carbon Tracker [Anthony et al. 2020] and Code Carbon, codecarbon.io]) and server-side tools to monitor usage in data centers (e.g., GA4HPC [Green Algorithms 2023], Amazon Web Services' dashboard on the cloud [AWS 2023]). The pros and cons of each approach are summarized in Lannelongue and Inouye (2023). In summary, task-agnostic, general-purpose calculators such as online tools can be used across all fields of computational science but require the user to manually input information such as runtime and memory usage. Task-specific tools integrate tightly with the existing code to avoid this (e.g., Python packages for machine learning [Anthony et al. 2020]; Code Carbon, codecarbon.io [Henderson et al. 2020]), but different tools need to be built for each task. The tools mentioned so far are all user-based (either integrated into the analysis pipeline or by inputting values in a calculator a posteriori). An alternative is to track usage from the server side (i.e., a calculator located on the computing platform and tracking usage continuously). Instead of being task-specific, this approach is platform-specific, particularly adapted to computations based in high-performance computing facilities. When feasible, it can address the limitations of the previously mentioned approaches. As part of the Green Algorithms Initiative, GreenAlgorithms4HPC (Green Algorithms 2023) is one example of such a tool, and some cloud providers integrate similar tools into their dashboards (AWS 2023). Recent studies investigated the accuracy and reliability of some of these tools and found that estimates were overall consistent and in line with real power consumption, particularly when training deep learning models (Bouza Heguerte et al. 2023; Jay et al. 2023).

About Experimental Work

Although not the focus of this work, computational tools rely heavily on data from experimental methods, either for training or validation. Similar to computations, part of the footprint of laboratories is from energy use. Estimates of the energy usage of laboratories range from three to ten times the energy needs of an equivalent size office, and the University of Oxford estimated that 60% of the university's GHG emissions are due to laboratory buildings (Royal Society of Chemistry 2022). For example, an ultralow temperature (ULT) freezer and a fume hood can use as much energy as one to four typical households (My Green Lab 2023; Nature Portfolio 2023), with 16–22 kWh/day for a ULT freezer (University of Exeter 2023) and ~75 kWh/day for a fume hood (Berkeley Lab 2023). For these simple actions such as closing the sashes of a fume hood or increasing a ULT freezer from −80°C to −70°C can result in substantial energy savings (more than 40% and 30%–40%, respectively) (Royal Society of Chemistry 2022). Laboratory work also comes with heavy usage of water and chemicals, as well as single-use plastic (Labconscious 2020). It is estimated that each year, research laboratories produce ~5.5 million tonnes of plastic waste (Urbina et al. 2015), an autoclave can use ~270 L of water per cycle and an ultrapure water purification system discards 80% of the water input (My Green Lab 2023).

ALGORITHMS AND PROTEIN SCIENCE

In this section, we give a series of examples of the carbon footprint of popular algorithms and machine learning applications in protein science (Fig. 2). The list is not exhaustive, so we would encourage readers to investigate the carbon footprint of models in their field of expertise that we may not have discussed here.

Cite this article as *Cold Spring Harb Perspect Biol* doi: 10.1101/cshperspect.a041473

Molecular Simulations

Computer simulations are key tools to understand how different components interact together, with structure-based drug discovery one of its successful applications. Molecular docking methods predict how compounds (e.g., proteins) are likely to bind to each other and, to do so, require significant computing power due to the complexity of the task. Grealey et al. (2022) used a benchmark that studied a one million ligand campaign from the Directory of Useful Decoys (DUDs) (Ruiz-Carmona et al. 2014). The DUD benchmark set contains 39 protein–ligand complexes with crystal structure, with 100 active ligands per complex on average, each with 36 decoys. When comparing three methods, rDock (Ruiz-Carmona et al. 2014), AutoDock Vina (Trott and Olson 2009), and Glide (Friesner et al. 2004), and using world average carbon intensities, Grealey et al. (2022) found carbon footprints of 13 $kgCO_2e$ for Glide, 154 $kgCO_2e$ for rDock, and 514 $kgCO_2e$ for AutoDock Vina (between 27 and 1082 kWh of energy). The latter runs for over 40,000 core hours, for example. While Glide seems to emit almost 40 times less GHGs than AutoDock Vina, it is worth noting that, in contrast to AutoDock Vina and rDock, Glide is not freely available.

Grealey et al. (2022) also looked at the computing requirements of simulating molecular dynamics of the Satellite Tobacco Mosaic Virus (one million atoms) for 100 nanoseconds and found the carbon footprint to be 18 $kgCO_2e$ with Amber (ambermd.org/index.php) (75 kWh) and 95 $kgCO_2e$ with NAMD (Phillips et al. 2020) (400 kWh); however, the two softwares use slightly different resolutions so a direct comparison is not straightforward.

Protein–Protein Interactions

In silico methods to predict protein–protein interactions have grown more popular in recent years. While earlier tools tended to rely on low-power machine learning algorithms, most methods now leverage deep learning and involve longer training times. A recent comparison of machine learning (random forest) and deep learning (recurrent neural networks) found that in some situations, the deep learning approach could emit 22,000 more GHGs for similar performance (Lannelongue and Inouye 2022). However, runtimes remain small and, in this case, training the deep learning model once had a carbon footprint of 356 gCO_2e and ~36 $kgCO_2e$ (75 kWh) when including fine-tuning the network.

Protein Structure Prediction: AlphaFold and ESMFold

Compared to the interaction prediction models above, other works on proteins rely on larger and more complex algorithms. Protein structure prediction from primary sequences has been a recent area of intense activity for deep learning networks in protein science. AlphaFold, released in 2021 by DeepMind (Google) (Jumper et al. 2021), was a significant leap forward toward solving the protein-folding problem. ESMFold was released a year later by Meta (Lin et al. 2023), claiming almost similar accuracy but faster inference. We will not assess the differences between the predicted structures but rather estimate the carbon footprints of training and predicting with these large neural networks based on numbers included in the original publications.

128 TPUv3 were used for 11 d to train and fine-tune AlphaFold once. Using the Green Algorithms calculator, this required 8.25 MWh of energy and would emit approximately 3.92 tCO_2e (using the average world carbon intensity and Google's best PUE of 1.11). Prediction runtimes, and therefore carbon footprints, vary greatly with protein length: it takes 9.2 min and 24 gCO_2e for a 384-residue protein but 3 $kgCO_2e$ for a protein with 2500 residues (using the ensemble model).

Three versions of ESMFold were released, with 700 million, 3 billion, and 15 billion parameters. 512 NVIDIA v100 GPUs were used to train these models once; it took 8 d for the smaller one (700M), 30 d for the 3B one, and 60 d for the larger model. These long runtimes come with proportionally large carbon footprints. Using the same PUE and carbon intensity as above, training the larger 15B model required 246 MWh and emitted 117 tCO_2e. Training the

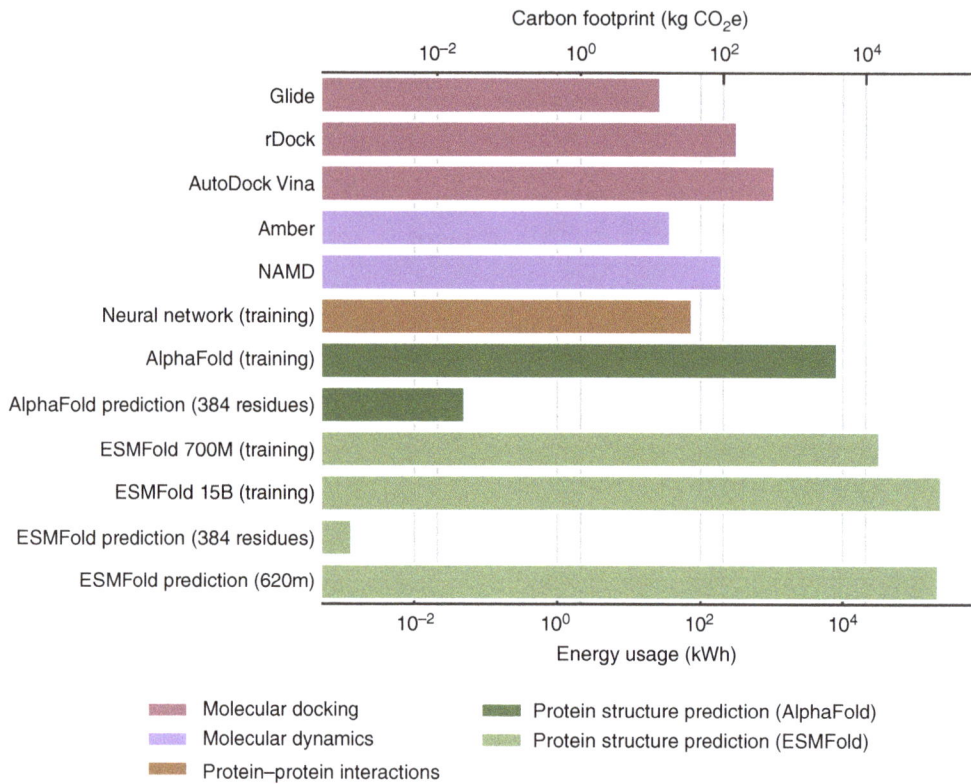

Figure 2. Energy usage and carbon footprint of the main use cases presented here (log scale). Carbon footprints are in kilograms of CO_2e and assume an average carbon intensity of 0.475 $kgCO_2e/kWh$.

700M and 3B models emitted, respectively, 16 tCO_2e (33 MWh) and 58 tCO_2e (123 MWh). ESMFold could predict a structure from a 384-residue protein in 14.2 sec on one GPU, which corresponds to just 0.6 gCO_2e. Interestingly, while the training cost of ESMFold is higher than AlphaFold, the inference cost is lower, hinting at some trade-offs that need to be considered. In their publication, the authors present predictions for 620 million sequences from the MGnify90 database, which took 28,000 GPU days to complete (2 wk with a cluster of 2000 GPUs), requiring 224 MWh of energy and emitting 106 tCO_2e.

CONCLUSION

The algorithms presented in this work are examples of protein models with noticeable environmental impacts, up to 117 tCO_2e to train one

of the ESMFold models. While many analyses not discussed here will have negligible carbon footprints, the existence of such complex models highlights the importance of monitoring, acknowledging, and reducing such impacts when possible.

There are a few mitigating factors to take into account when discussing these high-carbon algorithms. For example, training these large protein-folding models are meant as one-off costs; training the model once and making it available publicly alongside predicted structures prevents the unnecessary development of local models. Inference costs tend to then be significantly lower in most cases. For example, AlphaFold's team and EMBL's European Bioinformatic Institute have made 200 million structures available in a database and received over 700M API requests and 2.4M visitors in a year (Lannelongue et al. 2023). Pretrained foundation models like AlphaFold or ESMFold can also be downloaded and

 Cite this article as *Cold Spring Harb Perspect Biol* doi: 10.1101/cshperspect.a041473

fine-tuned to tackle slightly different problems for a fraction of the environmental cost of training the full model (Motmaen et al. 2023). Moreover, it can also be useful to compare computational models to the corresponding experiments (e.g., X-ray crystallography for protein structures [Bertoline et al. 2023]). Simulations are generally thought to have lower financial and environmental costs, which is likely to be true but would need to be assessed in a detailed manner moving forward. Such open-access models present undeniable benefits, both for scientific progress and the environment. However, the release of AlphaFold in 2021, followed by RoseTTAFold a few months later, and then ESMFold shortly after shows that there is a risk of engaging in a race for always bigger models, similar to what has been seen with large language models recently (Bender et al. 2021). If not carefully considered, the environmental costs of developing each model may negate the expected environmental benefits. Recent works focusing on achieving similar performance with smaller models (e.g., DR-BERT for protein region annotation [Nambiar et al. 2023]) can be a promising way forward.

Computational scientists can do a number of things to be more sustainable while still actively engaging in machine learning research (Lannelongue et al. 2021b). The first thing is to estimate and monitor the impact of the algorithms used. Such estimations should ideally be done before starting a project to include the figures in cost-benefit analyses, and afterward to acknowledge and track carbon impacts. Using the most efficient tool for a task can also have great impacts. Two examples from the field of genome-wide association studies (GWASs): updating from v1 of Bolt-LMM to v2.3 can reduce energy usage, and GHG emissions, by 72% (Grealey et al. 2022). Switching from SAIGE to REGENIE (two other GWAS softwares) reduces carbon footprints by 85% and saving 2.4 tCO_2e according to a study from the team behind the REGENIE tool (Mbatchou et al. 2021). There is also scope to reduce carbon footprints by running computations when carbon intensity is low. For example, it has been shown that delaying machine learning tasks by up to 24 h can reduce GHG emissions by 10%–40% in the United States (even 80% in some areas) (Dodge et al. 2022). Finally, these actions are part of a wider context of more sustainable computational science encapsulated by the GREENER principles (governance, responsibility, estimation, energy and embodied impacts, new collaborations, education, and research) (Lannelongue et al. 2023), and individual actions will need to be accompanied by broader institutional support to ensure that the societal benefits of protein models outweigh their environmental costs.

ACKNOWLEDGMENTS

L.L. was supported by the University of Cambridge MRC DTP (MR/S502443/1) and the BHF programme grant (RG/18/13/33946). M.I. was supported by the Munz Chair of Cardiovascular Prediction and Prevention and the NIHR Cambridge Biomedical Research Centre (BRC-1215-20014; NIHR203312). M.I. was also supported by the UK Economic and Social Research 878 Council (ES/T013192/1). This work was supported by core funding from the British Heart Foundation (RG/18/13/33946) and the NIHR Cambridge Biomedical Research Centre (BRC-1215-20014; NIHR20 3312). The views expressed are those of the author(s) and not necessarily those of the NIHR or the Department of Health and Social Care. This work was also supported by Health Data Research UK, which is funded by the UK Medical Research Council, Engineering and Physical Sciences Research Council, Economic and Social Research Council, Department of Health and Social Care (England), Chief Scientist Office of the Scottish Government Health and Social Care Directorates, Health and Social Care Research and Development Division (Welsh Government), Public Health Agency (Northern Ireland), and British Heart Foundation and Wellcome.

REFERENCES

ALLEA. 2022. *Towards climate sustainability of the academic system in Europe and beyond.* ALLEA, Berlin. doi:10 .26356/climate-sust-acad

Allen MR, Shine KP, Fuglestvedt JS, Millar RJ, Cain M, Frame DJ, Macey AH. 2018. A solution to the misrepresentations of CO2-equivalent emissions of short-lived climate pollutants under ambitious mitigation. *NPJ Clim Atmos Sci* **1**: 16. doi:10.1038/s41612-018-0026-8

Anthony LFW, Kanding B, Selvan R. 2020. Carbontracker: tracking and predicting the carbon footprint of training deep learning models. arXiv doi:10.48550/arXiv.2007.03051

Apple Environmental Reports. 2023. A first for Apple. https://www.apple.com/environment

Arias PA, Bellouin N, Coppola E, Jones RG, Krinner G, Marotzke J, Naik V, Palmer MD, Plattner GK, Rogelj J, et al. 2021. *Technical summary: in climate change 2021: the physical science basis. Contribution of working group I to the sixth assessment report of the intergovernmental panel on climate change*, pp. 33–144. Cambridge University Press, Cambridge. doi:10.1017/9781009157896.002

AWS. 2023. Customer carbon footprint tool. https://aws.amazon.com/aws-cost-management/aws-customer-carbon-footprint-tool

Bender EM, Gebru T, McMillan-Major A, Shmitchell S. 2021. On the dangers of stochastic parrots: can language models be too big? In *Proceedings of the 2021 ACM conference on fairness, accountability, and transparency, in FAccT '21*. Association for Computing Machinery, New York. doi:10.1145/3442188.3445922

Berkeley Lab. 2023. Laboratory fume hood energy modeler. https://fumehoodcalculator.lbl.gov

Bertoline LMF, Lima AN, Krieger JE, Teixeira SK. 2023. Before and after AlphaFold2: an overview of protein structure prediction. *Front Bioinform* **3**: 1120370. doi:10.3389/fbinf.2023.1120370

Bouza Heguerte L, Aurélie B, Lannelongue L. 2023. How to estimate carbon footprint when training deep learning models? A guide and review. *Environ Res Commun* doi:10.1088/2515-7620/acf81b

Carbon Footprint. 2023. 2022 Grid electricity emissions factors. https://www.carbonfootprint.com/docs/2023_02_emissions_factors_sources_for_2022_electricity_v10.pdf

Clément LPPVP, Jacquemotte QES, Hilty LM. 2020. Sources of variation in life cycle assessments of smartphones and tablet computers. *Environ Impact Assess Rev* **84**: 106416. doi:10.1016/j.eiar.2020.106416

Dell Technologies. 2023. Reducing our impact, driving progress. https://www.dell.com/en-uk/dt/corporate/social-impact/advancing-sustainability/sustainable-products-and-services/product-carbon-footprints.htm

Dodge J, Prewitt T, Tachet Des Combes R, Odmark E, Schwartz R, Strubell E, Luccioni AS, Smith NA, DeCario N, Buchanan W. 2022. Measuring the carbon intensity of AI in cloud instances. In *2022 ACM Conference on Fairness, Accountability, and Transparency, Seoul, Republic of Korea*. ACM, New York. doi:10.1145/3531146.3533234

Friesner RA, Banks JL, Murphy RB, Halgren TA, Klicic JJ, Mainz DT, Repasky MP, Knoll EH, Shelley M, Perry JK, et al. 2004. Glide: a new approach for rapid, accurate docking and scoring. 1: Method and assessment of docking accuracy. *J Med Chem* **47**: 1739–1749. doi:10.1021/jm0306430

Grealey J, Lannelongue L, Saw WY, Marten J, Méric G, Ruiz-Carmona S, Inouye M. 2022. The carbon footprint of bioinformatics. *Mol Biol Evol* **39**: msac034. doi:10.1093/molbev/msac034

Green Algorithms. 2023. The green algorithms project. https://github.com/GreenAlgorithms/GreenAlgorithms4HPC

Henderson P, Hu J, Romoff J, Brunskill E, Jurafsky D, Pineau J. 2020. Towards the systematic reporting of the energy and carbon footprints of machine learning. *J Mach Learn Res* **21**: 1–43.

Hill N, Bramwell R, Karagianni E, Jones L, MacCarthy J, Hinton S, Walker C, Harris B. 2020. 2020 Government greenhouse gas conversion factors for company reporting: methodology paper for conversion factors final report. Department for Business, Energy and Industrial Strategy, London.

IEA. 2019. Global energy & CO_2 status report 2019. https://www.iea.org/reports/global-energy-co2-status-report-2019/emissions

Jansen R, Yu H, Greenbaum D, Kluger Y, Krogan NJ, Chung S, Emili A, Snyder M, Greenblatt JF, Gerstein M. 2003. A Bayesian networks approach for predicting protein–protein interactions from genomic data. *Science* **302**: 449–453. doi:10.1126/science.1087361

Jay M, Ostapenco V, Lefèvre L, Trystram D, Orgerie AC, Fichel B. 2023. An experimental comparison of software-based power meters: focus on CPU and GPU. In *CCGrid 2023 - 23rd IEEE/ACM International Symposium on Cluster, Cloud and Internet Computing*. Bangalore, India.

Jumper J, Evans R, Pritzel A, Green T, Figurnov M, Ronneberger O, Tunyasuvunakool K, Bates R, Žídek A, Potapenko A, et al. 2021. Highly accurate protein structure prediction with AlphaFold. *Nature* **596**: 583–589. doi:10.1038/s41586-021-03819-2

Knödlseder J, Brau-Nogué S, Coriat M, Garnier P, Hughes A, Martin P, Tibaldo L. 2022. Estimate of the carbon footprint of astronomical research infrastructures. *Nat Astron* **6**: 503–513. doi:10.1038/s41550-022-01612-3

Labconscious. 2020. Going green in a wet lab. https://www.labconscious.com/blog/going-green-in-a-wet-lab-symbolic-ivs-high-impact-actions

Lacoste A, Luccioni A, Schmidt V, Dandres T. 2019. Quantifying the carbon emissions of machine learning. arXiv doi:10.48550/arXiv.1910.09700

Lannelongue L, Inouye M. 2022. Pitfalls of machine learning models for protein–protein interactions. bioRxiv doi:10.1101/2022.02.07.479382

Lannelongue L, Inouye M. 2023. Carbon footprint estimation for computational research. *Nat Rev Methods Primers* **3**: 9. doi:10.1038/s43586-023-00202-5

Lannelongue L, Grealey J, Inouye M. 2021a. Green algorithms: quantifying the carbon footprint of computation. *Adv Sci* **8**: 2100707. doi:10.1002/advs.202100707

Lannelongue L, Grealey J, Bateman A, Inouye M. 2021b. Ten simple rules to make your computing more environmentally sustainable. *PLoS Comput Biol* **17**: e1009324. doi:10.1371/journal.pcbi.1009324

Lannelongue L, Aronson HEG, Bateman A, Birney E, Caplan T, Juckes M, McEntyre J, Morris AD, Reilly G, Inouye M. 2023. GREENER principles for environmentally sustainable computational science. *Nat Comput Sci* **3**: 514–521. doi:10.1038/s43588-023-00461-y

Lin Z, Akin H, Rao R, Hie B, Zhu Z, Lu W, Smetanin N, Verkuil R, Kabeli O, Shmueli Y, et al. 2023. Evolutionary-scale prediction of atomic-level protein structure with a language model. *Science* **379:** 1123–1130. doi:10.1126/science.ade2574

Mbatchou J, Barnard L, Backman J, Marcketta A, Kosmicki JA, Ziyatdinov A, Benner C, O'Dushlaine C, Barber M, Boutkov B, et al. 2021. Computationally efficient whole-genome regression for quantitative and binary traits. *Nat Genet* **53:** 1097–1103. doi:10.1038/s41588-021-00870-7

Motmaen A, Dauparas J, Baek M, Abedi MH, Baker D, Bradley P. 2023. Peptide-binding specificity prediction using fine-tuned protein structure prediction networks. *Proc Natl Acad Sci* **120:** e2216697120. doi:10.1073/pnas.2216697120

My Green Lab. 2023. Energy. https://www.mygreenlab.org/energy.html

Myhre G, Shindell D, Bréon FM, Collins W, Fuglestvedt J, Huang J, Koch D, Lamarque JF, Lee D, Mendoza B, et al. 2013. *Anthropogenic and natural radiative forcing: in climate change 2013: the physical science basis. Contribution of working group I to the fifth assessment report of the intergovernmental panel on climate change.* Cambridge University Press, Cambridge.

Nambiar A, Forsyth JM, Liu S, Maslov S. 2023. DR-BERT: a protein language model to annotate disordered regions. bioRxiv doi:10.1101/2023.02.22.529574

Nature Portfolio. 2023. The demand for ultracold storage has soared. https://www.nature.com/articles/d42473-021-00361-7

Nguyen B, Sinistore J, Smith J, Arshi PS, Johnson LM, Kidman T, diCaprio TJ, Carmean D, Strauss K. 2020. Architecting datacenters for sustainability: greener data storage using synthetic DNA. In *IEEE electronics goes green 2020.* Fraunhofer Institute for Reliability and Microintegration IZM, Berlin.

Our World in Data. 2017. CO_2 and Greenhouse Gas Emissions. https://ourworldindata.org/co2-and-other-greenhouse-gas-emissions

Phillips JC, Hardy DJ, Maia JDC, Stone JE, Ribeiro JV, Bernardi RC, Buch R, Fiorin G, Hénin J, Jiang W, et al. 2020. Scalable molecular dynamics on CPU and GPU architectures with NAMD. *J Chem Phys* **153:** 044130. doi:10.1063/5.0014475

Royal Society of Chemistry. 2022. *Sustainable laboratories: a community-wide movement toward sustainable laboratory practices.* Royal Society of Chemistry, Cambridge.

Ruiz-Carmona S, Alvarez-Garcia D, Foloppe N, Garmendia-Doval AB, Juhos S, Schmidtke P, Barril X, Hubbard RE, Morley SD. 2014. Rdock: a fast, versatile and open source program for docking ligands to proteins and nucleic acids. *PLoS Comput Biol* **10:** e1003571. doi:10.1371/journal.pcbi.1003571

Schwartz R, Dodge J, Smith NA, Etzioni O. 2020. Green AI. *Commun ACM* **63:** 54–63. doi:10.1145/3381831

Seagate. 2023. Product sustainability. https://www.seagate.com/esg/planet/product-sustainability

Stevens ARH, Bellstedt S, Elahi PJ, Murphy MT. 2020. The imperative to reduce carbon emissions in astronomy. *Nat Astron* **4:** 843–851. doi:10.1038/s41550-020-1169-1

Trott O, Olson AJ. 2009. Autodock Vina: Improving the speed and accuracy of docking with a new scoring function, efficient optimization, and multithreading. *J Comput Chem* **31:** 455–461. doi:10.1002/jcc.21334

Urbina MA, Watts AJR, Reardon EE. 2015. Labs should cut plastic waste too. *Nature* **528:** 479–479. doi:10.1038/528479c

University of Exeter. 2023. Sustainable labs. http://www.exeter.ac.uk/about/sustainability/sustainablelabs/energy/ultfreezers

World Health Organization. 2021. *Children and digital dumpsites: E-waste exposure and child health.* World Health Organization, Geneva. https://apps.who.int/iris/handle/10665/341718

Index